Fabrication of Metallic Pressure Vessels

Wiley-ASME Press Series

Fabrication of Process Equipment
Owen Greulich, Maan H. Jawad

Engineering Practice with Oilfield and Drilling Applications
Donald W. Dareing

Flow-Induced Vibration Handbook for Nuclear and Process Equipment
Michel J. Pettigrew, Colette E. Taylor, Nigel J. Fisher

Vibrations of Linear Piezostructures
Andrew J. Kurdila, Pablo A. Tarazaga

Bearing Dynamic Coefficients in Rotordynamics: Computation Methods and Practical Applications
Lukasz Brenkacz

Advanced Multifunctional Lightweight Aerostructures: Design, Development, and Implementation
Kamran Behdinan, Rasool Moradi-Dastjerdi

Vibration Assisted Machining: Theory, Modelling and Applications
Li-Rong Zheng, Dr. Wanqun Chen, Dehong Huo

Two-Phase Heat Transfer
Mirza Mohammed Shah

Computer Vision for Structural Dynamics and Health Monitoring
Dongming Feng, Maria Q Feng

Theory of Solid-Propellant Nonsteady Combustion
Vasily B. Novozhilov, Boris V. Novozhilov

Introduction to Plastics Engineering
Vijay K. Stokes

Fundamentals of Heat Engines: Reciprocating and Gas Turbine Internal Combustion Engines
Jamil Ghojel

Offshore Compliant Platforms: Analysis, Design, and Experimental Studies
Srinivasan Chandrasekaran, R. Nagavinothini

Computer Aided Design and Manufacturing
Zhuming Bi, Xiaoqin Wang

Pumps and Compressors
Marc Borremans

Corrosion and Materials in Hydrocarbon Production: A Compendium of Operational and
Engineering Aspects
Bijan Kermani and Don Harrop

Design and Analysis of Centrifugal Compressors
Rene Van den Braembussche

Case Studies in Fluid Mechanics with Sensitivities to Governing Variables
M. Kemal Atesmen

The Monte Carlo Ray-Trace Method in Radiation Heat Transfer and Applied Optics
J. Robert Mahan

Dynamics of Particles and Rigid Bodies: A Self-Learning Approach
Mohammed F. Daqaq

Primer on Engineering Standards, Expanded Textbook Edition
Maan H. Jawad and Owen R. Greulich

Engineering Optimization: Applications, Methods and Analysis
R. Russell Rhinehart

Compact Heat Exchangers: Analysis, Design and Optimization using FEM and CFD Approach
C. Ranganayakulu and Kankanhalli N. Seetharamu

Robust Adaptive Control for Fractional-Order Systems with Disturbance and Saturation
Mou Chen, Shuyi Shao, and Peng Shi

Robot Manipulator Redundancy Resolution
Yunong Zhang and Long Jin

Stress in ASME Pressure Vessels, Boilers, and Nuclear Components
Maan H. Jawad

Combined Cooling, Heating, and Power Systems: Modeling, Optimization, and Operation
Yang Shi, Mingxi Liu, and Fang Fang

Applications of Mathematical Heat Transfer and Fluid Flow Models in Engineering and
Medicine
Abram S. Dorfman

Bioprocessing Piping and Equipment Design: A Companion Guide for the ASME BPE Standard
William M. (Bill) Huitt

Nonlinear Regression Modeling for Engineering Applications: Modeling, Model Validation,
and Enabling Design of Experiments
R. Russell Rhinehart

Geothermal Heat Pump and Heat Engine Systems: Theory and Practice
Andrew D. Chiasson

Fundamentals of Mechanical Vibrations
Liang-Wu Cai

Introduction to Dynamics and Control in Mechanical Engineering Systems
Cho W.S. To

Fabrication of Metallic Pressure Vessels

Owen R. Greulich
Consultant

Maan H. Jawad
Global Engineering & Technology, LLC

This Work is a co-publication between ASME Press and John Wiley & Sons Inc.

Registered Office
John Wiley & Sons, Inc., 111 River Street, Hoboken, NJ 07030, USA

Editorial Office
111 River Street, Hoboken, NJ 07030, USA

For details of our global editorial offices, customer services, and more information about Wiley products visit us at www.wiley.com.

Wiley also publishes its books in a variety of electronic formats and by print-on-demand. Some content that appears in standard print versions of this book may not be available in other formats.

Library of Congress Cataloging-in-Publication Data

Names: Greulich, Owen R., author. | Jawad, Maan H., author.
Title: Fabrication of metallic pressure vessels / Owen R. Greulich, Maan H. Jawad.
Description: First edition. | Hoboken, NJ : John Wiley & Sons, Inc., 2021. | Includes bibliographical references and index.
Identifiers: LCCN 2021035014 (print) | LCCN 2021035015 (ebook) | ISBN 9781119674863 (hardback) | ISBN 9781119674900 (adobe pdf) | ISBN 9781119674887 (epub)
Subjects: LCSH: Pressure vessels–Design and construction.
Classification: LCC TA660.T34 G74 2021 (print) | LCC TA660.T34 (ebook) | DDC 681/.76041–dc23
LC record available at https://lccn.loc.gov/2021035014
LC ebook record available at https://lccn.loc.gov/2021035015

Cover image: © Chris A. Cimarolli
Cover design by Wiley

Set in 9.5/12.5pt STIXTwoText by Straive, Pondicherry, India

SKY10030788_102221

To our wives

Cathy Greulich

Dixie Jawad

Contents

Preface

Pressure vessels are fabricated in thousands of facilities throughout the world. The fabrication processes differ from company to company, and even from plant to plant for the same company. Even within the same plant, construction of similar vessels will at times be performed in different ways for a variety of reasons.

Some companies produce large quantities of the same or essentially duplicate vessels. They typically develop designs that lend themselves to high production rates, as well as specialized tooling and processes to optimize production of those designs.

Other manufacturers specialize in pressure vessels for a particular function, such as heat exchanger vessels, and design their processes, tooling, and facilities around the type of product produced.

Still other fabricators make a specialty of constructing unique vessels. For these organizations, nearly every product is different from every other, covering a range of sizes, configurations, thicknesses, and purposes. Their business often comes from research organizations or from businesses that use very limited numbers of vessels for special applications, which cannot typically be obtained off the shelf. While using many of the same tools and machines as other manufacturers, fabrication of each vessel is planned as an individual project.

The volume of information that engineers need to absorb to work in the current environment has increased, and at the same time the opportunities for experience in manufacturing environments have in many cases decreased. The authors of this book, recognizing a dearth of readily available information in the field, felt that it would be useful to share their long experience in pressure vessel fabrication with a consolidated reference in this area.

The topics in this book cover various processes required in the fabrication of process equipment. This material will give the reader a broad understanding of the steps required in fabricating pressure vessels and includes such topics as cutting, forming, welding, machining, and testing. Each chapter presents a specific fabrication step and details its characteristics and requirements. Equations, charts, tables, figures, and other aids are presented, where appropriate, to help the reader implement the requirements in actual fabrication. Additional data is presented in the appendices at the end of the book as an aid to the user.

Acknowledgments

This book could not have been written without the help of many people.

Many thanks to Marks Brothers with the help of Nathan Marks and Dean Marleau, and to Harris Thermal with the help of Eric Groenweghe, Arnold Fuchs, Brice Parrow, Josh Thatch, and Jim Nylander for spending their time with the authors to access various pieces of equipment and machinery in their fabrication plants. Thanks is also given to Nooter Construction with the help of Chris Cimorelli, Mike Bytnar, and Steve Meierotto for providing many pressure vessel photographs.

Historical photographs were obtained with the help of Pat Hachanadel and Patrick Wayne of Los Alamos National Laboratory, Zhili Feng of Oak Ridge National Laboratory, and Nolan O'Brien of Lawrence Livermore National Laboratory.

Susumu Terada of Kobe Steel in Japan and David Anderson of Doosan Babcock in England helped with metric unit conversions. Sam Greulich lent his artistic talent to restoring some old photographs and Mike Kelly assisted in obtaining material cost comparisons. Bud Brust provided welding residual stress plots.

Many of the weld symbols in the book were obtained courtesy of the American Welding Society with the help of Peter Potela. Photographs of weld equipment were supplied by CB&I Storage Solutions with the help of Koray Kuscu and Dale Swanson. A photo of a pipe beveler was supplied by E. W. Wachs with the help of Keith Polifka.

Appendix I contains shackle dimensions obtained courtesy of Crosby Corporation with the assistance of Michael Campbell. Bigge Corporation with the assistance of Randy Smith supplied a photo of a heavy transporter.

Lane Barnholtz of Clemco and Gavin Gooden of Blast One gave permission for publishing a blast room and a paint room photo, respectively.

Special thanks are given to Gabriella Robles of Wiley and Mary Grace Stefanchik of ASME for their expert help, without which this book would not have been possible.

1

Introduction

1.1 Introduction

The fabrication of process equipment involves a straightforward but complex sequence of operations that is developed and refined by industry or by individual manufacturers over the years. Each successful manufacturer of such equipment will have its own ways of working and will differ from others in the details of how processes are performed and level at which documentation becomes formalized, but the essential elements remain the same.

Some fabricators of process equipment have a standard product line, either available off the shelf or made to order. Those that do not have a product line and that bid for individual jobs within their field(s) of expertise are referred to as job shops or custom fabricators. Whether fabricating a piece of equipment on a job shop basis or producing a standard product, the organization must develop a design, procure or produce the component parts, and assemble them, all the while ensuring quality and maintaining quality assurance documentation.

1.2 Fabrication Sequence

To provide a background for the remaining chapters, which delve into the details of each aspect of pressure vessel fabrication, consider a large pressure vessel for a process application. The fabrication process flow proceeds as follows:

The pressure vessel manufacturer receives a request for quotation from the procurement organization for a petro-chemical plant. A job file will be created and a project engineer or estimator will be assigned.

If the design of the pressure vessel is fully defined by the purchaser, including all dimensions, materials, interfaces, etc., then the bidding process will be straightforward. However, if just interfaces and process requirements are provided, then this will allow the fabricator leeway to use its particular experience, efficiency, or capability. Either way a job file will be created to document what is required and what has been accomplished. This allows keeping track of preliminary analyses, decisions, and details, and it ensures that work and research such as sourcing of unusual components done at the bidding stage does not have to be repeated if the company is successful in getting the job.

More sophisticated customers, such as oil refineries and larger chemical companies, may provide a fairly refined design and will often have their own design specification. Such company

Fabrication of Metallic Pressure Vessels, First Edition. Owen R. Greulich and Maan H. Jawad.

specifications usually include requirements that may increase the cost of fabrication over that of a minimal design. The further details are usually based on company experience indicating that long term overall costs are reduced by the additional requirements. Others will leave much of the design to the fabricator, just defining interfaces and process requirements such as temperature, pressure, volume and envelope dimensions, and chemical compatibilities. Or they may provide the design of a vessel that is being replaced but still allow some design and fabrication flexibility for the new vessel.

If only limited design information is received, then a preliminary design must be roughed out to produce a cost estimate. Even if the design is fully defined, the fabricator will still need to resolve items including many of the weld details, weld processes, and things such as whether a nozzle is fabricated using a pipe and a flange or a long welding neck (LWN) flange. Not every detail needs to be worked out at this stage, but there needs to be sufficient resolution of the design that a reasonable cost estimate can be produced. Accuracy should be precise enough that the company can be confident of making a profit on the job and at the same time be competitive on price and delivery. Extra time invested at this point can often find ways to keep overall fabrication costs down, resulting in a higher bid success rate and helping ensure that no unpleasant surprises occur after receipt of a contract.

This book will not address the details of developing a bid on process equipment except to note that accurate bidding involves a thorough understanding of what it takes to produce the required equipment, and enough clarity in the estimate to ensure that all aspects of the effort are covered. Fabricators with standard products may use sophisticated internal estimating programs to develop pricing information. Other fabricators rely on the background and experience of their estimators to put together material and labor costs for each and every job, and some use standard industry programs to assist.

Once the order is actually received, the design and process flow will be finalized and a quality assurance package begun.

If not already accomplished at the bid stage, trade-offs will be assessed, such as stronger material or additional inspection such as radiography or ultrasonic testing to allow increased joint efficiency to reduce vessel wall thickness. This can reduce total material weight and the amount of welding required. Some parts of the design may be decided based on shorter lead times for one option than for another. Some are based on the particular equipment and capabilities available within the company. Others are based simply on cost. After all aspects of the design have been defined, a detailed material list will be produced. This may be done using in-house or specialized industry software, or it may be done by hand. Any material not available from stock must be procured, and process flow may be adjusted accordingly.

It is usual to identify long lead time items and contract for them immediately. Typically, these include heads (if not made in-house), special valves, filters, and forgings, any mill orders, anything made of exotic materials, and anything else that was identified during the bid stage as requiring extra time. Some custom manufacturers may stockpile such items as exotic materials and exotic weld supplies in anticipation of future orders to minimize lead time and get an edge over their competitors.

Weld procedures may be developed at this time if they are not available, as coupons can then be produced and tested in parallel with the wait for materials and components without extending the overall schedule.

Also, at this time quality assurance personnel develop plans for the required inspections, tests, and hold points that will take place throughout fabrication. This will include a number of dimensional inspections, verification that reported test results are compliant with applicable requirements, verification of process control of welding and other processes, and review of radiographs and other nondestructive examination (NDE) results. Review by the Authorized Inspector will be included if the work involves an ASME code stamp, which for a pressure vessel it almost certainly does.

Additionally, this is when the layout department is likely to become involved. The layout department personnel have a thorough understanding of geometry, trigonometry, fabrication, and some of the behavioral characteristics of materials while they are being fabricated. They are trained in how to lay out intersections of such items as pipe or cone sections with heads or shells. The layout department will plan for efficient use of materials, produce detailed layouts for the heads and shell sections, mark locations and contour cutouts for nozzle installations, etc. The first part of this effort takes place in the flat, when shell sizes and weld bevels are prepared. Other parts occur throughout fabrication. Shell layout will include allowance for weld shrinkage.

As the material arrives, it will go through a receiving inspection and be checked for compliance with specifications, with material mill test reports and other documentation placed in the quality assurance file. Early arrivals are often stockpiled but segregated from non-code materials that have not gone through quality assurance acceptance until enough components are available to begin work and continue through the flow without unnecessary starts and stops. Even if shell plates are available from stock, it is usual to postpone cutting them until the heads arrive so that actual head dimensions can be measured, or to request a "taping" (a measurement of the circumference) from the head manufacturer prior to shipping. This allows the shell diameter to be adjusted if needed, from its nominal dimension to permit an optimal fit to the heads. This slight adjustment to the shell circumference is often necessary since it is difficult to bring the head circumference to a precise dimension during forming due to the three-dimensional nature of the head. The difference is usually fairly insignificant, but even an eighth of an inch (3.2 mm) of diameter can make fit up and welding more, or less, efficient.

The shell sections are rolled subsequent to cutting to size and beveling for welds. Their longitudinal joints (straight seams) will be tacked into alignment, and then welded. The heat and stresses of welding will cause a certain amount of shrinkage and distortion. This may be, to some extent, controlled by alternating weld passes on the inside and outside of the weld. However, if distortion is excessive after welding, the shells will be reworked with hydraulic rams or will be re-rolled to bring them back within tolerance. Working from the zero point on each shell course, nozzles and other appurtenances will be laid out full scale on the plates, with indications as to weld preps as needed.

The *fitter* assembles the shell courses, referred to as "courses" or "cans," to each other and to the heads. Circumferential shell welds are usually welded on positioning rolls to allow welding to be performed in the flat position, which is the preferred position because it is the easiest orientation for producing high volumes of high-quality weld. Next, nozzles will be fabricated, and reinforcing pads laid out, cut, and formed, then fit in place and welded, either preceded or followed by fitting and welding of supports.

Note that while this description looks simple, the work involves a high level of training and skill on the part of the layer out, fitter, and welder. The fitter has the job of fitting and tacking together the assembly within fairly tight tolerances and the welder must be able to produce hundreds of feet of weld with the least amount of rejectable indications.

Nozzles, for example, must be laid out and then fit accurately in the holes cut in the shell sections, correctly oriented, with the proper projection, and with bolt patterns on the flanges in the correct orientation. This is essential in order they fit correctly with piping that may already exist in the field or that may be assembled elsewhere. The fitter will typically install "spiders" and other braces to minimize distortion such as shells going out of round or nozzles sinking excessively during welding. The welder also has a part, controlling his welding within the parameters of the Welding Procedure Specification, in accordance with which he has already demonstrated the ability to produce top quality welds. The welder also maintains preheat and interpass temperatures and speed of progression, and makes in process adjustments as his experience dictates to maximize productivity, avoid weld defects, and control distortion.

After welding is completed, the welds will be inspected. Common inspection methods include the following:

1) Visual.
2) Magnetic particle for ferromagnetic materials such as steel.
3) Dye penetrant for either magnetic or nonmagnetic materials.
4) Dimensional inspections.
5) Radiography.
6) Ultrasonic examination.

It is usual to do these inspections before any required post weld heat treatment (PWHT), even if they are required after PWHT as well, so that any needed repairs can be completed prior to final heat treatment. This is because repair of a defect found after PWHT will normally require repair and an additional heat treatment. Such additional heat treatment can be costly as well as have the potential to reduce material mechanical properties.

After PWHT and required final NDE, any final machining that is needed takes place. Intermediate machining processes may already have taken place if thick welds require J-grooves or if unique machining is required because of special configurations. Also, for vessels such as heat exchangers requiring tubesheets or other special components, machining of these tubesheets and components is accomplished in parallel with other work on the vessel.

Next, the vessel will be pressure tested when inspections and NDE demonstrate compliance with all requirements. Pressure testing is done either by a hydrostatic test, which is preferred for safety reasons, or by a pneumatic test. Although failures are not anticipated, access is usually restricted during such tests due to the potentially high levels of stored energy. This is especially true during pneumatic tests, but even though water is considered an incompressible fluid, the energy stored by compression of water or other liquid and any trapped air and the stretch of the metallic shell can result in a significant hazard during such tests.

Once the pressure test is completed and all other quality requirements are verified, the vessel is ready for final cleaning and application of any required paint, conversion finishes, anodizing, or other surface treatments. A name plate describing various vessel parameters is then attached to the vessel, indicating compliance with the applicable code and other requirements.

Finally, with fabrication, inspection, NDE, and testing completion, and coatings applied, the pressure vessel is readied for shipment. Shipment may include low level pressurization with a clean, dry, inert gas, sometimes referred to as "pad pressure." It is used to ensure that nothing is sucked into the vessel on cold days and to prevent condensation. Shipment also includes blocking or cribbing, special supports, possible packaging, and tie-down on the truck, railcar or barge for shipping.

Once the product arrives at the customer's facility, it will often undergo further inspection to ensure that all of the requirements have been met and that there has been no damage during shipping. The Quality Assurance package, when supplied, will be reviewed in detail and placed on file. Only then can the vessel be installed and put into service.

1.3 Cost Considerations

The cost of a pressure vessel is a function of many parameters. In areas where labor is costly, it is often the biggest single factor, but many decisions by both the designer and fabricator influence overall cost. The most effective design from a cost standpoint will be one in which schedule, cost and availability of materials, cost and capability of labor, inspection options, and available equipment and tooling are all considered. In addition, short versus long term product cost considerations may need to be discussed with the customer.

It follows that the designer will either have some experience in all of these areas or will work closely with people who do. Similarly, the shop management will be familiar with a wide range of production techniques, including means of cutting and machining, forming, fixturing and fit up, welding, heat treatment, inspection and testing, cleaning, painting and other surface treatments, and packaging and shipping options and requirements.

If large numbers of vessels of the same or similar designs are fabricated, design and fabrication choices will be different from those involving fabrication of a single unit.

The particular capabilities of a vessel fabricator often make one variation of a design more cost effective than another, and if the designer is not directly associated with the fabricator, it makes sense for these two parties to discuss design options with an eye on cost reduction.

This book is not about fabrication cost estimating, and this chapter does not address actual product cost. It, however, addresses a number of considerations affecting the cost of an overall pressure vessel fabrication to help the user, designer, and fabricator make judicious choices regarding design and fabrication approaches.

1.3.1 Types of costs

For a business, one way of dividing costs is to separate them into either capital or operating costs. Capital costs are the one-time expenses such as purchase of land, construction of a plant, and major equipment purchases that are expected to last a long time. A small hand grinder, for example, would not be considered a capital cost, while the costs of constructing a building or purchasing a large forge would be. Operating costs are the other costs of being in business, including wages and salaries, real estate expenses (rent, taxes, etc.), materials, furniture, consumables, maintenance, etc.

This way of looking at expenses is useful in understanding what things cost overall, and it might be enough for a company with a single product line. For calculating and controlling costs of production of individual products in a job shop, it is usually easiest to work with burdened labor rates that represent the hourly cost of performing an operation, plus material and other direct costs of a particular job, plus capital costs. The burdened labor cost includes such items as direct wages, cost of vacations and holidays, social security and other tax cost, sick leave, and pension or 401k plans.

Some companies use a single rate for essentially all personnel whose time is charged to a job, while others charge a rate that varies by function or even by the individual assigned to the job. Sometimes costs are broken down further to identify and charge for specific assets outside of the burdened labor rate. This is most likely to occur in a case in which an asset of particularly high value is used only on some jobs. In such a case, dividing its cost among all jobs would subsidize those jobs that require this equipment at the expense of those that don't. The result would be extremely competitive prices on the jobs requiring this equipment, but a lack of competitiveness on those that don't need it.

Companies arrive at burdened labor and equipment rates in different ways, but the intent is to allocate costs in a way that allows bidding jobs, recovering costs, and making a reasonable profit. Because the fabrication environment is competitive, it is important to understand enough about the individual cost elements that (1) wise trade-offs between design approaches can be made to ensure competitiveness, and (2) accurate total cost of a particular fabrication can be identified for pricing purposes and to ensure a reasonable profit.

1.3.2 Design choices

1.3.2.1 Major cost decisions – long term choices

Some design choices must typically involve the customer because they involve significant product cost differences that can only be amortized over the long run. An example of this occurs with a vessel that will contain a corrosive medium. In this case, material choices may make a significant difference in short term vessel costs. A vessel might be fabricated with a corrosion allowance, anticipating that at the end of some term (approximately five years, for example) the vessel will simple be replaced. Another approach would be to fabricate it entirely of a material that does not undergo corrosion in its particular internal and external environments, or to clad it with such a corrosion-resistant material. The cost of fabricating a pressure vessel of high alloy steel or other material may be significantly greater – perhaps double or more – than that of a fabrication using steel. If a more expensive product allows essentially unlimited life versus five years for the steel pressure vessel, then amortizing the cost of the single vessel versus initial vessel purchase plus replacements, and downtime and labor for the replacement, can make the farsighted decision attractive. Whichever way this decision goes, all other cost issues still apply.

1.3.2.2 Labor–material trade-offs

Some choices regarding materials simply minimize material costs. Others have the additional advantage of reducing labor. A third category reduces costs by eliminating whole operations. A fourth category is to increase labor in situations where labor cost is minimal and material cost can be reduced without a comparable increase in cost of labor.

1.3.2.3 Selecting a less expensive material

Cost reduction by minimizing material cost is represented by a situation in which two different metal alloys of different costs (per unit weight) result in the same wall thickness. This occurs when either the wall is fixed (for example, when a minimum wall is required for handling or for stiffness reasons), or when rounding from the required minimum wall to the next stock thickness results in the same fabricated thickness for both. If there is no other operational reason to use a more expensive material (SA 516-70, for example, rather than SA-36), then the obvious choice is a less expensive one.

1.3.2.4 Selection of a material with a higher allowable stress

In a given class of materials, using one with a higher allowable stress is beneficial in pressure vessels with high pressures and larger diameters. For example, use of SA 516-70 rather than SA 285C reduces wall thickness. The cost of material may remain about the same, since SA 516-70 is more expensive per pound than SA 285C, a fact somewhat balanced by the lesser amount of material used. The reduction in wall thickness reduces cost in multiple ways, however. The time needed for rolling the vessel shell and forming the heads is less. The thickness of welds and therefore weld volume and welding time are diminished. Handling costs may be less. And depending on the fluid medium, the reduction in vessel weight might lead to smaller or thinner supports or saddles.

1.3.2.5 Component selection to eliminate operations

Design changes to eliminate whole operations are options to be considered. This category includes selection of vessel diameter to coincide with standard pipe sizes and the use of integrally reinforced designs. Figure 1.1 shows a detail of a hemispherical head where additional material on the outside is left in place to minimize machining cost.

If a shop rolled and welded vessel shell or nozzle can be replaced by a piece of off-the-shelf pipe, whether seamless or welded, then the costs for layout, crimp, and individually rolling that shell are all included in the pipe cost, which is typically produced in a dedicated facility that only produces pipe, but does it very efficiently. When this can be done, the material cost of the completed shell section is often little more than the material cost of the unrolled shell plate. A further benefit is that standard caps may then be available for use as vessel heads. These, too, being mass produced, will likely be significantly less expensive that custom-formed heads.

When nozzles beyond a certain size penetrate a vessel wall, reinforcement is required to take the pressure loads that would otherwise have been transmitted through the material cut from the shell or head for placement of the nozzle. The ASME code puts limits on what material may be counted as contributing to this load carrying capability. Simple area replacement is typically used, provided that the reinforcing material is of strength equal to or greater than that of the material it is replacing. There are numerous ways of providing this material. Because the code allows essentially any material within a certain distance to be counted, any excess material in the shell itself, the nozzle wall, the weld, added reinforcing pads, or shell inserts, may be considered.

The best means of providing reinforcing beyond that inherent in the design is often fairly obvious, but in some cases a cost estimate for more than one approach may be needed to evaluate the trade-off.

If a vessel has a limited number of penetrations requiring reinforcement, accepting the labor and material cost of providing reinforcement on a few nozzles may be inexpensive compared to providing a heavier shell that results in an integrally reinforced design. When a vessel has many nozzle

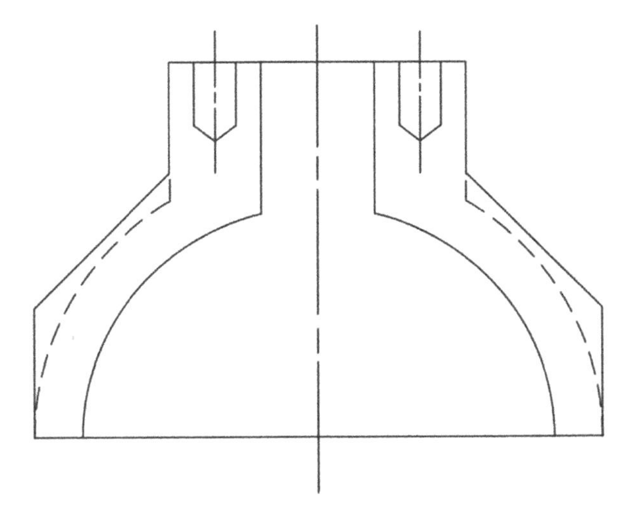

Figure 1.1 Outside machining of a hemispherical head

penetrations requiring reinforcement, however, the labor associated with providing that reinforcement may far exceed the additional cost of a heavier shell wall and thicker shell and nozzle to shell welds. If most or all of the nozzles requiring reinforcement are located in the same area, it may make sense to make one shell course thicker than the others to provide integral reinforcement.

Another way of providing additional nozzle reinforcement when a flanged nozzle is required is the use of LWN flanges. If the nozzle protrusion is not excessive, then unless the cost of labor is extremely low or the cost of material extremely high (e.g., high nickel materials for their corrosion resistance or high temperature strength), it will almost always be more economical to use an LWN flange than to add a reinforcing pad. The neck of an LWN flange normally has an outside diameter equal to the hub diameter of a slip-on flange, and it may be ordered in a variety of lengths. Thus, particularly if it is acceptable to allow the nozzle to protrude into the vessel, an LWN flange can almost always fulfill the need for additional reinforcement. While the cost of an LWN flange is significantly greater than that of either a slip-on or a welding neck flange, it has the advantage of eliminating the following costs: flange to nozzle weld, reinforcing (or insert) plate, reinforcing or insert plate layout, forming of reinforcing plate, drilling and tapping of reinforcing plate vent hole, fit up of reinforcing plate, and welding of reinforcing plate both to the shell and to the nozzle.

1.3.2.6 Enhanced inspection for higher joint efficiency

Enhanced inspection to increase joint efficiency can result in a significant reduction in wall thickness on a heavy wall vessel. This, in some cases, sufficiently reduces the wall thickness to allow the use of the next smaller stock thickness, thereby reducing material and other fabrication costs. When inspection has not been performed to allow 100% joint efficiency of shell longitudinal or head welds, however, then locating shell longitudinal or head welds so that welds aren't included in zones used for reinforcement may, in some cases, be enough to eliminate the need for extra reinforcement, since the excess material in the shell can be counted based on the 100% joint efficiency of the parent material.

Major considerations in deciding whether to perform inspections to reduce other costs include the cost of the inspection, the anticipated labor and material cost savings, and the level of confidence that the welds will pass inspection the first time. If inspection shows that weld repairs are required, all savings in labor and material may be wiped out by the cost of repairs and reinspection, resulting in no benefit to the fabricator and a loss in terms of schedule, and tolerances may be affected as well.

Example 1.1 This example illustrates an actual vessel for which the design approach eliminated a large number of operations as well as fabrication risk by using a much heavier wall than originally specified.

The heavy wall vessel shown in Figure 1.2 is 16 ft long, 30 in. diameter, 4 in. nominal wall, with flat bolted heads, 88 total penetrations, and the full shell length machined inside. Figure 1.3 shows the side views of the same vessel. The vessel might have been fabricated of much thinner material, but was fabricated this way to reduce cost.

The vessel was originally designed using a 1-3/4 in. thick shell with a number of heavier shell plate inserts with blind drilled and tapped holes for attaching instrumentation. The original design also had an added heavy section at each end with drilled and tapped holes for installation of cover flanges. The fabricator evaluated four approaches before making a proposal. Each approach included the large nozzles welded into fabricated shell sections. The four approaches were (1) as designed originally; (2) a centrifugal casting with flats machined and drilled and tapped for small

Figure 1.2 A vessel fabricated with a heavy wall to minimize cost (*Source:* Los Alamos National Laboratory)

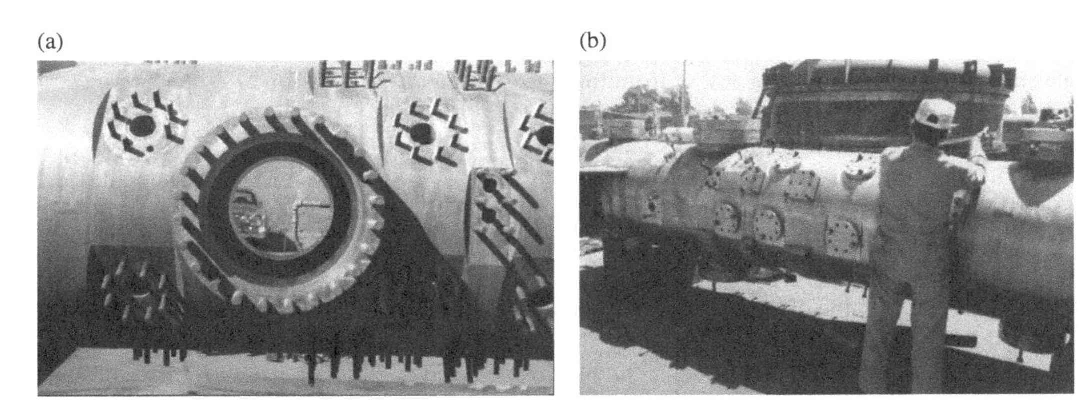

Figure 1.3 Side views of the vessel in Figure 1.2

penetrations, with the internal surface machined after insertion of the large nozzles so as to meet internal tolerance requirements; (3) a single piece, trepanned, heavy forged cylinder with the same approach as (2) for nozzle penetrations, and (4) a rolled and welded heavy plate wall shell with the same approach as (2) for nozzle penetrations.

The rolled and welded design proved to be the least costly. All of the heavy walled designs eliminated two circumferential welds at the ends, as well as the cost of layout and cutting of holes, and welding in the plates for the small openings. They also reduced the risk of distortion by minimizing the amount of welding required. The centrifugal cast and forged designs had higher costs for the basic cylinder than did the rolled and welded design.

When the user recognized the costs and the tolerance risks associated with welding a large number of nozzle plates, the rolled and welded design was accepted. The rolled and welded shell was produced by a pipe fabricating shop, helping to manage costs.

1.3.2.7 Process choices

Often decisions about cost of fabrication depend on the quantity of product being produced. Vessels will be more economical to produce if processes are optimized, but sometimes the cost of optimization is not warranted. For a single vessel, or even a small number of vessels, the cost of procuring forming equipment and optimal welding equipment and costs of developing tooling will likely exceed any profit on the job. Sometimes even setting up existing tooling for a vessel will not pay for itself, and it will be less expensive to fabricate the product using less efficient means but with essentially no initial setup cost.

1.3.2.8 Forming

Vessel fabricators will usually use one of four ways of making vessel shells.

First, as noted in Section 1.3.2.3.3, the least expensive way of producing a vessel shell is almost always to purchase a standard size of pipe, if it is available. This is usually true even for rolled and welded pipe.

Second, if large quantities are to be produced, is to develop dies and form shell sections using a large press. The cost of this tooling, even ignoring the cost of a press large enough to perform this type of work, is high, and it will only be justified by large quantities of product. For large quantities, however, this approach allows the production of shell sections (usually halves or thirds) with a single stroke of a press. Even the cost of installing the dies may be fairly high, and may not be cost effective for single vessels.

Third, rolled shell sections may be produced using forming rolls as described in Section 4.3, followed by placement of longitudinal welds. This technique is especially useful for diameters and shell lengths that can be rolled out of a single piece, since it efficiently produces cylinders requiring only a single longitudinal weld.

Finally, shells can be produced on a press brake. This is usually more labor intensive than either forming rolls or forming dies, but for small quantities of shell courses or if control of all aspects of the production is needed, it can make sense. For a company possessing a press brake but not forming rolls, rolling of pressure vessel shells can be accomplished in either of two ways: first, the shell can be "bump rolled" on the press, usually in sections, and second, the company can either buy the finished product or send shell material to a fabricator possessing a set of rolls for rolling. If the first approach is taken, the labor cost for bump-rolling itself is probably greater than that for a product produced using forming rolls, but the cost of extra layout, pre-crimping or cutting off extra material allowed in place of crimping (thicker sections), and shipping the product both directions are eliminated. Thus, for a single product, bump-rolling may be adopted, while the second approach is likely if a number of shells are required and the roll setup costs can be better distributed over the number of shells produced.

If a very large quantity of the product is to be produced, particularly if it is to be produced on an ongoing basis, then a company may invest in a set of forming rolls. The cost of the rolls is then amortized over the life of the product line, costs go down, and the company has a new capability.

1.3.2.9 Fixturing

As in the case of tooling for rolling a shell course, the value of fixturing for fit-up and assembly is often limited for production of a single or low volume product, but as production rises the cost of fixturing may remain constant while the benefits increase.

For a single shell, tack-welded lugs, wedges, and clamps are often all that are needed and used for alignment, though in some cases hydraulic rams may be used. Lugs will be flame cut out of stock plate and welded in place – number of lugs, thickness of lugs, and amount of weld vary depending on how much "persuasion" the fit up is anticipated to require. After pushing shell edges (for example) into alignment so that they can be tacked together for welding, the wedges are removed, the lugs are ground free or knocked off, and any damage to the plate surface is repaired and ground flush.

Compare this scenario to the shop that has many shells of either the same size or a small number of predictable sizes to be produced. In this case, design and construction of fixtures to accomplish the same thing can cut individual shell fit-up time significantly. Once shell fit-up fixtures are constructed, the following might take place: The rolled plate section is placed on the fixture. Portions of the fixture will be swung into place and pinned. Hydraulic rams will push the sides and ends of the shell to bring the longitudinal joint into rough alignment. Other rams are used to bring the edges into the same plane. The side rams may be further adjusted to provide the proper root opening. The longitudinal seam is tacked, the rams are released, the fixture arms are moved back to provide space to remove the shell course, and the next shell is brought in.

There is often a sizable investment in a fixture such as this, and any such equipment that is developed will occupy shop space, so it shouldn't be done without consideration of the returns. However, this investment can cut what may be an 8–16 (or even more) hour job to a matter of an hour or so.

1.3.2.10 Welding

A similar situation arises in the case of manual versus automated welding. A number of different processes may be used to produce welds. Each has its benefits and drawbacks. Chapter 7 discusses welding in detail and provides a comparison of various welding processes, including deposition rates. Items to consider include equipment and setup costs versus the benefits of more efficient placement of welds, design for production runs rather than individual fabrications, and weld configurations, such as narrow welds to minimize weld metal required and residual stresses.

1.3.2.11 Hydrotesting

Pressure testing is most often performed using water or other comparably incompressible fluids. Hydrotest of a single vessel is usually accomplished by filling it, pumping to pressure, holding, and draining the vessel. For single vessels, the water is usually dumped after use, and pumping is accomplished using a small positive displacement pump.

If the quantity of vessels produced in accordance with a particular vessel design is such that multiple vessels are tested daily, then it is common to set up test fixtures and to salvage and recycle the water. As with other means employed to reduce per unit cost, the savings must be weighed against the up-front cost of fixturing, constructing a reservoir, etc.

1.3.3 Shipping

For most pressure vessels, the cost of shipping is not more than a few percent of the total cost, yet even that is enough that it should be considered in the price of the product. For products that are extremely large or extremely heavy, however, that percentage may increase.

The size and load capacity of standard rail cars facilitate shipment of many vessels that might require permits as wide, long, or heavy loads if shipped by truck. Rail rates (per pound) are often much less than truck rates. Barges even more so, if the size of the product justifies them. This is especially the case if permits or special routing are needed for trucking. Rail shipments often take longer than trucks, however, due to the way that rail traffic is routed.

In any case, unless the estimator is confident of knowing shipping costs with a good degree of accuracy, it would be good to verify costs with shippers prior to bidding a job. See Chapter 11 for more information regarding shipping.

1.3.4 General approach to cost control

Effective management of cost involves making trades based on actual costs of the delivered product. It therefore requires assessment not only of material and labor costs but also the cost of shipping. This will be especially important for shipment of large and/or heavy fabrications. These are likely to require permits and may require special equipment and routing, raising costs far above the usual cost per pound for shipping.

A general rule, with some exceptions, is that labor costs outweigh material costs and that labor is therefore the area most ripe for cost reductions.

If, for example, material represents 10% of the cost of a product, then any reduction in material costs must clearly be less than 10% of the cost of the overall product. This could be the case for a carbon steel vessel with complex fit-up. For this, vessel reductions in labor likely do not increase material costs significantly and should be considered as ways of reducing overall costs.

A vessel fabricated of certain nickel alloys, titanium, or zirconium, on the other hand, will have very high material costs. In this situation, reducing material costs may be effective in reducing overall costs.

Seeking only the lowest hourly rates risks, at times, finding the lowest productivity, but where skilled labor is acquired cheaply, overall product costs may be low.

Thus, it is important to assess the overall cost of a delivered product. When a design change is made, whether or not with the intent of reducing costs, overall costs must be reassessed. It will sometimes be found that the change results in even greater savings than anticipated, but it will also sometimes be found that the savings are eaten up by increases in other areas.

1.4 Fabrication of Nonnuclear Versus Nuclear Pressure Vessels

The fabrication of nuclear components such as vessels, pumps, valves, piping, and storage tanks in the United States must meet the requirements of Section III Division 1 of the ASME Boiler and Pressure Vessel Code as well as the rules of the U.S. Nuclear Regulatory Commission (NRC). This book is written for nonnuclear applications. While the general fabrication processes such as forming, machining, and welding are the same for both nonnuclear and nuclear components, the quality control process is different regarding the details of these operations.

Nuclear components constructed in accordance with the ASME code are considered in “classes” that are used to construct pressure equipment in accordance to its relative importance to safety. The three most common classes are Class 1, 2, and 3.

Class 1 components, including vessels such as reactor vessels, pressurizers and the primary side (tubes) of steam generators are exposed to radioactive coolant fluid, and they consequently are considered to bear the highest importance to safety. Class 1 vessels therefore require the most stringent levels of quality control for the various fabrication operations, compared to Class 2 and 3 vessels. Class 2 vessels generally resemble ASME VIII-2 (editions prior to 2007) requirements, while Class 3 components are generally similar to VIII-1 requirements.

All three classes of nuclear components are subject to strict quality control during construction. Quality assurance requirements for nuclear applications are provided in Article 4000 of the ASME Nuclear Code Subsection NCA (General Requirements for Division 1 and Division 2), along with its references to ASME NQA-1 (Quality Assurance Requirements for Nuclear Facility Applications, Part I and Part II).

Some of the details required during fabrication of Class 1 nuclear vessels, pumps, valves, and piping are as follows:

1) All materials must be provided with documentation showing Certified Material Mill Test Reports. The location of the test specimens taken from the mill plate must be identified for traceability. When a piece of the mill plate is cut out for use as a vessel part, its location in the mill plate is recorded and identified. Depending on the material, the method of removing the plate, such as machining or burning then grinding, may need to be recorded. The type and identification number of the grinding wheel may also need to be recorded. These requirements must be met for each piece of material in the pressure vessel.
2) Hot forming during fabrication must be qualified and documented. The effect of hot forming on material properties and final thickness for some materials may have to be recorded.
3) All weld electrodes and wires must be identified, including recording of heat numbers, location where used, and properties, with full documentation and traceability.
4) Each weld in the vessel must be identified with regard to the location as well as WPS and PQR.
5) All weld repairs on materials and welds during fabrication must be identified and documented. Such documentation must include weld procedures and their effect on properties such as strength and impact values.
6) Examination of welds and components must be documented.
7) Records of operators, equipment used, calibration of equipment, and results of tests must be maintained.

This limited sample of Class 1 requirements illustrates the extent to which quality control and documentation required for nuclear components exceed those for nonnuclear components.

1.5 Units and Abbreviations

This section describes the unit conventions used in this book. Consistency of units is important in communicating technical data. While throughout the engineering field there is generally good comprehension of units, conversions, and their use, various industries and companies have their own conventions. This can result in confusion, mistakes, and accidents. An example is when NASA lost a 125-million-dollar Mars Climate Orbiter when the navigation team at JPL used the customary NASA metric units in its calculations of acceleration readings while Lockheed Martin that built the spacecraft provided the vital acceleration data in the English units. And as the Los Angeles Times put it, “In a sense, the spacecraft was lost in translation.”

Rates will use abbreviations, not followed by a period, and a slash rather than “per.” Example: “inches per minute” will be written as “in./min”, and in metric “mm/min.”

While in machining the cutting speed is generally referred to as "surface feet per minute," for consistency with other sections this will be written as "ft/min" with no periods at the ends of the abbreviations.

"Micro" will be represented by the lowercase Greek letter mu, written as μ.

English units will be used, followed by the metric equivalent. Dimensions will generally be rounded when converted, and will show no more than three significant figures. For example, "6 in." becomes "150 mm." This convention will be followed unless something is clearly intended as an exact dimension or for nominal sizes that are clearly not rounded measurements of the specified dimension. Thus, 6 in. plate is written as 150 mm, and 6 in. pipe is written as DN 150. 3/4 in. plate is written as 19 mm plate.

Appendix A provides a list of abbreviations and conversions commonly used in the pressure vessel industry.

1.6 Summary

The description in this chapter represents a brief summary of the many tasks that must be accomplished in the fabrication and preparation of a pressure vessel, as well as some discussion of aspects of cost reduction. This description and associated cost control techniques can be applied both to less and more complex vessels. The same issues occur and the same processes are applied. The following chapters provide a more detailed look at each aspect of this process, and the appendices provide a compendium of information that is often needed during the design or production of pressure vessels.

2

Materials of Construction

2.1 Introduction

Pressure vessel components are normally fabricated from ferrous alloys, nonferrous alloys, or composite material. The type of material selected for construction depends mostly on the type of service and the construction code being used. Pressure vessels operating above 15 psi (100 kPa) internal pressure are generally required to be built in accordance with the ASME Boiler and Pressure Vessel Code (BPVC) [1] by most jurisdictions in the United States. Accordingly, these vessels must be constructed of materials approved by the ASME code. Vessels operating below 15 psi (100 kPa) do not need to comply with the ASME rules and the materials of construction are not necessarily approved by the ASME code. However, other rules specified by the jurisdiction may apply.

Specifications for ferrous alloys in the United States are designated by ASTM [2] with a letter A such as A-516 material, while specifications for nonferrous alloys are designated by a letter B such as B-168 material. The ASME pressure vessel code adds a letter S to the specification such as SA-516 and SB-168 to indicate the material is approved by ASME for use in pressure vessel construction. The majority of the SA and SB specifications in the ASME code are identical to their corresponding A and B specifications in ASTM although there are some cases in which the ASME standards are more stringent. Accordingly, the fabricator of a pressure vessel built in accordance with the ASME code needs to specify SA or SB materials for construction.

All ASTM materials used in manufacturing pressure vessels are also designated by a Unified Numbering System (UNS) number. The UNS number consists of a prefix and a five-digit number. The UNS prefix designations for commonly used materials and alloys in pressure vessels are given in Table 2.1.

Hence, carbon steel plate SA-516 Grade 70 is designated as UNS K02700 and nickel alloy plate SB-575 type Hastelloy C-276 as UNS N10276.

Almost all of the pressure vessels fabricated in accordance with the ASME code are constructed of one of three ferrous material categories or one of five nonferrous material categories. The ASME BPVC lists over 1500 ferrous alloys and over 1000 nonferrous alloys that can be used in the construction of pressure vessels. The properties of these materials and their categories are explained in the remainder of this chapter.

Fabrication of Metallic Pressure Vessels, First Edition. Owen R. Greulich and Maan H. Jawad.

Table 2.1 Unified Numbering System (UNS) designations

UNS designation	Alloy type
A00001 to A99999	Aluminum and aluminum alloys
C00001 to C99999	Copper and copper alloys
F00001 to F99999	Cast irons
G00001 to G99999	AISC and SAE carbon and alloy steels
J00001 to J99999	Cast steels
K00001 to K99999	Miscellaneous steels and ferrous alloys
N00001 to N99999	Nickel alloys
R00001 to R99999	Reactive metals and alloys R05xxx – tantalum alloys R5xxxx – titanium alloys R6xxxx – zirconium alloys
S00001 to S99999	Stainless steels
W00001 to W99999	Welding filler metals

2.2 Ferrous Alloys

Ferrous alloys used in pressure vessel manufacturing are divided into three categories [3]:

1) Carbon steel.
2) Low alloy steel (Cr–Mo steels).
3) High alloy steel (stainless steels).

The properties of the steel are affected by the chemical composition and heat treatment during production of the steel as well during fabrication.

2.2.1 Carbon steels (Mild steels)

The majority of vessels manufactured worldwide are made of carbon steels, also referred to as mild steels. Carbon steels are magnetic and their average density is about 0.284 lb/in.^{3} (7.85 g/cm^3). Carbon steels have a carbon content of less than 2% and they easily oxidize in air at room temperature, Figure 2.1. The chemical composition of regular carbon steel consists of iron with small amounts of a few added elements such as carbon, manganese, and silicon for enhanced strength and workability. Phosphorous and sulfur are also controlled. Table 2.2 lists the chemical composition for a typical carbon steel plate, SA-516 Grade 70.

When corrosion is not a factor and temperatures are not excessive, mild steel is typically an economical and practical choice. This is true from both labor and material cost standpoints.

Carbon: Added to increase the strength of steel. However, the toughness of steel tends to decrease with an increase in the carbon content.
Manganese: Increases the toughness of steel by lowering the transition temperature.
Phosphorous: Added to increase the strength and hardness of steel, but is controlled to avoid embrittlement that occurs at higher levels.

Figure 2.1 Carbon and low alloy steels

Table 2.2 Chemical composition of SA-516 Grade 70 plates

Element, product form	Composition, %
Carbon, for $t \leq ½$ in., max.	0.27
Manganese, for $t \leq ½$ in.	0.79–1.30
Phosphorous, max.	0.035
Sulfur, max.	0.035
Silicon	0.13–0.45
Iron	Remainder

Silicon: Used as a deoxidizer and to produce more uniform grain distribution and it improves toughness.
Sulfur: Increases the machinability of steel but tends to reduce toughness.

2.2.1.1 Applications

Vessels constructed of carbon steel normally operate at temperatures of up to around 650°F (345°C). Above this temperature, their strength drops rapidly and it is more economical to use other types of steels. Some typical products are as follows:

Plates: SA-285 and SA-516.
Forgings: SA-181 and SA-350.
Pipes: SA-524 and SA-671.

2.2.2 Low alloy steels (Cr–Mo steels)

These steels contain, in addition to the elements listed for low carbon steels, chromium, molybdenum, and nickel to enhance strength at elevated temperature. Their density is essentially the same as that of the carbon steels. The percentage of alloying elements is generally kept below 10%. Common products are 1.25Cr–0.5Mo, 2.25Cr–1Mo, and 9Cr–1Mo steels.

Low alloy steels are typically selected for elevated temperatures when corrosion is not a major consideration. Their higher cost is justified by their ability to withstand higher temperatures.

Commercially produced low alloy steels are supplied in the annealed condition, the normalized-and-tempered, or the quenched-and-tempered conditions. Since chromium and molybdenum increase the strength at elevated temperatures, give better toughness, and improve corrosion resistance, these steels are often used in pressure vessels operating in the range of 750°F (400°C) to 1200°F (650°C). The complexity of Cr–Mo steels requires the mill to specify additional parameters such as annealing, normalizing, tempering, and quenching to produce a product with the appropriate properties. The following list indicates the effects of various alloying elements and heat treatments with respect to low alloy steels:

Chromium: Increases the high temperature strength and corrosion properties of steel. It also increases the toughness.

Molybdenum: Generally used in conjunction with chromium to increase hardness of steel at high temperatures. It also decreases the temper embrittlement of steel. It is more expensive than chromium.

Nickel: Increases ductility and toughness of steel.

Annealing (solution annealing): Consists of heating the steel and then furnace cooling down to room temperature. The slow cooling process results in a refined grain structure and soft material. The annealing temperature depends on the type of steel. Carbon and low alloy steels are usually annealed at about 1380°F (750°C), while high alloy stainless steels are solution annealed at 1900°F (1040°C).

Normalizing: Consists of heating the steel to about 1430°F (775°C) and then air cooling it to room temperature. The purpose of normalizing is to increase the strength and hardness of steel. It also refines the grain and homogenizes the structure.

Quenching: Is a rapid cooling of steel, usually in water or oil, to increase hardness.

Tempering: Consists of additional heating of the product to reduce hardness of normalized or quenched steels and to increase toughness. An example of tempering is the ASTM requirement of a tempering temperature of 1250°F (675°C) for SA-387 Grade 22 material (2.25Cr–1Mo).

The process of producing the correct physical properties for a particular low alloy steel using a combination of annealing, normalizing, tempering, and quenching can be extremely complicated. Only steel mills specializing in the production of such steels can adequately produce the steel. The process requires coordination between the steel mill, fabricator, and user. The fabricator needs to inform the mill regarding how many times the steel will be subjected to heat treating during the fabrication process, and how many heat treatments the steel might be subjected to during the life of the vessel. The mill will use the information from the fabricator and user to decide how to produce the steel so the strength is maintained after all of the specified heat treatments. This process is complicated and requires careful coordination between the mill, fabricator, and user.

2.2.3 High alloy steels (stainless steels)

Stainless steels have a density of about 0.289 lb/in.^{3} (8.00 g/cm^3) and do not oxidize in air at room temperature, Figure 2.2. They contain a larger percentage of chromium and molybdenum than low alloy steels. Some stainless steels also include nickel in their composition.

High alloy steels are used for both elevated temperatures and for corrosive environments.

Stainless steels are generally divided into three categories:

1) Martensitic stainless steel.
2) Ferritic stainless steel.
3) Austenitic stainless steel.

2.2.3.1 Martensitic stainless steel

These steels are magnetic and contain about 12% chromium with a small amount of aluminum and other alloying elements. The chromium content gives them strength at elevated temperatures and good corrosion resistance. Type 410 is one of the most commonly used of these steels.

2.2.3.2 Ferritic stainless steel

These steels are magnetic and contain about 13% chromium. They have good strength at elevated temperatures and better corrosion resistance than martensitic steels. One of the most commonly used is type 405.

2.2.3.3 Austenitic stainless steel

These steels are nonmagnetic, have good corrosion resistance, and contain nickel plus a higher content of chromium compared to types 410 and 405. Type 304 contains 18Cr–8Ni and can be used up to 1500°F (815°C) in Section VIII, Division 1, but use of a stabilized grade such as 321 or 347 is

Figure 2.2 Stainless steel

generally wise above about 850°F (455°C) to avoid carbide precipitation. Type 316 contains 16Cr–12Ni–2Mo and has better strength than type 304. It can be used up to 1550 °F (845 °C), though avoiding use in the range of 850°F (455°C) to 1200°F (650°C) is again wise.

Stainless steel products are sometimes pickled subsequent to mill production or fabrication.

Table 2.3 Cost of various steels

Material	Cost index per lb
Carbon steel	1.0
Low alloy steel	2.0
High alloy steel	3.0–8.0

Pickling: Pickling is a chemical surface treatment of stainless steels to remove such items as stains, rust, scales, grease, oils, or embedded iron particles, and to improve the surface finish. Pickling typically has an additional favorable effect of leaving the surface of the steel in a less reactive state and is sometimes referred to as "passivating." The pickling solution usually consists of a hydrofluoric and nitric acid mixture.

2.2.4 Cost of ferrous alloys

The approximate cost ratios, per pound of weight, for the three categories of steel discussed in this section are given in Table 2.3.

2.3 Nonferrous Alloys

The ASME BPVC lists alloys of five nonferrous metals: aluminum, copper, nickel, titanium, and zirconium. The characteristics of these alloys are briefly described in the following sections.

2.3.1 Aluminum alloys

Aluminum alloys are nonmagnetic. Their average density is about 0.10 lb/in.^{3} (2.76 g/cm^3) which is about one-third that of steel. Aluminum alloys generally develop a passive oxide layer on the surface, and do not oxidize further in air at room temperature, Figure 2.3. The maximum design temperature for aluminum alloys listed by the ASME code is typically 400°F (205°C) or less. Aluminum can be alloyed with various elements to increase strength and corrosion resistance.

The Aluminum Association designations for aluminum alloys with additional elements are shown in Table 2.4.

The fourth decimal, on the far left, in Table 2.4 designates the type of added alloy. The third decimal indicates a modification to the original alloy limit or impurity. The first and second digits in groups 2000–8000 identify the different aluminum alloys within the group. Hence, in alloy 5254, the 5 on the far left

Figure 2.3 Aluminum alloy

Table 2.4 Designation of aluminum with additional elements

Alloy group	Group number
Pure aluminum (>99% AL)	1000
Aluminum plus copper	2000
Aluminum plus manganese	3000
Aluminum plus silicon	4000
Aluminum plus magnesium	5000
Aluminum plus magnesium and silicon	6000
Aluminum plus zinc	7000
Aluminum plus other elements	8000

indicates it is aluminum with a magnesium alloy. The 2 indicates the second modification to the magnesium limit in this group, and 54 indicate the alloy number within this group. The first and second digits in group 1000 identify the purity of aluminum. Hence, 1060 indicates 99.60% aluminum.

The ASME code lists four tempers of aluminum as shown in Table 2.5.

Hence, from Table 2.1 and Tables 2.4 through 2.7 product SB-209 A95652 H32 is a plate material (SB-209) of aluminum alloy (A9XXXX) with magnesium as the main alloying element (5XXX). This is the sixth revision (X6XX) to original specification number (XX52). The material is strain hardened (HXX) using strain hardening and stabilizing (X3X) process with temper control 2 (XX2).

Similarly, material SB-211 A96061 T651 is a bar material (SB-211) of aluminum alloy (A9XXXX) with magnesium and silicon as the main alloying elements (6XXX). This is the zero revision (X0XX) to the original specification number (XX61). The material is thermally treated (TXXX) using a solution heat treating and artificial aging (X6XX) process with type 51 stress relieving (X51).

Aluminum alloy and temper selections are generally based on material compatibility and structural strength. This selection process is addressed extensively in other references. Jawad and Farr [4] list the effects of various acids and salts on the corrosion of aluminum.

Table 2.5 Various tempers of aluminum

Temper	Description
F	As fabricated
H	Strain hardened. This temper is followed by two digits to indicate the degree of strain hardening as shown in Table 2.6. The first digit indicates the operation and the second digit the degree of hardening
O	Annealed. Produces soft grade
T	Thermally treated. This temper is followed by one or two digits to indicate the degree of treatment as shown in Table 2.7

Table 2.6 Levels of strain hardening

Hardness number	Operation
H1	Strain hardened only
H2	Strain hardened and partially annealed
H3	Strain hardened and stabilized
H4	Strain hardened and painted
HX2	Quarter hard
HX4	Half hard
HX6	Three quarter hard
HX8	Full hard
HX9	Extra hard
HXX1	The third digit indicates the degree of control of temper or identifies special mechanical properties

Table 2.7 Level of thermal treatment

Thermal treatment number	Operation
T1	Naturally aged after cooling from elevated temperatures
T2	Cold worked after cooling from elevated temperatures
T3	Solution heat treated, cold worked, and naturally aged
T4	Solution heat treated and naturally aged
T5	Artificially aged after cooling from elevated temperatures
T6	Solution heat treated and artificially aged
T7	Solution heat treated and stabilized
T8	Solution heat treated, cold worked, and artificially aged
T9	Solution heat treated, artificially aged, and cold worked
T10	Cold worked after cooling from elevated temperatures and then artificially aged
TX11 or TXX11	Additional digits that indicate type of stress relief

2.3.2 Copper alloys

Copper alloys are nonmagnetic and their average density is 0.324 lb/in.^{3} (8.97 g/cm^3), about 14% heavier than steel. Most copper alloys oxidize in air at room temperature, Figure 2.4. Exceptions are the copper nickel alloys that tend to keep their original color in air. The most common alloys of copper are as follows:

1) Brass – alloy of copper and zinc.
2) Bronze – alloy of copper and tin.
3) Copper nickel – alloy of copper and up to 49.9% nickel.

Figure 2.4 Copper alloy

The maximum design temperature for most copper alloys listed by the ASME code is about 400°F (205°C), but it can go as high as 700°F (370°C) for copper nickel alloys. Copper is reddish-brown in color, turning to brown when oxidized, and finally a green color with further weathering. However, the color changes as the percentage of alloys increases. Copper nickel alloys have the same color as stainless steels. The numbering system for alloys is shown in Table 2.8.

2.3.2.1 Tempers of copper alloys

Copper alloys have a multitude of temper designations listed in ASTM B-601. Hardness values refer to the percent reduction in area in accordance with Table 2.9.

Table 2.8 Copper alloy UNS numbering system

Type of copper	Wrought	Cast
Copper	C10100 to C13000	C80100 to C81200
Brass	C20500 to C28580	C83300 to C85800
Tin Brass	C40400 to C48600	C83300 to C84800
Aluminum Bronze	C50100 to C52400	C90200 to C91700
Phosphor Bronze	C60800 to C64210	C95200 to C95900
Silicon Bronze	C64700 to C66100	C87000 to C87999
Silicon Red Brass	C69400 to C69710	C87300 to C87900
Copper Nickel	C70100 to C72950	C96200 to C96900

Table 2.9 Hardness reductions in area

Hardness	Percent reduction in cross-sectional area
1/8	5
1/4	11
1/2	21
3/4	29
Hard	37
Extra hard	50
Spring	60
Extra spring	68

Table 2.10 Temper designations

Temper designation	Description
O	Processed to produce specific mechanical properties
OS	Processed to produce specific grain size
H	Cold worked
HR	Cold worked and stress relieved
M	As manufactured
T	Heat treated
W	Welded

Some temper designations are provided in Table 2.10. These designations may be followed by additional letters to further describe the processing.

Some of the copper alloy tempers are listed below.

Annealed tempers

Some annealed tempers, designated by a letter O, are listed in Table 2.11.

Annealed tempers specified by average grain size

Some annealed tempers specified by average grain size, designated OS, are listed in Table 2.12.

Cold worked tempers

Some cold worked tempers, designated H, are listed in Table 2.13.

Table 2.11 Annealed conditions

Annealed temper O	Process
O10	Cast and annealed
O11	Cast and precipitation heat treated
O20	Hot forged and annealed
O25	Hot rolled and annealed
O30	Hot extruded and annealed
O31	Extruded and precipitation heat treated
O32	Hot extruded and temper annealed
O40	Hot pierced and annealed
O50	Light annealed
O60	Soft annealed
O61	Annealed
O65	Drawing anneal
O68	Deep drawing anneal
O70	Dead soft anneal
O80	Annealed and 1/8 hard cold worked
O81	Annealed and 1/4 hard cold worked
O82	Annealed and 1/2 hard cold worked

Table 2.12 Some tempers by grain size

Temper OS	Average grain size, mm
OS005	0.005
OS010	0.010
OS015	0.015
OS025	0.025
OS035	0.035
OS045	0.045
OS050	0.050
OS060	0.060
OS065	0.065
OS070	0.070
OS100	0.100
OS120	0.120
OS150	0.150
OS200	0.200

Table 2.13 Some cold worked tempers

Cold worked tempers H	Process
H00	1/8 hard cold worked
H01	1/4 hard
H02	1/2 hard
H03	3/4 hard
H04	Hard
H06	Extra hard
H08	Spring
H10	Extra spring
H50	Extruded and drawn
H52	Pierced and drawn
H55	Light drawn and light cold worked
H58	Drawn, general purpose
H60	Cold heading, forming
H66	Bolts
H70	Bending
H80	Hard drawn
H90	As finned

Cold worked and stress relieved tempers

Some cold worked and stress relieved tempers, designated HR, are listed in Table 2.14.

Table 2.14 Some cold worked and stress relieved tempers

Cold worked and stress relieved tempers	Process
HR01	1/4 hard cold worked and stress relieved
HR02	1/2 hard cold worked and stress relieved
HR04	Hard cold worked and stress relieved
HR06	Extra hard cold worked and stress relieved
HR08	Spring cold worked and stress relieved
HR10	Extra spring cold worked and stress relieved
HR12	Special spring cold worked and stress relieved
HR20	As finished
HR50	Drawn and stress relieved
HT04	Hard temper and treated
HT08	Spring temper and treated
HE80	Hard drawn and end annealed

As-manufactured tempers

Some as-manufactured tempers, designated M, are listed in Table 2.15.

Table 2.15 Some as-manufactured tempers

As-manufactured tempers	Process
M01	Sand cast
M02	Centrifugal cast
M03	Plaster cast
M04	Pressure die cast
M05	Permanent mold cast
M06	Investment cast
M07	Continuous cast
M10	Hot forged then air cooled
M11	Forged then quenched
M20	Hot rolled
M25	Hot rolled then rerolled
M30	Hot extruded
M40	Hot pierced
M45	Hot pierced then rerolled

Heat treated tempers

Some heat treated tempers, designated T, are shown in Table 2.16.

Some additional heat treated tempers are shown in Table 2.17.

Table 2.16 Some heat treated tempers

Heat treated tempers	Process
TQ00	Quench hardened
TQ30	Quench hardened and tempered
TQ50	Quench hardened and temper annealed
TQ55	Quench hardened and temper annealed then cold drawn and stress relieved
TQ75	Interrupted quench
TB00	Solution heat treated
TD00	Solution heat treated and 1/8 hard cold worked
TD01	Solution heat treated and 1/4 hard cold worked
TD02	Solution heat treated and 1/2 hard cold worked
TD03	Solution heat treated and 3/4 hard cold worked
TD04	Solution heat treated and hard cold worked
TF00	Precipitation hardened
TF01	Precipitation heat treated plate – low hardness

(Continued)

Table 2.16 (Continued)

Heat treated tempers	Process
TF02	Precipitation heat treated plate – high hardness
TX00	Spinodal hardened
TH01	1/4 hard cold worked and precipitation heat treated
TH02	1/2 hard cold worked and precipitation heat treated
TH03	3/4 hard cold worked and precipitation heat treated
TH04	Hard cold worked and precipitation heat treated

Table 2.17 Some specialized heat treated tempers

Additional heat treated tempers	Process
TS00	1/8 hard cold worked and Spinodal* hardened
TS01	1/4 hard cold worked and Spinodal hardened
TS02	1/2 hard cold worked and Spinodal hardened
TS03	3/4 hard cold worked and Spinodal hardened
TS04	Hard cold worked and Spinodal hardened
TS06	Extra hard cold worked and Spinodal hardened
TS08	Spring cold worked and Spinodal hardened
TS10	Extra spring cold worked and Spinodal hardened
TS12	Special spring cold worked and Spinodal hardened
TS13	Ultra spring cold worked and Spinodal hardened
TS14	Super spring cold worked and Spinodal hardened
TM00	Mill hardened AM
TM01	Mill hardened 1/4 HM
TM02	Mill hardened 1/2 HM
TM03	Mill hardened 3/4 HM
TM04	Mill hardened HM
TM05	Mill hardened SHM
TM06	Mill hardened XHM
TM08	Mill hardened XHMS
TL00	Precipitation heat treated or spinodal heat treated and 1/8 hard cold worked
TL01	Precipitation heat treated or spinodal heat treated and 1/4 hard cold worked
TL02	Precipitation heat treated or spinodal heat treated and 1/2 hard cold worked
TL04	Precipitation heat treated or spinodal heat treated and hard cold worked
TL08	Precipitation heat treated or spinodal heat treated and spring cold worked
TL10	Precipitation heat treated or spinodal heat treated and extra spring cold worked
TR01	Precipitation heat treated or spinodal heat treated and 1/4 hard cold worked and then stress relieved
TR02	Precipitation heat treated or spinodal heat treated and 1/2 hard cold worked and then stress relieved
TR04	Precipitation heat treated or spinodal heat treated and hard cold worked and then stress relieved

* Spinodal decomposition is a controlled thermal treatment to achieve high strength.

Welded tube tempers

Some welded tube tempers, designated W, are listed in Table 2.18.

An example of a copper alloy specification is B-152 Alloy C11000 H03. B-152 is an ASTM specification for copper alloy plates. The last three digits "110" in UNS C11000 indicate copper with no additional alloying elements. H03 indicates that the plate is furnished as 3/4 hard, which is a 29.4% reduction in the original slab thickness.

Table 2.18 Some welded tube tempers

Some welded tube tempers	Process
WM00	As welded from 1/8 hard cold worked annealed strip
WM01	As welded from 1/4 hard cold worked strip
WM02	As welded from 1/2 hard cold worked strip
WM03	As welded from 3/4 hard cold worked strip
WM04	As welded from hard cold worked strip
WM06	As welded from extra hard cold worked strip
WM08	As welded from spring cold worked strip
WM10	As welded from extra spring cold worked strip
WM15	As welded from annealed strip and thermal stress relieved
WM20	As welded from 1/8 hard cold worked strip and then thermal stress relieved
WM21	As welded from 1/4 hard cold worked strip and then thermal stress relieved
WM22	As welded from 1/2 hard cold worked strip and then thermal stress relieved
WM50	As welded from annealed strip
WO50	Welded and light annealed
WO60	Welded and soft annealed
WO61	Welded and annealed
WC55	Welded and light cold worked
WH00	Welded and drawn 1/8 hard cold drawn
WH01	Welded and drawn 1/4 hard cold drawn
WH02	Welded and drawn 1/2 hard cold drawn
WH03	Welded and drawn 3/4 hard cold drawn
WH04	Welded and drawn hard cold drawn
WH06	Welded and drawn extra hard cold drawn
WH55	Welded and cold reduced or light drawn
WH58	Welded and cold reduced or drawn, general purpose
WH80	Welded and reduced or hard drawn
WR00	Welded, drawn, and stress relieved from 1/8 hard drawn
WR01	Welded, drawn, and stress relieved from 1/4 hard drawn
WR02	Welded, drawn, and stress relieved from 1/2 hard drawn
WR03	Welded, drawn, and stress relieved from 3/4 hard drawn
WR04	Welded, drawn, and stress relieved from hard drawn
WR06	Welded, drawn, and stress relieved from extra hard drawn

2.3.3 Nickel alloys

Nickel material is highly magnetic, but some nickel alloys are nonmagnetic. The average density of nickel alloys is about 0.306 lb/in.^{3} (8.47 g/cm^3), slightly denser than steel. Nickel alloys do not oxidize in air at room temperature, Figure 2.5.

Nickel alloys are used extensively in high temperature applications with corrosive environments. Various producers of nickel alloys use commercial names to identify their products. Table 2.19 shows some commonly used nickel alloys and their generic composition and provides a cross reference between commercial name, alloy designation, UNS number, and major composition.

Nickel alloys are normally suited for high temperature applications. Table 2.19 also provides the maximum design temperature listed in the ASME Boiler and Pressure Vessel Code for the listed alloys.

2.3.4 Titanium alloys

Titanium alloys are reactive, combining readily with oxygen at elevated temperatures. Hence, they pose challenges in welding and cutting during the fabrication process. They are nonmagnetic and their average density is about 0.163 lb/in.^{3} (4.50 g/cm^3), or a little less than 60% that of mild steel. Titanium alloys do not oxidize in air at room temperature, Figure 2.6.

Their strength-to-weight ratio makes them excellent material for pressure vessels in severe environments where corrosion is a problem. The disadvantages of using titanium alloys are their cost and the temperature limit of around 600°F (315°C).

2.3.5 Zirconium alloys

Zirconium alloys are nonmagnetic, reactive at elevated temperatures, and have an average density of about 0.240 lb/in.^{3} (6.64 g/cm^3). Zirconium alloys do not oxidize in air at room temperature,

Figure 2.5 Nickel alloy

Table 2.19 Some nickel alloy cross references and temperature limits

Commercial name	Alloy designation	UNS	Major composition	ASME Spec Number	Max temp*
Nickel 200	200	N02200	Ni	SB-160, 161, 162 163 366	600°F (315°C)
Nickel 201	201	N02201	Ni–low C	SB-160, 161, 162 163 366	600°F–1200°F (315°C–650°C)
Monel 400	400	N04400	Ni–Cu	SB-1276, 163, 164, 165, 366, 564	500°F–900°F (260°C–480°C)
Monel 405	405	N04405	Ni–Cu	SB-164	900°F (480°C)
Inconel 600	600	N06600	Ni–Cr–Fe	SB163, 166, 167, 168, 516, 517, 564	1200°F (650°C)
Inconel 625	625	N06625	Ni–Cr–Mo–Cb	SB-366, 443, 444, 446, 564, 704, 705	1200°F–1600°F (650°C–870°C)
Inconel 690	690	N06690	Ni–Cr–Fe	SB-166, 167, 168	850°F (455°C)
Incoloy 800	800	N08800	Ni–Fe–Cr	SB-163, 366, 407, 408, 409, 514, 515, 564	1500°F (815°C)
Incoloy 800H	800H	N08810	Ni–Fe–Cr	SB-163, 407, 408, 409, 514, 515, 564	1650°F–1800°F (900°C–980°C)
Incoloy 825	825	N08825	Ni–Fe–Cr–Mo–Cu	SB-163, 366, 423, 424, 425, 564, 704, 705	1000°F (540°C)
Hastelloy B2	B2	N10665	Ni–Mo–Fe	SB-333, 335, 366, 462, 564, 619, 622, 626	800°F (425°C)
Hastelloy B3	B3	N10675	Ni–Mo–Fe	SB-333, 335, 366, 462, 564, 619, 622, 626	800°F (425°C)
Hastelloy C4	C4	N06445	Ni–Mo–Cr	SB-366, 574, 575, 619, 622, 626	800°F (425°C)
Hastelloy C276	C276	N10276	Ni–Mo–Cr	SB-574, 575, 619, 622, 626	1250°F (675°C)
Hastelloy G	G	N06007	Ni–Cr–Fe–Mo	SB-366, 581, 582, 619, 622, 626	1000°F (540°C)
Hastelloy G2	G2	N06975	Ni–Cr–Fe–Mo	SB-582, 619, 622, 626	800°F (425°C)
Hastelloy G3	G3	N06985	Ni–Cr–Fe–Mo	SB-366, 581, 582, 619, 622, 626	800°F (425°C)
Hastelloy G30	G30	N06030	Ni–Cr–Fe–Mo	SB-366, 462, 581, 582, 619, 622, 626	800°F (425°C)
Hastelloy G35	G35	N06035	Ni–Cr–Mo	SB-366, 462, 564, 574, 575, 619, 622, 626	800°F (425°C)
Carpenter 20	20Cb	N08020	Ni–Fe–Cr–Cb	SB-366, 462, 463, 464, 468, 473, 729	800°F (425°C)
Rolled alloy 330	330	N08330	Ni–Cr–Si	SB-366, 511, 535, 536, 710	1650°F (900°C)
Alloy 904L	904L	N08904	Fe–Ni–Cr–Mo	SA-182, 240, 249, 312, 403, SB-649, 677	700°F (370°C)

* Temperatures are maximums for ASME Section VIII, Division 1 applications.

Figure 2.6 Titanium alloy

Figure 2.7 Zirconium alloy

Figure 2.7. The maximum design temperature for zirconium listed by the ASME code is about 700°F (370°C). They are used for pressure vessels under many severely corrosive environments. Also, their excellent radioactive neutron absorption property makes them ideal as isotope supports in nuclear components.

2.3.6 Tantalum alloys

Tantalum alloys are nonmagnetic, reactive at elevated temperatures, and have an average density of about 0.602 $lb/in.^3$ (16.7 g/cm^3). Tantalum alloys do not oxidize in air at room temperature, Figure 2.8. They are not listed in the ASME BPVC as a material of construction but are used as

Figure 2.8 Tantalum alloy

liners for pressure vessels with severely corrosive contents such as nitric and sulfuric acids, as a substitute for glass-lined vessels. Their high density is about 50% greater than that of lead. Their low yield strength and high elongation make them easy to form during fabrication. Their corrosion resistance is similar to that of silver and their cost is about the same.

2.3.7 Price of nonferrous alloys

The approximate cost ratio per pound of weight of some nonferrous alloys [4] compared to carbon steel is shown in Table 2.20. Note that a high strength to density ratio in some cases mitigates the high cost per pound of material.

Table 2.20 Approximate cost of various nonferrous alloys compared to carbon steel

Material	Cost index per lb
Carbon steel	1.0
Aluminum alloys	5.0
Copper alloys	6.0
Nickel alloys	16.0–50.0
Titanium alloys	55.0
Zirconium alloys	65.0
Tantalum alloys	330.0

2.4 Density of Some Ferrous and Nonferrous Alloys

The cost of material is determined by multiplying the calculated weight by the cost/unit weight for the material. The weight is obtained by multiplying the volume of required material by its density. Table 2.21 lists approximate densities of some of the materials used in the ASME BPVC for pressure vessels.

Example 2.1

What is the approximate weight of a 2 : 1 carbon steel ellipsoidal head with an inside diameter of 48 in. and a minimum wall thickness of 1.25 in.?

Solution

The forming process results in thinning in the knuckle region. This inside radius is referred to as the "inside corner radius" or ICR, when ordering. A usual allowance for thinning in this region is about 6%. Hence, $t = (1.25)(1.06) = 1.33$ in. The fabricator will start with a nominal $t = 1.375$ in.

With a nominal wall thickness of 1.375 in. and a 48 in. ID, the mean diameter will be

$$D = 48 + 1.375 = 49.375 \text{ in.}$$

From Table B.4, the surface area of the ellipse = $1.086D^2$.

Volume of the head material

$$V = (1.375)(1.086)\left(49.375^2\right) = 3640 \text{ in.}^3$$

The approximate weight of the ellipsoid, using the density of steel from Table 2.21, is

$$\text{Weight} = (3640)(0.284) \approx 1034 \text{ lb.}$$

In addition, in the United States, heads such as this are furnished with about a 2 in. straight cylindrical flange ("straight flange"). The weight of the straight flange is given by

$$\text{Weight of the cylinder} = (\pi)(49.375)(1.375)(2)(0.284) = 121 \text{ lb.}$$

Hence, the total weight of the head is $1034 + 121 = 1155$ lb.

The fabricator may also add a small amount of material at the end of the cylindrical shell for trimming.

Table 2.21 Approximate density of various alloys

Material	lb/ft^3	lb/in.^{3}	kg/m^3	g/cm^3
Carbon steel	490	0.284	7820	7.85
Stainless steel	470–500	0.272–0.296	7500–8000	7.53–8.19
Aluminum	170	0.098	2700	2.72
Copper	560	0.324	8900	8.97
Nickel	500	0.289	8000	8.01
Titanium	280	0.162	4500	4.49
Zirconium	400	0.231	6400	6.41
Tantalum	1,040	0.602	16,600	16.7

2.5 Nonmetallic Vessels

Stationary composite pressure vessels in the United States are normally designed in accordance with the ASME Section X Code and comply with ASTM D4097 – Standard Specification of Contact-Molded Glass-Fiber-Reinforced Thermoset Resin Corrosion-Resistant Tanks. ASME Section X limits the pressure to 150 psi (1 MPa) for bag-molded, centrifugally cast, and contact-molded vessels; 1500 psi (10 MPa) for filament-wound vessels; and 3000 psi (20.7 MPa) for filament-wound vessels with polar boss openings.

The US Department of Transportation provides requirements for various types of pressure vessels, including nonmetallic ones, in 49 CFR 173.301 General requirements for shipment of compressed gases and other hazardous materials in cylinders [5], UN pressure receptacles, and spherical pressure vessels.

Aerospace pressure vessels are addressed by the American Institute of Aeronautics and Astronautics (AIAA) standards S-080 Standard: Space Systems-Metallic Pressure Vessels, Pressurized Structures, and Pressure Components [6] and S-081 Standard: Space Systems-Composite Overwrapped Pressure Vessels [7], as well as various international standards.

While not considered pressure vessels, atmospheric storage tanks have the potential to have significant pressure at their bases because of the static head of the contents. Some of these vessels are constructed of polymeric or composite materials. Vessels that hold liquids without surface pressure are frequently made from polyethylene material in accordance with ASTM D1998 – Standard Specification for Polyethylene Upright Storage Tanks. The American Petroleum Institute uses API 12P (Specification for Fiberglass Reinforced Plastic Tanks) as a guide for fabricating unpressurized fiberglass vessels.

2.6 Forms and Documentation

The fabricator is responsible for ordering not only the material of construction but also for specifying any supplemental requirements needed for fabrication of the material or requested by the end user. There are numerous requirements needed for a given material that may or may not be part of the ASME/ASTM published specification of the material. A typical example is a limitation on various alloying or tramp elements. A sample of a requisition form for ordering material from the mill or warehouse including such supplemental requirements is shown in Table 2.22.

Each heat of material supplied by the mill to a fabricator of a pressure vessel must include a mill test report (MTR). The MTR lists mechanical and chemical properties of the material, other specified characteristics of the product such as Charpy V-notch values, and any additional restrictions or additions to the composition or heat treating of the product. ASTM only requires that MTRs furnished by steel mills list the elements included in the ASTM specification, such as the ones shown in Table 2.2. ASTM does not require reporting of other elements that may be present. These unreported elements, such as vanadium, may have an effect on weldability, hardenability, or toughness of the steel. Accordingly, some manufacturers and especially those fabricating thick-wall vessels or vessels operating in extreme environments may require the steel mills to report additional elements present in the steel they purchase.

Table 2.22 Sample material requisition form

MATERIAL REQUISITION FORM

ASME/ASTM Material Specification Number:
UNS number:
Dimension and number of pieces:
Supplemental requirements:

Additional tension tests:
Carbon equivalent (CE) value:

$$\text{Maximum CE} = C + (Mn)/6 + (Cr + Mo + V)/5 + (Ni + Cu)/15$$

Charpy V-Notch impact test:
Corrosion test:
Drop Weight Test:
Hardness test:
High temperature tension test:
Lateral expansion:
Liquid penetrant examination:
Magnetic particle examination:
Marking:
Product analysis:
Quench and temper heat treatment:
Radiographic examination:
Simulated PWHT of mechanical test coupons:
Ultrasonic examination:
Vacuum carbon-deoxidized steel:
Vacuum treatment:
Additional tests:

A sample MTR for a stainless steel plate is shown in Figure 2.9. Some of the commercial information in this table has been deleted from the MTR due to its irrelevance to the technical content discussed herein. Some of the items in the MTR are given as follows:

- Specification
  - The plate was hot rolled, annealed, and pickled.
  - The plate meets both the ASTM A240 and the ASME SA240 plate specifications.
  - Two other ASTM/ASME specifications are also met. The first is A-480/SA-480 "Specification for General Requirements for Flat-Rolled Stainless and Heat-Resisting Steel Plate, Sheet, and Strip." The second specification is A-666/SA-666 "Specification for Annealed or Cold-Worked Austenitic Stainless Steel Sheet, Strip, Plate, and Flat Bar."
  - The plate meets the QQS766D-A, a no longer active US federal specification for stainless steel and heat resisting alloys, plate, sheet, and strip.
  - The plate meets AMS 5507F a Society of Automotive Engineers (SAE) specification for corrosion and heat resistant steel (SAE30316L) sheet, strip, and plate, solution heat treated, and AMS 5524K, a second SAE specification for corrosion and heat resistant sheet, strip, and plate.
  - The plate meets SA-240 type 316 and type 316L specifications.
  - The corrosion test requirements of ASTM A-262 were met.
  - A 180° bend test was performed.
  - The plate is 1.25 in. × 60 in. × 240 in. (32 mm × 1500 mm × 6100 mm).

- Mechanical tests
  - A Rockwell B hardness test was conducted.
- Chemical composition
  - Copper content is reported, though not specified in the standard for 316 or 316L.
  - All other alloying elements are reported and are within the specified limits for SA240-316/316L.
- General information section
  - The material is free of mercury contamination.
  - No weld repairs were made on the plate during production.

MATERIAL TEST REPORT

From: North American Stainless
6870 Highway 42 East

Page : 1
Heat/Lot : 0CJ8
TIN : 050CJ8CB

Sold To: RYERSON
41 CENTURY DR
AMBRIDGE (PPC) , PA 15003-2513

Ship To: 41 CENTURY DRIVE
AMBRIDGE
PA15003

Vendor Information

Cust PO #: 540959 Cust PO DT: 04/18/06

Specifications

STAINLESS STEEL PLATE, HOT ROLLED, ANNEALED AND PICKLED.
ASTMA240/06a,480/06,666/03,ASMESA240/04-A05,480/04-A05,SA666/04
(X GRAIN), QQS766D-A X MG PERM,AMS5507F/AMS5524K X MRK
CORROSION: ASTM A262/02aE;180Bend-OK
Buyer Part: 75812700
316/316L HRAP 1.25 IN 60 IN 240 IN
Country of Origin: UNITED STATES

Mechanical Tests

		UOM	L	O	UOM	L	O
Yield Strength - 0.2% Offset	42.05	KSI	F	TRANSV			
Tensile Strength (UTS)	80.97	KSI	F	TRANSV			
Rockwell B	83		F	TRANSV			
Pct Uniform Elongation 2"/50mm	56.48	PCT	F	TRANSV			

Chemical Composition

C	Carbon-Pct	.024	CR	Chromium-Pct	17.190
CU	Copper-Pct	.447	MN	Manganese-Pct	1.698
MO	Molybdenum-Pct	2.163	N	Nitrogen-Pct	.050
NI	Nickel-Pct	10.422	P	Phosphorus-Pct	.033
S	Sulfur-Pct	.004	SI	Silicon-Pct	.271

Comments

Figure 2.9 Sample mill test report (*Source:* Harris Thermal Transfer Products)

MATERIAL TEST REPORT

From: North American Stainless
6870 Highway 42 East

Page : 2
Heat/Lot : 0CJ8
TIN : 050CJ8CB

Material free from mercury contamination. No weld repairs.
EN 10204 3.1 PED 97/23/EC Annex1, Para. 4.3 QQS763F Cond A
Material is RoHS-Compliant
MINIMUM SOLUTION ANNEAL TEMPERATURE 1950 F.
- Melted & Manufactured in the USA
This document certifies the material has been tested in
accordance with applicable specifications described herein
and has met those requirements.

QA by ERIC HESS 07/19/2006

Miscellaneous Data

Ghent , KY410459615
Control #: 84A60835 Date: 08/31/06
S.O. Number :773 486335 3
DC316L-HRAP PMP 1.250 X 60 X 240 Customer P.O.: 49354

Figure 2.9 (Continued)

- The British EN 10204 standard for statements of compliance for metallic products was met.
- The no longer active (US) federal specification QQS763F for corrosion resistant steel bars, wire, shapes, and forgings, since superseded by SAE-AMS-QQ-S-763 was met.
- The requirement of the European Union's RoHS "Restriction of Hazardous Substances" is met.
- The material was solution annealed at a temperature of 1950°F (1065°C) or greater.
- The material was melted and manufactured in the United States of America.

2.7 Miscellaneous Materials

Fabricated pressure vessels or their associated piping and other components often contain materials other than those discussed earlier. Some of these materials are discussed in the following sections.

2.7.1 Cast iron

Cast iron is similar to steel but contains carbon in the range of 2–6%. This carbon level makes the material very brittle and mostly unweldable. Cast iron is used in elbows, tees, and other connections in piping and flanges. The ASME code lists some cast iron specifications used for elbow, bends, and other such components.

2.7.2 Gaskets

2.7.2.1 Gasket types

A wide range of gasket types are in common use in pressure vessels and piping. For many applications, especially for low pressures in benign environments, the choice is not critical and is typically made based on cost. Higher pressure applications and aggressive environments are more demanding. In such environments, systems can be plagued by leaks if a suitable means of sealing is not identified. Extra tightening of bolts is occasionally effective, but if the wrong type of gasket has been selected, often the only permanent solution is replacement with the correct one. This can be costly, sometimes involving replacement of flanges or other connectors as well. Below is a list of some of the types of gaskets in common use.

Gaskets, Figure 2.10, are extensively used to seal flange connections at pressure vessel interfaces and in piping systems. A wide range of materials is used in gaskets [8]. However, gasket materials are not governed by the rules of the ASME code. The gasket materials vary and are selected based on temperature, pressure, and chemical compatibility.

Rubber o-rings, Figure 2.10(a), are commonly used at temperatures up to around 250°F (120°C) and relatively low pressures, although some specialty rubber o-rings are used up to a few thousand psi pressure. The surface finish for seating these gaskets is around 32 μin. (0.8 μm) rms.

Metallic C-rings, Figure 2.10(b), are considered self-energized and are used in high pressure and high temperature applications and constructed of stainless steel or nickel alloy materials. The surface finish for seating these gaskets is 32 μin. (0.8 μm) rms. Metallic o-rings, Figure 2.10(c), are also

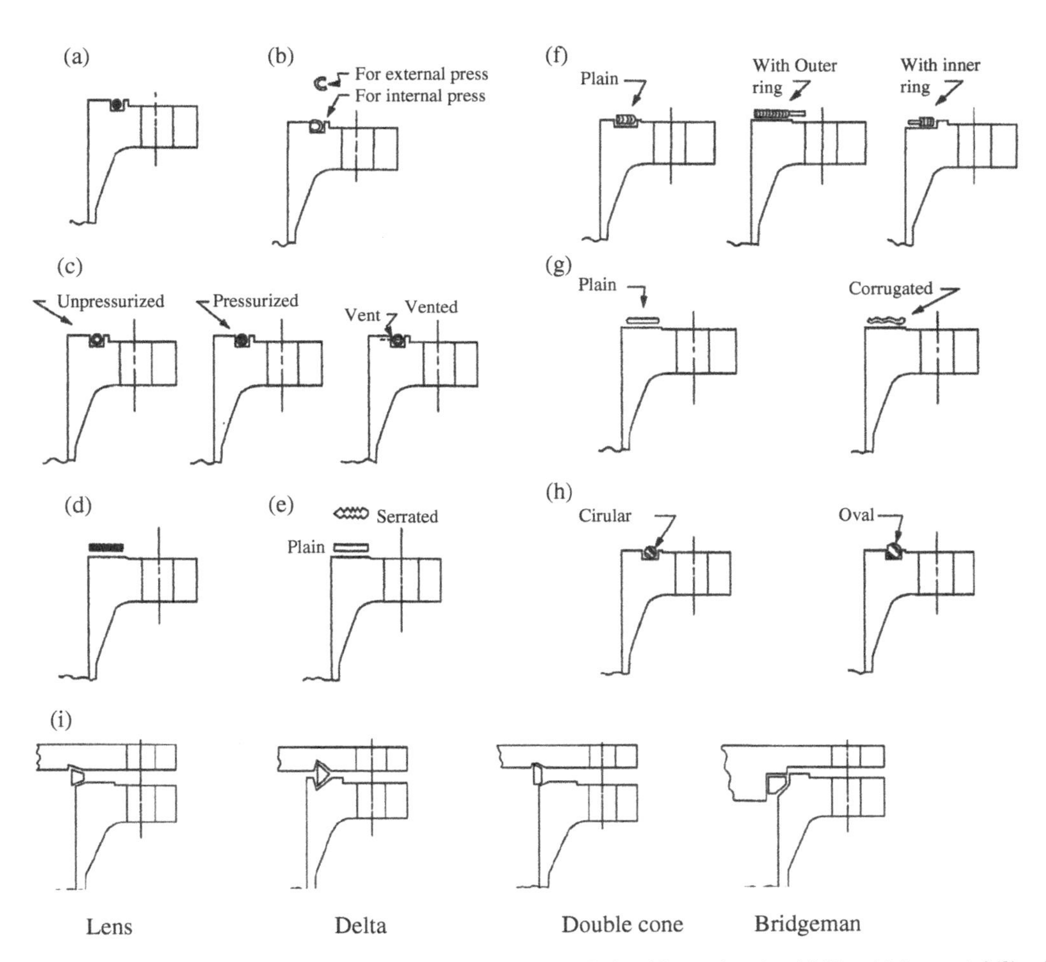

Figure 2.10 Samples of gasket types: (a) rubber o-ring (b) C-ring (c) metal o-ring (d) fiber (e) flat metal (f) spiral wound (g) jacketed (h) metal ring (i) high pressure (*Source:* Jawad and Farr [4], John Wiley & Sons)

common in high pressure applications due to their low seating stress. They also are constructed of stainless steel or nickel alloy materials.

Fiber gaskets, Figure 2.10(d), are very common in relatively low pressure applications, especially where the seating surface has complicated geometry, such as in heat exchangers with pass-partitions and oval or rectangular flanges. The gaskets are cut from flat sheets of fiber to any desired configuration. The material consists of about 70% fiber, 20% rubber, and 10% filler material and curative.

Figure 2.10(e) shows plain or serrated flat metal gaskets. These gaskets are used for high pressure applications and are normally constructed of stainless steel. They can be covered with rubber or graphite to improve sealing.

Spiral wound gaskets, Figure 2.10(f), are very popular in applications with moderate temperature and pressure. They consist of an inner composite fiber material wound in a spiral with thin stainless steel strips to hold them together and stabilize them. Such gaskets tend to be flimsy in large diameters so in many cases they are encased in an outside compression ring that provides stability. The outside compression ring also helps avoid extrusion of the gasket and provides a well-defined limit to prevent over-compression of the gasket during bolt tightening. Occasionally, an additional inner compression ring is also added for further stability.

Jacketed gaskets, Figure 2.10(g), are made of composite filler material encased with an outer metallic strip. The metallic strip is normally made of stainless steel. The gasket seating surface finish is around 63 μin. (1.6 μm) rms.

Solid metal ring gaskets, Figure 2.10(h), are commonly used for high pressure applications due to their compact sizes and their ability to seal high pressures. The material of construction ranges from soft iron to stainless steel to nickel alloys. They are furnished in round, oval, or octagonal cross section. While sometimes difficult to seat, once seated they provide a very tight and reliable seal.

Figure 2.10(i) shows some gaskets used in high pressure applications. The material of construction is normally soft iron. The seating stress is small and they rely on the applied pressure for sealing.

Most of the metal gaskets can be plated with silver or gold to provide easier sealing and to resist the corrosive contents of the vessel.

2.7.2.2 Gasket containment

An important consideration in gasket design is how the gasket is contained. Most gasket types, while working effectively as seals in the appropriate application, have little or no capability to resist pressure stress on their own. Thus, a number of means have been developed for their containment. ASME B16.5 [9] describes a number of flange facings that provide support or containment for gaskets. The following case illustrates why this is necessary.

Consider the o-gaskets shown in Figure 2.10 sketches (a), (b), (c), and (h). They can be represented by Figure 2.11(a).

The stress in the o-ring due to internal pressure, if it is unsupported, can be approximated by the equation

$$S = PDH/2A \tag{2.1}$$

where

A = cross-sectional area of the gasket
D = diameter subjected to pressure
H = height of the gasket subjected to pressure

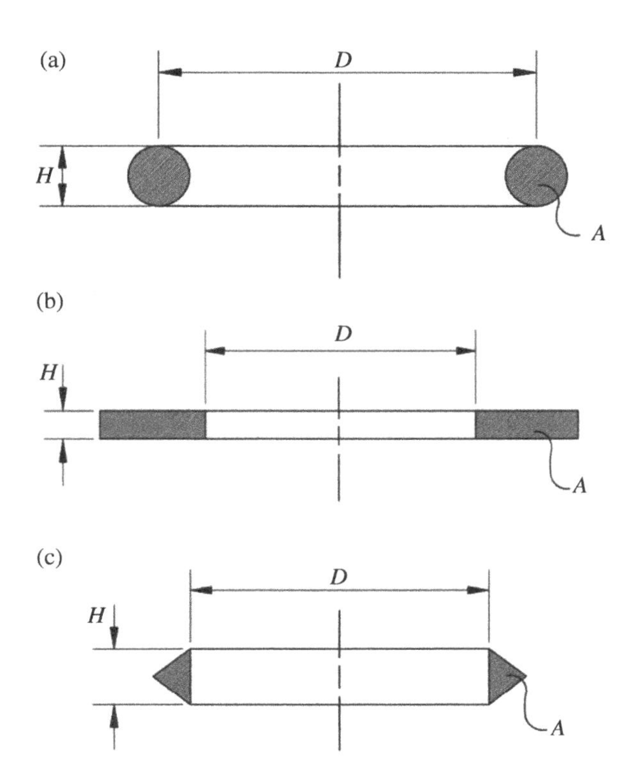

Figure 2.11 Gasket dimensions. (a) o-ring, (b) Flat gasket, (c) Lens ring

P = internal pressure
S = stress in the gasket

Equation (2.1) can be written in terms of pressure as

$$P = 2AS/DH \tag{2.2}$$

Example 2.2
A metallic o-ring gasket per Figure 2.10(h) has a yield stress $S = 12{,}000$ psi, diameter $D = 24$ in., and a cross-sectional diameter $H = 0.1875$ in. What is the maximum internal pressure that can be applied without extruding the gasket?

Solution

The cross-sectional area of the gasket is

$$A = \pi H/4 = \pi(0.1875)^2/4$$
$$= 0.0276 \text{ in.}^2$$

The maximum internal pressure that can be applied without extruding the gasket is obtained from Eq. (2.2) as

$$P = [2(0.0276)(12,000)]/[(24)(0.1875)]$$
$$P = 147 \text{ psi}$$

Hence, for pressures over 147 psi, there is a high likelihood of extruding the gasket unless it is contained. For o-rings, this containment is typically provided by a groove as shown in Figure 2.10(h).

For o-rings placed in a rectangular flange, it is often difficult to achieve a sufficiently fine finish for sealing on the bottom of the groove. For such flanges, the groove is typically produced using an endmill, which leaves machining marks across the groove, providing a potential leak path. These grooves are particularly difficult to machine because they are generally fairly narrow, and at the same time they often are machined with a dovetail to help retain the o-ring. For some applications, an alternative to laborious hand polishing of the bottom of a narrow groove is to polish the flat surface of the flange, then create the effect of a groove by tack welding keystock to the surface to contain the o-ring, Figure 2.12. This is particularly effective for polymer o-rings. For low pressure applications, only small welds are required to hold the stock in place, effectively fabricating a polished o-ring groove rather than machining it. This technique works for rectangular, circular, and other shapes of flanges, and creates an effect similar to that of the compression ring of a spiral wound gasket.

Note that in addition to providing the containment, either the groove or the keystock also defines the amount of compression of the o-ring, which may vary from a few percent for a solid metal o-ring to about 30% for some polymeric ones.

Equations (2.1) and (2.2) may also be used for other configurations. Hence, flat and jacketed gaskets shown in Figure 2.10 sketches (d), (e), and (g) can be represented as shown in Figure 2.11(b). Similarly, the delta gasket shown in Figure 2.10(i) can be represented as shown in Figure 2.11(c).

Spiral wound gaskets in Figure 2.10(f) are normally furnished with compression rings. The cross-sectional area of these rings can be checked for adequacy against Eqs. (2.1) and (2.2).

One of the most common containment methods for gaskets is that used on many raised faced flanges, Figure 2.10(d): a serrated finish as specified in ASME B16.5. In this case, a smooth surface is not desirable because it provides less "bite" for the gasket. Thus, ASME B16.5 specifies that the finish is to be "either a serrated concentric or serrated spiral finish having a resultant surface finish from 125 μin. to 250 μin. (3.2 μm to 6.3 μm)". It also provides further guidance regarding the radius of the tool to be used to create the finish, groove spacing, etc. Such a surface effectively retains a fiber gasket having essentially no tensile strength and makes an effective seal for compatible liquids and gases at pressures of hundreds of psi.

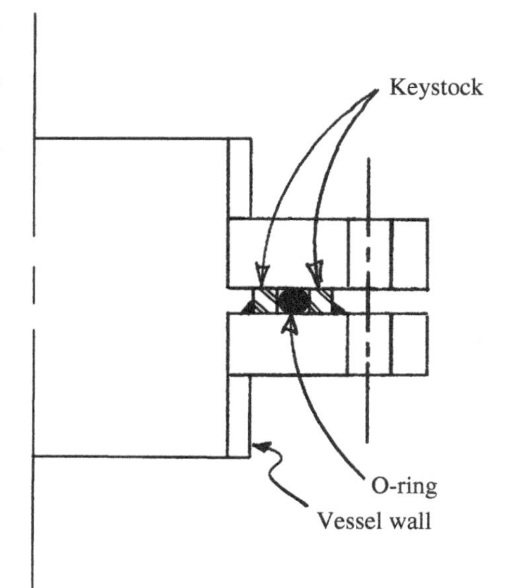

Figure 2.12 Use of keystock to create support for an o-ring

References

1 ASME. 2021. "*Boiler and Pressure Vessel Code, Section VIII, Division 1*". American Society of Mechanical Engineers, New York.

2 ASTM. "*Section 1 – Iron and Steel Products and Section 2 – Nonferrous Metal Products*". ASTM International, West Conshohocken, PA.

3 S. R. Avner. 2nd Ed, 1974. "*Introduction to Physical Metallurgy*". John Wiley Publishing, New York.

4 M. Jawad and J. Farr. 3rd Ed, 2019. "*Structural Analysis and Design of Process Equipment*". John Wiley Publishing, New York.

5 Code of Federal Regulations (CFR) (2011). 49 CFR 173.301 General requirements for shipment of compressed gases and other hazardous materials in cylinders, UN pressure receptacles, and spherical pressure vessels.

6 AIAA (2018). S-080A-2018, Standard: Space Systems-Metallic Pressure Vessels, Pressurized Structures, and Pressure Components, American Institute of Aeronautics and Astronautics, Reston, VA.

7 AIAA (2018). S-081B-2018, Standard: Space Systems-Composite Overwrapped Pressure Vessels, American Institute of Aeronautics and Astronautics, Reston, VA.

8 D. E. Czernik. 1996. "*Gaskets Design, Selection, and Testing*". McGraw Hill, New York.

9 ASME, 2017, "*Pipe Flanges and Flanged Fittings: NPS 1/2 Through NPS 24 Metric/Inch Standard B16.5-2017*". American Society of Mechanical Engineers, New York.

3

Layout

3.1 Introduction

Layout consists of the processes and procedures for measuring and marking material for cutting, drilling, forming, fit up, and welding. It is the beginning of the process of transforming dimensional information plus raw plate, pipe, and other material into pressure vessels.

In the past, the dimensional information for a pressure vessel was all on drawings. Today it may be on drawings, but its basic location is more likely inside a computer, with drawings merely an output. This situation means that this field, like many others, has evolved significantly over the past thirty years.

Competent layout personnel are still needed in the fabrication business, but the distribution of work has changed. Earlier, much time was spent on using solid geometry with projections, triangulation, and other techniques to produce the contours used to cut parts for forming and fit up. It is sometimes still needed, but much of this work is now being done numerically, with the results either fed directly to numerically controlled cutting equipment or produced as full scale plots for use in the shop.

Unless the design has been created in a system that (1) takes into account fabrication practices and processes and (2) operates seamlessly with equipment to produce all contoured plate, tubular, and other products, the layer out is still needed. Even if the aforementioned two criteria are met, the layer out may be the best option to use software to translate the design into a manufacturable form. In some cases, for one reason or another, layout by computer may not be the best option.

Sometimes work must be done at a time or location that precludes taking advantage of computing capabilities for layout. Also, some configurations can be dealt with more easily by traditional layout techniques than by providing data in standard input configurations due to unique shapes, thickness combinations, weld prep requirements, etc. There is also additional work that does not get done by the computer: For example, once a plate has been cut to contours, rolling or other forming instructions still need to be marked, and rolling templates provided, functions which are described in Section 3.7.

3.2 Applications

Common work performed by a layer out includes the following:

- Determining plate sizes for rolling shell courses.
- Development of layout for cones, transitions, and intersections.

Fabrication of Metallic Pressure Vessels, First Edition. Owen R. Greulich and Maan H. Jawad.

- Layout of reinforcing pads.
- Marking of plates for rolling or other forming, such as for work on a press brake.
- Marking for drilling, if not performed as part of machining.
- Provision of alignment marks for use by the fitter in assembling components.
- Marking of welding to be performed (e.g., size of fillet welds, intervals for intermittent welds).

In the production of a standard product, some or all of these items may be unnecessary (or be done only once for the lot), but in the production of one-off products or small lots, some or all of these will typically be required.

3.3 Tools and Their Use

A well-stocked layout department will have the following tools and equipment available:

- Computer – used to create design, develop contour layouts, optimize plate usage, estimate material costs, send information to a plotter, provide coordinates for hand layout.
- Plotter – produces full scale plots for direct usage, or scaled plots to assist the layer out in developing full scale plots for cutting.
- Interface to computer numerical control (CNC) machines such as mills, lathes, and cutting machines.
- Calculator – used for trigonometric and other calculations on basic layouts, or programmed to provide coordinates for layout.
- Levels – assist in field layout of intersections, bolt patterns, etc.
- Protractor – assist in layout for triangulation.
- Straight edges – marking, triangulation, etc.
- Measuring tapes – measuring existing parts and laying out new ones.
- Pi tapes® – determine true size of heads, pipes, etc. based on measurement of circumference using a tape on which one unit is pi inches (cm for the metric tape) long.
- Scales – accurate measurements.
- Calipers – thickness verification.
- Squares – layout of projections, use in triangulation.
- Dividers – transferring and repeating measurements.
- Trammel points – layout of large radii, transfer of large dimensions.
- Prick punch – accurate and indelible marking of critical locations.
- Center punches – accurate and indelible marking of critical locations.
- Ball peen hammers – use with center punches for marking, also for peening away incorrect center punch or prick punch marks.
- Chalk lines, pencils, crayons, marking paint – marking.
- Layout blue – make fine layout and marking easier and more visible.

3.4 Layout Basics

Most layouts are done in the flat. Whether done manually or by a computer, the operations are essentially the same. Contours for cutting are usually developed by projection and/or triangulation. A layout performed by the computer may use a mathematical representation of these same operations, or may, in some cases, use more sophisticated approaches to accomplish the same. Also, note that a forming layout must also be based on the neutral axis of the surface being produced.

Layout is a specialty field requiring special knowledge and training, and about which books have been written. Therefore, only the basics are presented this chapter, leaving the details of implementation for other works.

3.4.1 Projection

The flat shape that will roll up to form a cylindrical piece for a mitered elbow uses simple projection.

For purposes of this operation, consider a cylinder oriented vertically. The bottom view of the cylinder is drawn, along with a line at the angle at which the cylinder will be cut. The circle of the bottom view is divided into equal segments, small enough that when projected they provide enough points to produce a smooth curve. Above and to the right, a series of vertical lines is drawn, equally spaced so as to make a full circumference of the circle.

The dividing points are projected up onto the line, and from there to the right onto the respective vertical lines. Connecting the points produces the contour.

This technique works for any angle of miter that may be selected. From an analytical standpoint, the contour that will be projected on the right will always be a sine wave, with a factor equal to the sine of the angle of projection. This makes computer development of this contour a simple procedure.

In Figure 3.1(a), there is a plane of projection representing the angle of the miter cut. This projection provides the shape for a pipe or tube intersecting a flat plate or meeting another pipe or tube of the same size at an angle. The same technique is used to find the contour for a cylinder intersecting another cylinder of a different size, whether on center (Figure 3.1(b)) or offset, and whatever the angle. It provides the contours of intersection (cut) for both the run cylinder and the intersecting cylinder.

3.4.2 Triangulation

When the piece to be cut is shaped so that it cannot be presented in a view in which all dimensions are true length, triangulation must generally be used. This occurs if the shape to be cut or intersected is not of a continuous cross section.

A cone or its frustum is easily laid out if it is symmetric because the true length of measurements does not change around the perimeter. Therefore, the contour is a portion of a circle. The triangulation is simple because the angled side in an elevation drawing of the cone provides the critical dimensions. The included angle of the flat plate layout is given as θ and the outside radius by R_O. Figure 3.2 shows the associated analytical approach. The correlation between the various dimensions is given by

$$L = \left((R_2 - R_1)^2 + h^2\right)^{0.5} \tag{3.1}$$

$$R_o = \frac{R_2}{R_2 - R_1}\left((R_2 - R_1)^2 + h^2\right)^{0.5} = \frac{R_2 L}{R_2 - R_1} \tag{3.2}$$

$$\theta = \frac{360(R_2 - R_1)}{\left((R_2 - R_1)^2 + h^2\right)^{0.5}} = \frac{360(R_2 - R_1)}{L} \tag{3.3}$$

where

h = height of cone
R_1 = small radius
R_2 = large radius

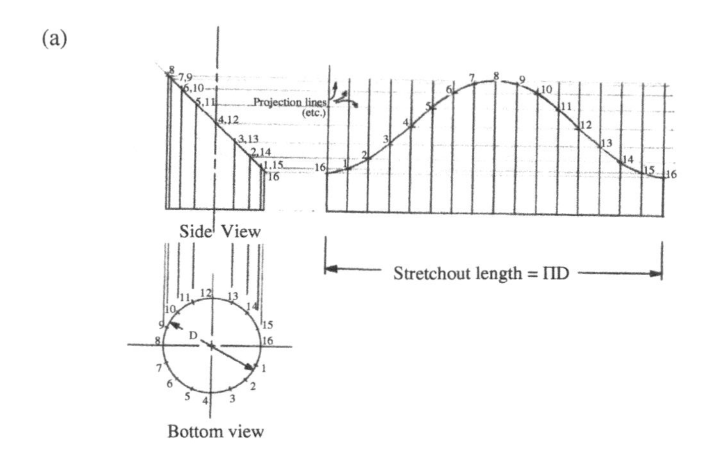

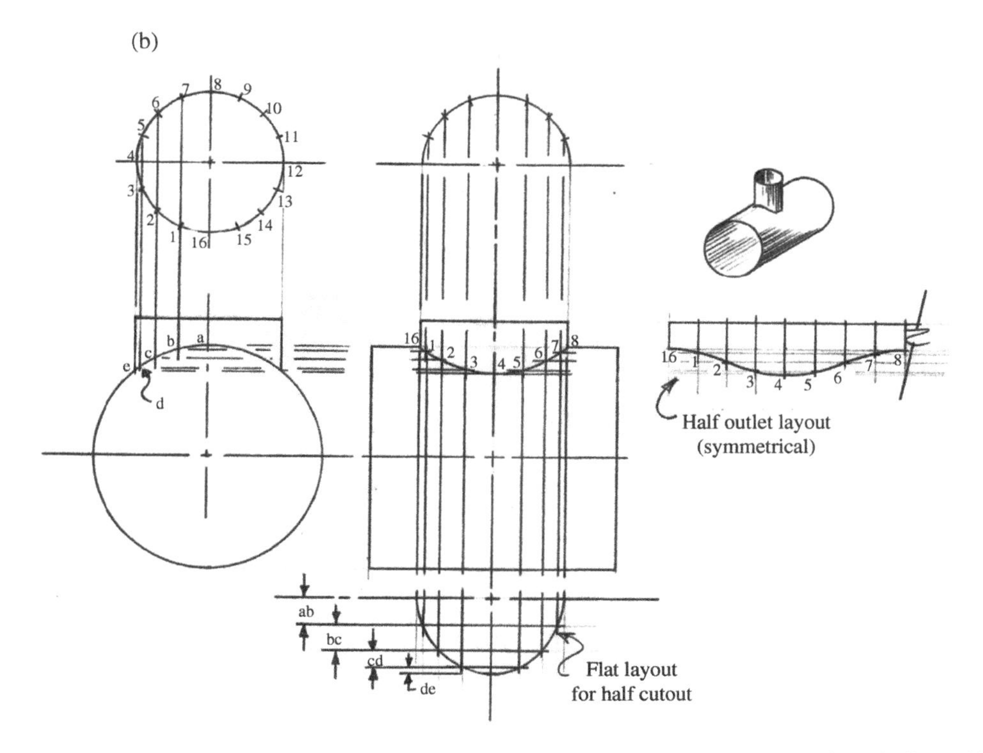

Figure 3.1 Projections of a cylinder. (a) Cylinder cut at an angle for creating a mitered elbow. (b) Two intersecting cylinders of different sizes, making a reducing tee

An eccentric cone occurs in offset reducers, and four quarters of eccentric cones occur in each rectangular-to-round transition. In the special case of a concentric square-to-round transition, all of the quarters are the same, Figure 3.3. In this case, the layout for each formed corner section is the same. The same procedure is used for each individual corner of a rectangular-to-round symmetric or asymmetric transition. The difference is that a concentric rectangular-to-round transition will have two each, of two different eccentric cone quarters, and an eccentric square- or rectangular-to-round transition will have four eccentric cone quarters, all being different. Figure 3.3 illustrates the triangulation required to produce this layout.

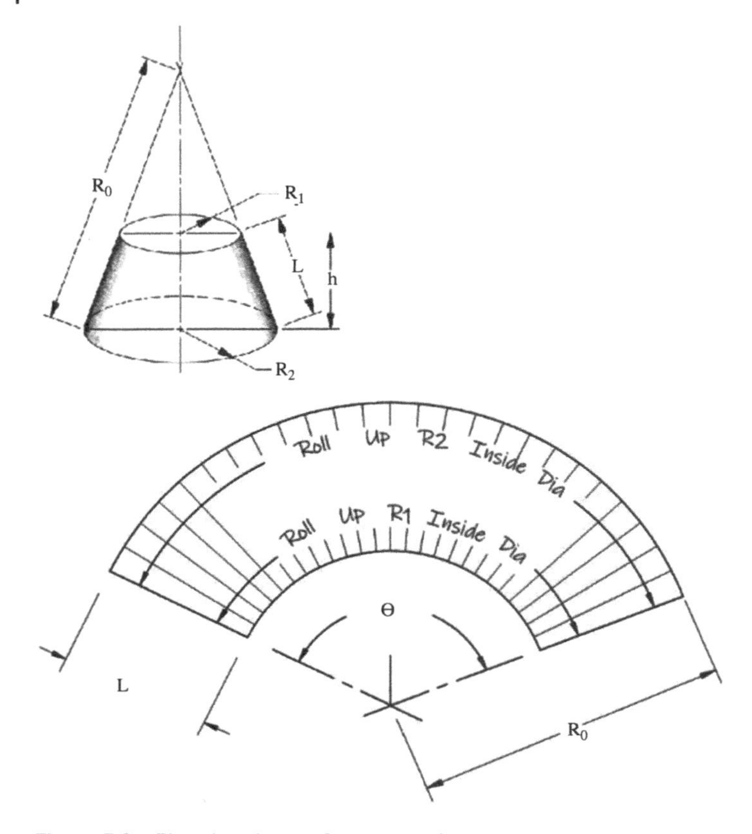

Figure 3.2 Flat plate layout for concentric cone

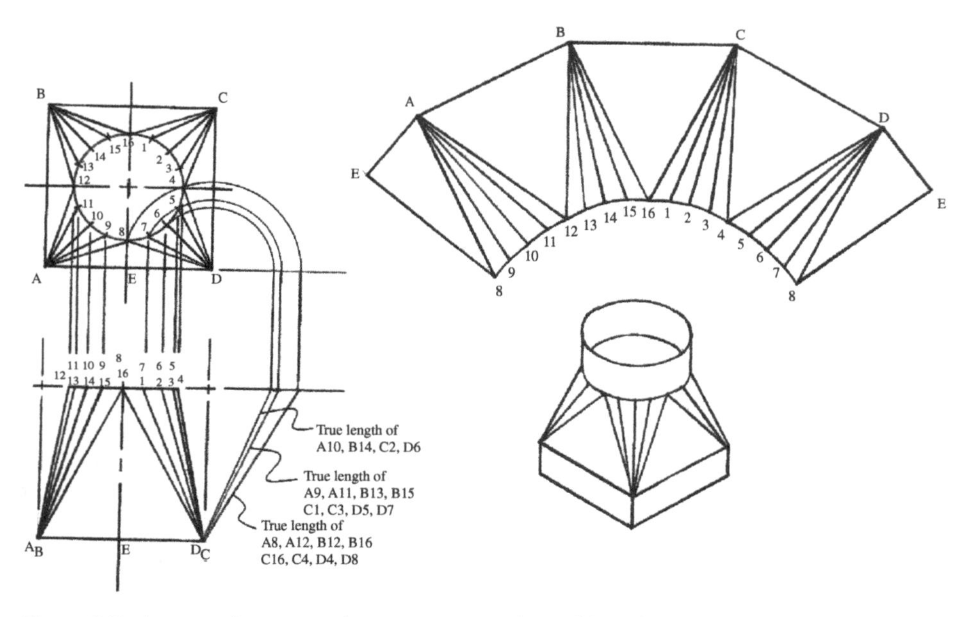

Figure 3.3 Layout of a concentric square-to-round transition piece

In the layout of the flat shape of sections, triangulation is used to find the true length of the lines, and again to develop the flat layout of the pieces. This is illustrated in Figure 3.3. In this construction, the only dimensions that can be picked directly from the drawing are E8 and the comparable dimensions on the other four sides. To obtain each other dimension in the flat layout from the plan and elevation, a triangle is constructed using the height of the transition and the distance in the plan view from point A, B, C, or D to the point in question. For example, to find D4 or D8, or any of the comparable dimensions with respect to the other corners, an arc is struck in the plan view with D as its center and D4 (and D8, by symmetry) in the plan view as its length. This length is carried to the horizontal line at the top of the elevation, and the true length is read from that point to D in the elevation.

Although this example shows a concentric square-to-round transition piece, the same technique using triangulation to find the true length of lines is used for rectangular and eccentric transitions. The difference is that more lengths must be triangulated because of the lack of symmetry.

3.5 Material Thickness and Bending Allowance

The same techniques are used for sheet metal layout as for plates, but because the sheet material is relatively thin a thickness allowance is often not needed. Because pressure vessels are fabricated using plates, usually 1/4 in. (6 mm) or greater, thickness must be considered. The neutral axis of the plate is the plane in the mid-section of a plate that is neither compressed nor stretched when the plate is bent. It is therefore the location that is considered in creating the flat layout of a plate that will later be formed. It can also be described as the plane in the plate at which there are neither tensile nor compressive stresses. When a plate is rolled to a large radius, it is generally safe to assume the neutral axis to be at the middle of the plate. For a tight radius, however, the neutral axis tends to move toward the inside of the bend. Opinions among experts and operators differ as to the location of the neutral axis in various situations. Example 3.2 illustrates three different methods that are used in developing the plate width required to form an angle.

Since fabrication drawings normally provide either inside or outside dimensions, and since bends are not of zero radius, a bend allowance is usually provided to ensure accurate forming.

Example 3.1

A plate of thickness $t = 1$ in. will be formed to create a cylinder with a 60 in. outside diameter D_o. What length of plate is required?

Solution

Since t/D_o is small, the neutral axis is assumed to be at the midpoint of the wall. Thus, the diameter of the neutral axis is given by

$$D_n = 60 - 2 \times 0.5 = 59$$

The circumference of the neutral axis is thus

$$C_n = \pi \times 59 = 185.35$$

This will be rounded to 185-3/8 in. This is the net length of the shell plate in the circumferential direction. If it is determined that allowance for flats at the end of the cylinder is not required, then this is also the cut length of the shell. If the forming rolls to be used do not have pinch (crimp) capability, then either re-rolling may be required after welding or extra material can be left to allow for pre-bending ("pinching" or "crimping") in a press brake. The amount of extra material to be allowed will be determined by the size of the bottom die to be used in the press, usually about $10t$ wide. Thus, 5 in. plus enough material to overhang the die during forming must be left at each end, for a total initial cut length of about 198 in. The flat layout will include marking for cutting off the excess material after pre-forming the ends. Also, after pre-forming the ends, the ends will be beveled for welding prior to rolling.

3.6 Angles and Channels

Standard structural steel sections are typically less expensive per pound than the cost of plate plus labor for bending of a formed channel. They are therefore used for supports and other attachments when possible. Vessel saddles for large horizontal vessels are much wider than typically available structural products, however, so formed channels are often used. Also, the range of available materials for structural sections is limited.

Saddles can also be produced by welding plates together and this is often done, but for a shop with a press brake of sufficient capacity, it is often much less expensive to form a large channel. Layout costs are about the same, but there is less cutting and much welding is eliminated. Also, formed channels and angles can be produced in standard plate thicknesses and of any formable alloy. On the other hand, a formed channel will almost always have a larger inside bend radius than is found in structural sections and a significant outside radius. This is usually not a concern, but should be considered in the decision-making process. Also, it should be noted that, if available, a standard steel shape requires only cutting to length and contour and may therefore be less expensive than a formed angle or channel.

For flat layouts, there is agreement that using the neutral axis of the plate will yield a correct layout. There is less agreement as to the location of the neutral axis. The most common assumption is that the neutral axis is at the midpoint of the plate, but various sources assume different locations. The following example illustrates two common approaches to calculate the developed length of a piece to be formed to create an angle.

Example 3.2

Consider a plate of $t = 1$ in. to be formed to make a 6 in. × 6 in. angle as in Figure 3.4. Assume the inside radius is equal to the thickness, in this case 1 in. An inside radius equal to the thickness is a commonly accepted limit on forming mild steel to avoid cracking on the outside of the bend. Calculate the length of the material, L, that must be added to the straight sections in order to produce this angle. This length is commonly referred to as the bend allowance.

Solution

Various approaches are used to determine the forming allowance for angles. Fortunately, the tolerances on such angles are not usually critical and the differences in results are not great.

Method 1: One of the most common approaches is to assume the neutral axis to be at the midpoint of the thickness and calculate the length of the arc. The bend allowance, or developed length L of the bend, is then given by

$$L = \frac{1.5\pi}{2}$$

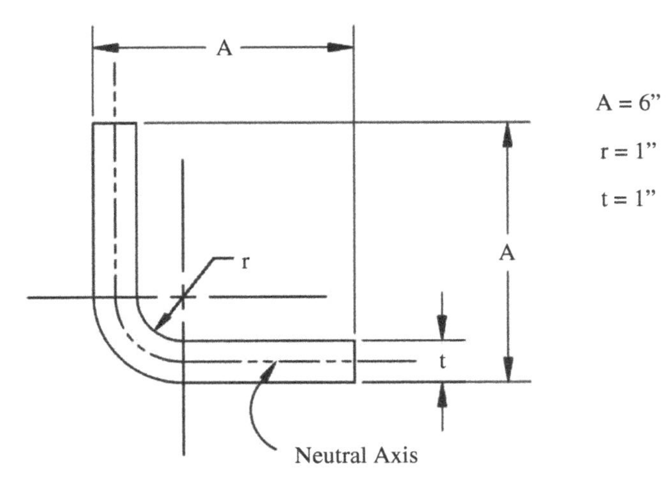

Figure 3.4 Formed angle

or,

$$L = 2.36 \text{ in.}$$

Method 2: Machinery's Handbook [1], a commonly used reference in the metalworking industries, provides three formulas for the bend allowance for 90° bends, depending on the material. For soft brass and soft copper, it offers

$$L = 0.55t + 1.57r$$

For the case above, with $t = r = 1$ in., this yields

$$L = 2.12 \text{ in.}$$

For half-hard copper and brass, soft steel, and aluminum, the formula is

$$L = 0.64t + 1.57r$$

or,

$$L = 2.21 \text{ in.}$$

For bronze, hard copper, cold-rolled steel, and spring steel,

$$L = 0.71t + 1.57r$$

or,

$$L = 2.28 \text{ in.}$$

Method 3: This method is used by some fabrication facilities for the case of an inside bend radius equal to the thickness, generally recognized as a minimum bend radius. Calculations are simplified by just using the sum of the inside dimensions of the angle. Thus,

$$L = 2t$$

and

$$L = 2 \text{ in.}$$

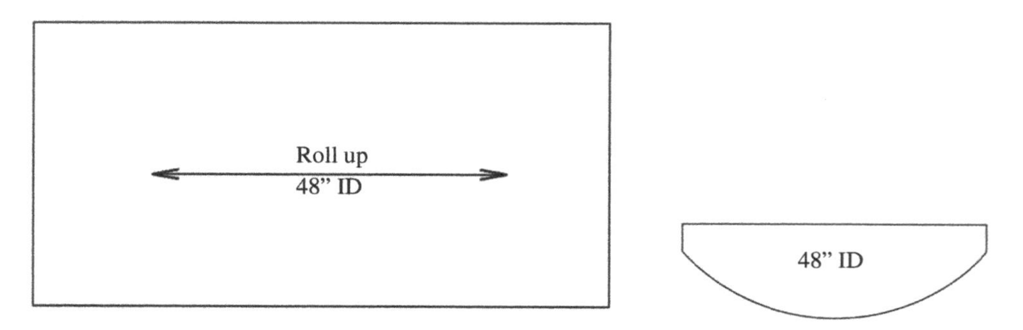

Figure 3.5 Plate layout for rolling, with rolling template

Method 1 places the neutral axis midway through the thickness of the material. Method 2 places it, for the case above, 28%, 40%, or 45% of the thickness from the inner wall. Method 3 places it at about 27% out from the inner wall, yielding almost exactly the same value as that given by the first formula in Method 2.

For tight bends, there is some validity to using a value of the neutral axis that is somewhat toward the center of the radius from the midpoint of the material, due to Poisson expansion of the inner wall. The differences in length, compared to the size of the formed product, are not great.

To form the above angle in a press brake, it would be usual to use a bottom die not less than $10t$, or 10 in. wide. Because the legs of the angle must be large enough to span the die when fully formed, either a smaller die would have to be used to finish the forming, or both legs would need to be longer than 7 in. (the 6 in. angle described above could not be formed without substituting a second bottom die to finish the forming operation). On an angle of this size, the differences cited above are not likely to be significant. On thinner plates or sheets, the differences in the calculated bend allowances would be proportionately smaller.

Different fabricators use different bend allowances, and all seem to be successful. If a large number of parts is to be produced, then making test products and optimizing the process are valid, while for most applications related to the fabrication of pressure vessels, optimization is probably not justified.

3.7 Marking Conventions

As noted in Section 3.4, flat plate layouts can be developed manually or using software. Manually developed layouts are plotted full size for use with pattern burning machines or for direct marking onto the plates to be cut. Layouts developed using software can either be plotted or be sent electronically to plate cutting machines. Flame, plasma, laser, and water jet cutting are all available options, all with full automated capabilities.

Once a plate is cut, further layout helps communicate information to shop personnel for follow-on operations. Most plates used in pressure vessel fabrication are not used without forming. This communication with shop personnel is critical to ensuring proper forming and fit up. A skilled roll or brake operator can form a cylinder or other shape as needed, but only if the need is communicated.

Sheet metal and forming handbooks provide guidance on how to transfer information effectively from the layout loft to the shop floor, and some companies have their own internal standards and conventions. While these conventions often vary to some extent, they also have some commonality.

Marking usually uses paint sticks, marking crayons, soapstone, or felt tip markers, with the addition of center punch marks when there is a need for particular accuracy. Caution should be exercised when using paint sticks or marking crayons on unpainted surfaces as the residue may react with the fluid in the vessel and cause local corrosion.

A plate to be rolled will normally be marked on the inside with a long arrow in the direction of rolling and a notation "Roll up XX ID," with XX indicating the inside diameter. A forming template of sheet metal, Masonite, or other rigid material will be provided to verify the contour as the rolling progresses. The layout department often maintains a number of templates for commonly formed radii or shapes.

For brake work, brake marks will normally be indicated at least at the two edges of the plate, and, if not extending the full width of the plate, will often be numbered to allow operators working at opposite sides of the plate to verify easily that both are working to the same mark. Brake marks are further differentiated from other markings by the addition of symbols centered on the line. A diamond symbol is common, and a circle symbol is sometimes used.

For 90° (or other angle) bends, there will be an indication "UP XX" or Down XX," with up or down providing he direction, and XX designating the included angle in degrees, Figure 3.7.

Plates to be "bump rolled" will have continuous lines across or a series of brake marks along each edge, numbered, and with an indication of the radius to be attained, Figure 3.8.

If a cone is being formed, a template will be provided for both the large and the small radii. Note that the cone layout shown in Figure 3.2 is for the full 360° layout. If the cone cannot be formed and

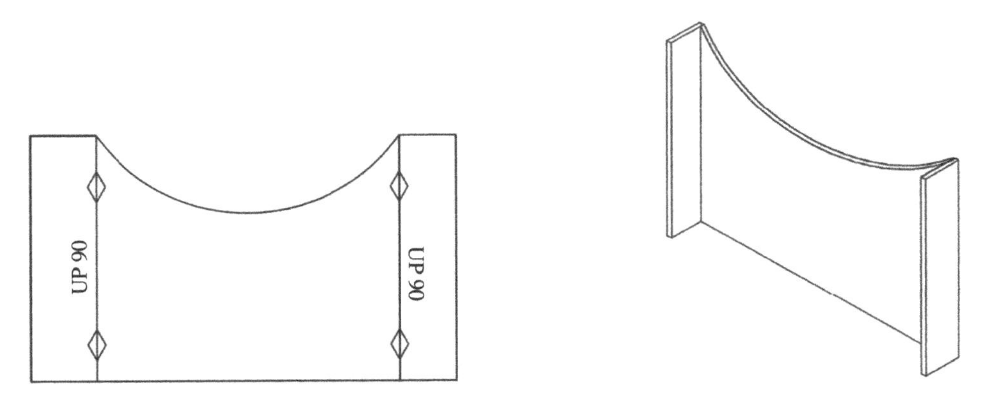

Figure 3.6 Plate layout for formed channel for vessel saddle

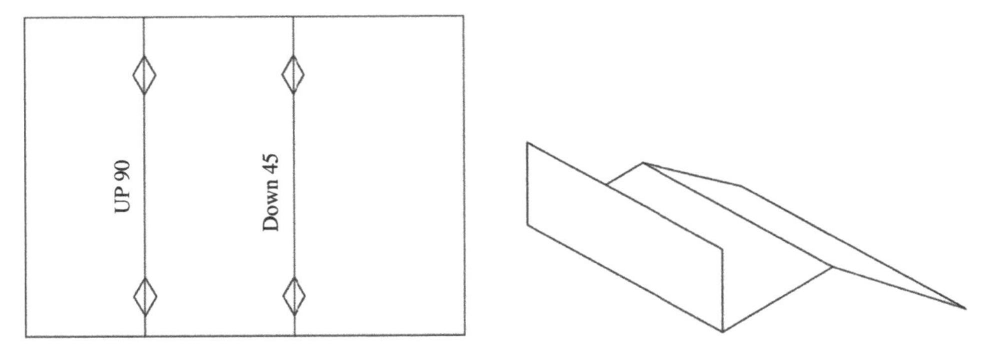

Figure 3.7 Plate layout illustrating opposite direction bends

Figure 3.8 Plate layout for bump rolling ends of plate

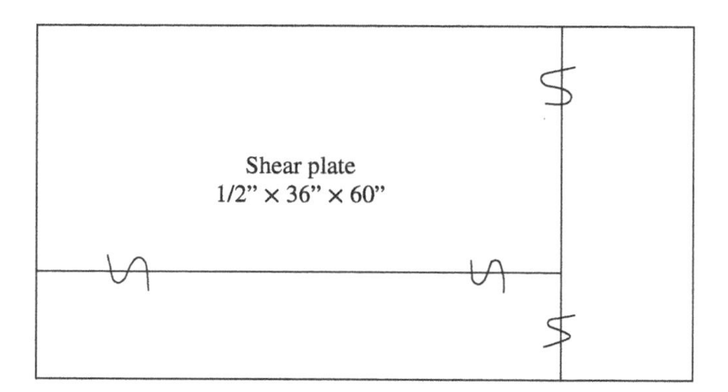

Figure 3.9 Plate layout for shearing plate rectangle from larger plate

slid off the end of the brake ram, it will have to be formed in two pieces and will have two longitudinal welds instead of one.

Plates to be further cut will have cut lines marked. Shear cuts are often indicated by a letter "S" across the lines, Figure 3.9.

3.8 Future of Plate Layout

As noted at the beginning of this chapter, the layout profession has evolved significantly in recent years. As software becomes further integrated with material cutting and forming equipment, the evolution will continue. As obsolete equipment is phased out, the changes will accelerate. However, human expertise will always be needed to develop and interpret the software and its products, and there will always be a need for someone to address the special cases not covered within the software.

Reference

1 E. Oberg et al., 1996, "*Machinery's Handbook, 25th Edition*". Industrial Press, South Norwalk, CT.

4

Material Forming

4.1 Introduction

Forming is the general term referring to the bending and stretching operations performed to shape components from plate, pipe, tube, or structural shapes through plastic deformation without the addition or removal of material.

The operations performed range from simple bending of plate to make an angle, through three-dimensional forming of heads, flues, and other components. While specialized terminology is often used for various operations, the terms “forming” and “bending” are both broadly applied. Thus, for example, a head may be said to be dished, spun, pressed, drawn, and in some cases even forged, depending on the specific process used. But it is always formed or bent.

This chapter describes forming of plates as well as other products such as pipes and tubes. It covers both two-dimensional bending, such as forming a plate into an angle or a cylinder, and three-dimensional forming, as in producing a head. After a brief description of various issues applicable to all metal forming, the most commonly used forming tools are discussed, along with their advantages and disadvantages.

4.1.1 Bending versus three-dimensional forming

Forming uses tensile, compressive, or combined tensile and compressive loads to produce desired material shapes. These loads are applied in a variety of ways.

For simple bending, the most common tools are forming rolls and press brakes. For three-dimensional forming, large C-presses and various progressive forming machines are used.

In both cases, the forming process involves straining the material such that permanent deformation occurs. In simple bending such as forming an angle or bending a pipe to form an elbow, a moment is normally used to induce uniaxial strain that varies across the section. A longer moment arm produces greater moment, allowing forming of heavier sections. When performing three-dimensional forming, the metal must typically be stretched or compressed biaxially, as when dishing a head. Thus, through section yielding usually occurs in both directions. The loads required for this are generally significantly higher, consequently requiring larger and heavier equipment.

4.1.2 Other issues

When forming plates or bars, spreading may occur at the edges along the inner surface of the bend and cracking at the outer surface. This spreading is often referred to as “mushrooming.” It

Fabrication of Metallic Pressure Vessels, First Edition. Owen R. Greulich and Maan H. Jawad.

Figure 4.1 Test sample showing change of geometry due to bending

is common to grind the plate edges prior to forming to reduce stress concentrations and the likelihood of cracking. A severely bent piece of plate or bar might suffer wrinkling of the inner surface and texturing of the outer surface. Figure 4.1 shows spreading of the inside fibers, contraction of the outside, curvature on the outside of the bend, and texturing of the surface on the outside surface, reflective of the varying material properties. A small crack or delamination is also visible.

In tubes or pipes, thickening of the wall at the inside of the bend and thinning at the outside are expected due to Poisson's effect.

Occasionally, axial compression is applied during bending, for example, during forming of tight tube and pipe bends, which causes the neutral axis of the component to move toward the outside of the bend during bending. It increases thickening of the pipe wall on the inside of the bend and reduces thinning on the outside of the bend. The reduced thinning of the outer wall of the bend has the benefit of reducing the likelihood of cracking or overstress on the outside of the bend. The application of axial loads complicates the forming process and requires special dies. It is a practice used only when production quantities justify the additional cost.

When a flanged and dished head is formed, thickening is anticipated in the knuckle and the flange area due to the compression involved in reducing the diameter of the flat plate as it is rolled down into the knuckle and straight flange. Thinning is expected in the crown region due to biaxial stretching.

Three-dimensional forming involves a complex stress state involving tensile, compressive, and shear stresses and strains and it requires special dies, higher tonnage, and skilled operators.

4.1.3 Plastic Theory

This section provides a brief introduction to the portion of plastic theory most likely to be useful to fabricators. Analysis of stresses in simple bending is relatively manageable without complex models or special modeling software. Forming of three-dimensional products such as heads is much more

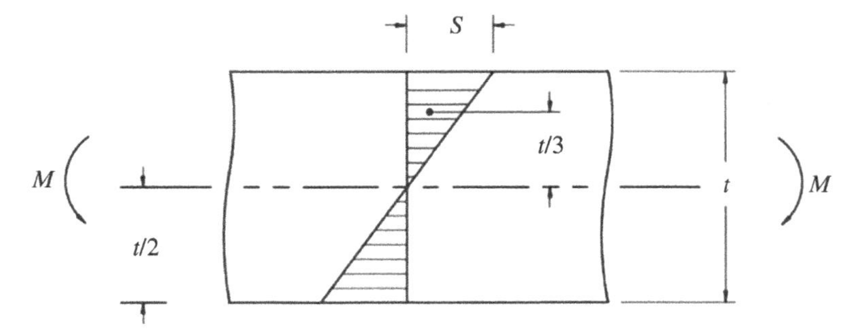

Figure 4.2 Stress distribution up until extreme fiber yield

difficult than two dimensional forming such as bending, and an adequate treatment of this topic requires far more space than can be allocated to it in this book. Therefore, only simple bending is addressed here.

Equipment designed to bend a plate to produce an angle or a pipe to form a bend, accomplishes it by inducing a moment. In this way, one side of the material is subjected to compression and the other side to tension, Figure 4.2. As the strain increases, the outer fibers yield, and yielding progresses inward, Figure 4.3. At some point, the full section has yielded, Figure 4.4. After the material has yielded sufficiently, the moment is removed and the product springs back to the point at which stresses are balanced through the material. This results in a more complex

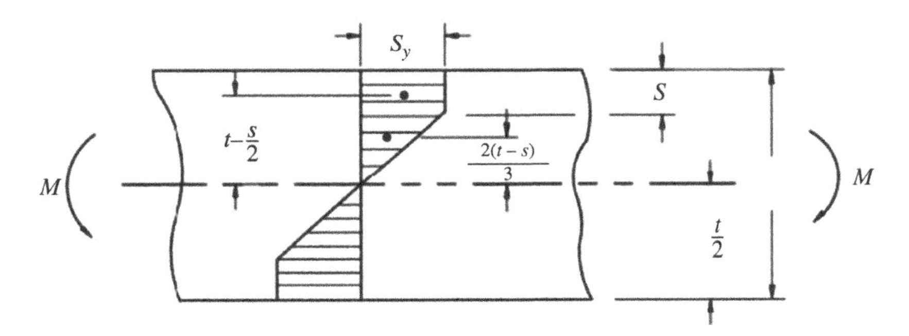

Figure 4.3 Stress distribution after extreme fiber yield and before full section yield

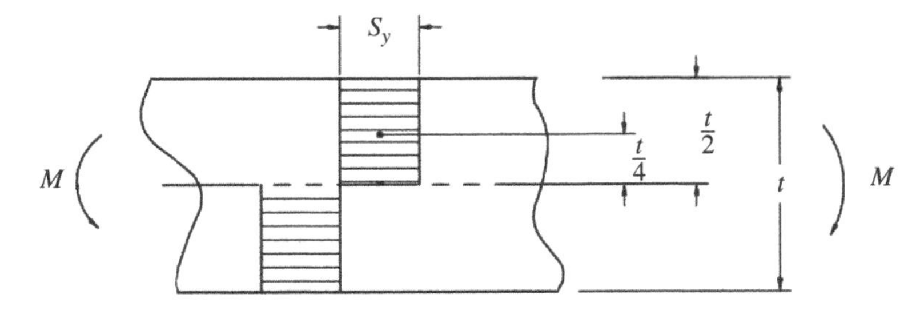

Figure 4.4 Stress distribution after full section yielding

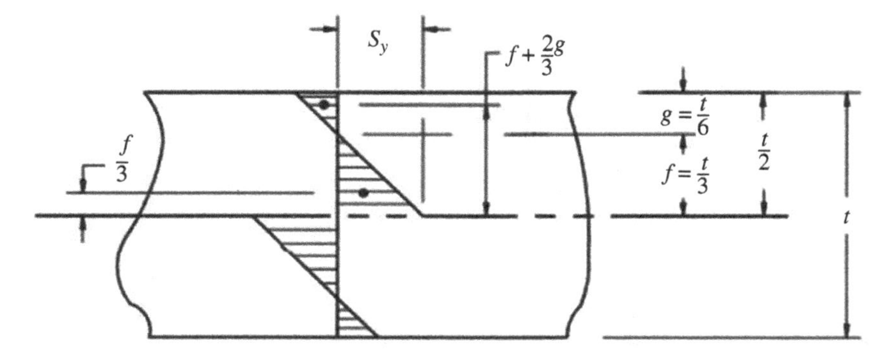

Figure 4.5 Stress distribution after full section yielding and removal of bending moment

stress distribution shown in Figure 4.5. For ease of modeling, we will consider a unit width of the plate. The analysis that follows could be performed equally well and more generally using integrals.

Assuming a linear elastic perfectly plastic material, a simple model of through section stress for a plate in bending is illustrated in Figures 4.2 through 4.5.

These simplifying assumptions are approximations, but are useful in visualizing and understanding the stresses and loads associated with metal forming.

Using the simplification in Figure 4.2, the load–moment relationship for a plate of width b can be expressed as

$$M = FLb \tag{4.1}$$

where

F = the load in tension (above the neutral axis) or compression (below)
L = the moment arm from the neutral axis to the centroid of the load
b = the width of the plate being bent

It can be seen that

$$F = \left(\frac{1}{2}\right)\left(\frac{t}{2}\right)\text{Sb} \tag{4.2}$$

and

$$L = \left(\frac{2}{3}\right)\left(\frac{t}{2}\right) \tag{4.3}$$

Thus, the total resisting moment for a given value of stress is equal to

$$M = 2\left(\frac{1}{2}\right)\left(\frac{t}{2}\right)\text{Sb}\left(\frac{2}{3}\right)\left(\frac{t}{2}\right) \tag{4.4}$$

or

$$M = \frac{St^2b}{6} \tag{4.5}$$

And if the point of incipient yield is desired, S is set equal to the yield stress S_y, yielding

$$M = \frac{S_y t^2 b}{6} \quad (4.6)$$

This is the moment required to begin permanently deforming a plate or bar in bending. It should be remembered that S_y in Eq. (4.6) is the actual yield stress of the material in question, not the specification value. Thus, for example, for SA-36 steel plate, with a minimum yield stress of 36 ksi (248 MPa), the value to be used in this equation is more likely to be in the range of 42 ksi (290 MPa). This is because actual material yield stress nearly always exceeds the minimum by a significant amount. For SA-36, experience indicates 42 ksi (290 MPa) as a typical value.

Similarly, the moment required for full yielding across the section of a plate is given by the same moment formula:

$$M = 2FLb \quad (4.7)$$

But with F and L values reflecting the different stress condition and moment arm. Thus,

F = the load in tension (above the neutral axis) or compression (below)

$$F = \left(\frac{t}{2}\right) S_y b \quad (4.8)$$

L = the moment arm from the neutral axis to the centroid of the load

$$L = \left(\frac{1}{2}\right)\left(\frac{t}{2}\right) \quad (4.9)$$

Therefore, the moment to fully yield a plate or bar is represented by

$$M_y = 2\left(\frac{t}{2}\right) S_y b \left(\frac{1}{2}\right)\left(\frac{t}{2}\right) \quad (4.10)$$

or

$$M_y = \frac{t^2 S_y b}{4} \quad (4.11)$$

Again, the actual material properties must be used. For mild steels, the yield provides a fairly good representation of required load. For high strength and work hardening metals, the ultimate strength may better reflect actual loads.

Finally, consider the stress state of a formed plate or bar after full yielding, which springs back elastically upon release of the forming load. This condition is represented in Figure 4.5. As shown in Figure 4.4, the material has fully yielded, but the loads are unbalanced if the moment is released. Superposing the spring back stress (which resembles the reverse of the stress shown in Figure 4.2) onto the forming stress in Figure 4.4 results in the stress distribution shown in Figure 4.5. For ease of manipulation, the two triangles have vertical legs dimensioned as "b" for the inner triangle and "a" for the outer.

Setting the moments of the two triangles equal to each other yields

$$\left(\frac{1}{2}\right) f S_y \left(\frac{f}{3}\right) = \left(\frac{1}{2}\right)\left(\frac{g}{f} S_y g\right)\left(f + \frac{2g}{3}\right) \quad (4.12)$$

Simplifying yields

$$\frac{f^2}{3} = \left(\frac{g^2}{f}\right)\left(f + \frac{2g}{3}\right) \tag{4.13}$$

Further simplifying and letting $g = 1$ gives

$$f + \frac{2}{3} = \frac{f^3}{3} \tag{4.14}$$

The solutions to Eq. (4.14) are $f = 2, -1$, and -1.

The negative solutions can be discarded because they represent loads on the other side of the neutral axis, which is not what we are looking for.

Thus,

$$f = 2g \tag{4.15}$$

so

$$f = \frac{t}{3} \text{ and } g = \frac{t}{6} \tag{4.16}$$

And the residual stresses are approximated as shown in Figure 4.5.

These approximations apply whether the moment is achieved through the use of a press or rolls, but only as approximations. In actual metalworking, many factors such as material strength (including variability), material thickness, rolling (grain) direction, work hardening, friction over the die, and die parameters are involved in producing the actual bending force required. Thus, while knowledge of the calculated moment is helpful, it alone is not sufficient for successful metal forming.

Manufacturers of metal forming equipment typically provide forming load charts showing load per unit length for various types and thicknesses of material and various widths of bottom die. Such charts can be produced for any material of known properties. It is common, however, to use a single chart and to apply a formability factor for other materials. Table 4.2 shows a typical forming load chart.

Press brakes and forming rolls are typical tools used to induce the required bending moments.

Through thickness yielding normally occurs in both transverse directions in three-dimensional forming. In the crown of a head, the yielding is tensile in both directions. In flanging, however, there is a combination of compressive circumferential stresses and a bending moment in the radial direction. Developing sufficient force to produce this through section yielding is often difficult. Sometimes large presses are used to press the material into a die. In other cases, progressive forming is used, as in spinning or rolling a head.

When the required loads become too high, or when equipment that is sufficiently heavy is not available, the material may be heated to facilitate forming. A sufficient increase in metal temperature will greatly reduce the required forming loads. Heating is also used when the material properties would be negatively affected by the amount of deformation or when the material would simply fail during processing if not heated.

The above analysis can be used to calculate forming loads to determine whether an article can be formed on a particular piece of forming equipment. See the following example:

Example 4.1

A high strength steel plate is inserted in a pyramid roll as shown in Figure 4.6. The roll has a working capacity of $P = 175$ tons (350,000 lb) and the spacing of the lower rollers $L = 24$ in. Thickness of the plate is $t = 1.25$ in. and the width of the plate being bent is $b = 96$ in. The yield stresses at various temperatures, taken from actual plate samples, are given as follows.

Temperature, °F	100	200	300	400	500
Yield stress S_y, ksi	75	66.3	60.7	56.3	53.1

Determine whether the plate can be formed at room temperature or whether it requires heating.

Solution

The maximum bending moment in the plate is $M = PL/4$

$$M = \frac{FL}{4} \tag{4.17}$$

Substituting this value into Eq. (4.11) and rearranging the terms give

$$F = \frac{bt^2 S_y}{L} \tag{4.18}$$

Substituting the actual parameters yields

$$F = \frac{(96)(1.25^2)(75,000)}{24} \tag{4.19}$$

or

$$F = 468,750 \text{ lb} \tag{4.20}$$

Since this value is greater than the roll capacity of 350,000 lb, the plate must be heated if it is to be formed on this piece of equipment.

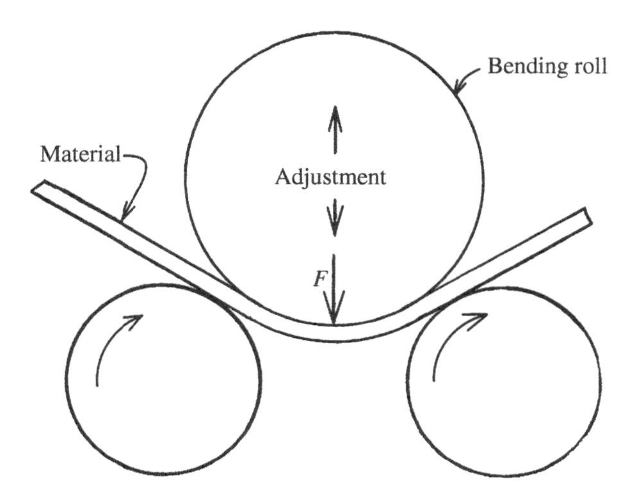

Figure 4.6 Plate inserted in a pyramid roll forming machine

Reformatting Eq. (4.18) gives the formula for the maximum yield stress for these conditions, or

$$S_y = \frac{FL}{bt^2} \tag{4.21}$$

Thus,

$$S_y = \frac{(350,000)(24)}{(96)(1.25^2)} \tag{4.22}$$

$$= 56,000 \text{ psi} \tag{4.23}$$

It can be seen from the table provided with this example that the plate must be heated to a minimum of 400°F in order to be rolled on this machine.

4.1.4 Forming limits

Both direct physical considerations and limits placed by codes govern the extent to which material can be formed. The limits in the codes are, of course, just codifications of experience harkening back to the direct physical and engineering considerations. They are typically for categories of material that have unique limits and forming requirements.

For some materials and applications, the only forming limits are those imposed by the materials themselves. The ASME Boiler and Pressure Vessel Code (BPVC) specifies limits for the following specific applications.

4.1.4.1 For carbon steels

Paragraph UCS-79 of ASME Section VIII, Div. 1 has many requirements related to the extent of forming. It specifies that carbon steel pressure parts fabricated by cold forming shall be heat treated subsequent to forming that results in extreme fiber elongation exceeding 5% from the supplied condition except for P-No. 1, Group Nos. 1 and 2; and P-No. 15E materials (Chapter 8 defines P materials). Practically, it may be noted that this limitation means, for example, that 1 in. (25 mm) plate may not be formed to less than a 10 in. (250 mm) mean radius without heat treatment.

For P-No. 1, Group Nos. 1 and 2; and P-No. 15E materials, heat treatment is required when extreme fiber elongation exceeds 40%, or for various specific conditions specified in UCS-79(d)(1) through UCS-79(d)(5).

4.1.4.2 For high alloy steels

If forming exceeds certain limits depending on temperature and alloy, then if material is worked below the temperature specified in Table UHA-44 of ASME Section VIII, Div. 1, solution annealing is required.

4.1.4.3 For ferritic steels with tensile properties enhanced by heat treatment

For ferritic steels with tensile properties enhanced by heat treatment, Paragraph UHT-79 of ASME Section VIII, Div. 1 specifies that pieces formed after heat treatment at a temperature lower than the final tempering shall be heat treated in accordance with Table UHT-56 when extreme fiber elongation exceeds 5%.

4.1.4.4 Calculation of forming limits

The following ASME BPVC table provides formulas for calculating forming strains.

The following example illustrates the calculation of strain to determine whether heat treatment is required after forming.

Example 4.2

The 1.25 in. thick high strength steel plate in Example 4.1 is rolled to a 40 in. mean diameter. Does it require heat treatment after forming?

Solution

The strain after forming is calculated using the appropriate formula in Table 4.1 as follows:

$$\varepsilon_f = \left(\frac{50t}{R_f}\right)\left(1 - \frac{R_f}{R_o}\right) \tag{4.24}$$

Thus,

$$\varepsilon_f = \left(\frac{(50)(1.25)}{20}\right)\left(1 - \frac{20}{\infty}\right) \tag{4.25}$$

$$= 3.1\% \tag{4.26}$$

Since this value is less than 5%, no heat treatment is required after forming.

4.1.4.5 Other factors affecting formability

Factors in material formability include surface finish of both the material to be formed and of the forming dies, metallurgical cleanliness (less tramp elements and less inclusions leading to greater forming success), forming technique, die selection, and lubrication. Forming using a poorly maintained, unlubricated dies, for example, increases the tensile stress in the extreme outside fiber over that for a smooth lubricated die, thereby increasing the likelihood of failure. This is exacerbated by stress concentrations in the material if the surface is rough or heavily scaled.

Table 4.1 ASME-1 [1] equations for calculating forming strains

Type of part being formed	Forming strain
Cylinders formed from plate	$\varepsilon_f = \left(\frac{50t}{R_f}\right)\left(1 - \frac{R_f}{R_o}\right)$
For double curvature (e.g., heads)	$\varepsilon_f = \left(\frac{75t}{R_f}\right)\left(1 - \frac{R_f}{R_o}\right)$
Tube and pipe bends	$\varepsilon_f = \frac{100r}{R}$

General note:

ε_f = calculated forming strain or extreme fiber elongation
R = nominal bending radius to the centerline of the pipe or tube
R_f = final mean radius
R_o = original mean radius, equal to infinity for a flat plate
r = nominal outside radius of the pipe or tube
t = nominal thickness of the plate, pipe, or tube before forming

Source: ASME [1].

4.1.5 Grain direction

It is recognized that whenever possible it is best to form material across (perpendicular to) the direction of the grain (the mill rolling direction). For some materials, this is simply a "nice-to-have," but for others it is critical, particularly for tight bends.

Many carbon and high alloy steels can be formed to an inside radius equal to the material thickness regardless of forming direction. Aluminum, magnesium, titanium, and other alloys are typically less forgiving. Even in the mill rolling direction, forming of these materials may require a larger radius or heating for a successful bend.

4.1.6 Cold versus hot forming

The practice of hot forming metals can greatly reduce the loads on equipment as well as the likelihood of material failure during the forming process. If the forming temperature is high enough, it may also eliminate the need for post-forming heat treatment. In addition, it almost eliminates the problem of spring back.

The temperatures required to significantly reduce forming loads or to forgo heat treatment, however, are so high as to make material handling difficult and dangerous. For example, the temperatures specified in Paragraph UHA-44 and Table UHA-44 of Section VIII, Div. 1 are in the range of 1900°F–2000°F (1040°C–1095°C). Also, for creep-strength-enhanced ferritic steels, "cold forming' is defined as any forming performed below 1300°F (705°C).

At such temperatures, there is no question of material being manipulated by hand, even using gloves and other protective equipment. Special tooling and handling equipment are therefore needed. In addition, working materials at these temperatures is very hard on the equipment itself, potentially inducing thermal stresses in the equipment and affecting its strength as well.

For thick plates and thick-walled or large-diameter pipes that would otherwise be beyond the capability of available equipment, however, hot forming is a necessity. Certain wall-thickness-to-diameter ratios would be unattainable by cold forming. This is sometimes caused by limitations of equipment, and sometimes by the inability of the material to sustain the required deformation at ambient temperatures.

4.1.7 Spring back

Whenever a material is cold formed, it can be expected to spring back a small amount toward its original shape. This happens much less, or not at all, with hot forming. How much a plate springs back depends on material, thickness, degree of bend, die selection, etc. Calculations of spring back are possible, but accuracy is limited due to a variety of parameters. Generally, the amount that a plate deflects before any permanent deformation occurs provides an approximation. An experienced press worker is often the best judge of how much spring back to expect.

Spring back is much reduced when the material is "coined," or deformed in compression by the bottoming of the die on it. Unfortunately, this is rarely possible with plate forming due to the extremely high loads required and the extreme loads on the dies. As a result, most plate bending is air bending and accurate work requires continual checking of contour with a template.

A consequence of spring back and its variability is that it is often difficult to bend a product directly to its final form. Instead, it is usual to "sneak up" on the required bend radius because over-bending is often harder to correct than under-bending. In rare cases, the extent of spring back may be such that a die is not deep enough to produce the required bend.

An unrelated phenomenon that is often dealt with in the same way as spring back is the tendency for plates to be very slightly thicker toward the center of the plate than at the edges. This is because when the material is being rolled at the mill, the rolls may deflect slightly. The consequence of a plate being thicker at the middle than at the edges varies, depending on how the plate is being formed.

The stiffness of a press brake ram is typically extremely great because it consists of a large heavy plate on edge. If the dies are sufficiently stiff, and they usually are, the thicker material in the center of the plate may result in a tighter bend in that region. On the other hand, because most forming is done in the center of the brake, dies often get worn in this area, with the reverse effect. Either way, variations in forming across the width of a plate are often compensated for by checking contour, then inserting a piece of thin sheet between the top die and the article being formed, in the region where the bend is too loose. Bringing the die to the same location will not further form the portion of the plate without the additional sheet. The thin sheet, however, has the same effect as bringing the die lower, tightening the bend locally.

When forming heavy plate on a roll forming machine, the effect can be the opposite. The slightly higher section modulus combined with roll deflection can result in a looser bend in the center of the plate. To compensate for this tendency, the top roll is often machined to a slightly larger diameter in the center ("crowned") than at the ends. If this is not sufficient to correct the problem, or if the particular machine does not have crowned rolls, then a piece of sheet can be inserted to counter this effect.

While spring back is rarely a critical problem, it is a phenomenon that must be considered and that often must be corrected.

4.2 Brake Forming (Angles, Bump-Forming)

Press brakes provide the capability of making long bends across steel sections in a wide range of thicknesses. Many press brakes are mechanical and provide high production rates in the forming of sheet metal for cabinets and other applications. Brakes used for heavy plate forming are more often hydraulic, with ratings typically ranging in the hundreds of tons. Figure 4.7 illustrates the general configuration of a typical press brake.

A lower platen accepts a bottom die, while the ram above accepts a top die. Usually, the female die is located on the platen and the male die is attached to the ram, though this may be reversed for special applications. If vee dies are used, both dies are usually made with an angle of less than 90° so that a 90° angle can be bent, allowing for spring back, without bottoming the dies. Die stresses go up immensely when bottoming (also referred to as "coining" the metal) occurs.

Bending of metal without bottoming the dies is often referred to as air bending, Figure 4.8. This is the most common way of forming plate material in a press brake.

While press brakes are often used to put single bends in plates, such as in forming an angle, they are much more versatile than this.

By placing a series of small bends at equal intervals in a plate, a very close approximation to a true radius can be formed. Rolling a plate in this manner is more time consuming than rolling it on a set of forming rolls, but may at times have advantages. Some shops maintain a press brake but no rolls. Maintaining equipment that is only used intermittently is expensive and takes up shop space, and rolling capability may be available from suppliers in the vicinity.

Figure 4.7 Press brake general configuration (*Source:* Marks Brothers, Inc., ASME Fabricator)

Figure 4.8 Hydraulic press with four-way bottom die, forming without contacting the bottom of the vee in the bottom die (*Source:* Harris Thermal Transfer Products)

Figure 4.9 Forming cone sections on a brake (*Source:* Marks Brothers, Inc., ASME Fabricator)

Because forming rolls make a cylinder much more quickly than does a press brake, often a significant amount of time is saved by shipping the material to a supplier with rolls. However, it is accompanied by a loss of control and often a cost in schedule. Also, if the available rolls do not have the capability of pinching the ends of the plate, then a brake can be used to crimp the ends prior to rolling, and for extremely heavy sections the ends may be crimped and remaining flats cut off.

When brake rolling, a large bottom die is generally used. Use of a large bottom die not only results in a larger bend radius but also reduces forming loads significantly. Further, while it seems counterintuitive, the size of the bottom die often has a greater effect on the resulting radius than does the top die. Thus, if a top die with a large radius is used and the bottom die is not significantly larger, a gap often develops between the plate being formed and the center of the top die. A small top die in a large bottom die can result in a large radius in the formed plate, but too small a top die can result in indentations in the plate.

Many forming rolls are very limited in their ability to roll cones because of limited adjustability. Forming a series of radial bends in a plate by using a press brake results in a section of a cone, Figure 4.9. By proper layout the cone can be made either concentric or eccentric. This is a convenient way of producing nonstandard reducers and shell sections. In addition, three-dimensional dies can later be used to provide a knuckle radius at the end of the cone for efficient distribution of stress.

Press brakes can produce eccentric cone segments, in combination with flat triangular sections, to create round-to-rectangular transitions of various shapes and offsets.

4.2.1 Types of dies

Forming dies are available in a multitude of configurations. Companies typically stock a number of dies applicable to the type of work that they most commonly perform. New dies are designed and fabricated as needed for new applications. Vee dies are common for forming thin materials, with a four-way bottom die as a common configuration. To maximize flexibility, a thicker material is usually formed in a heavy rectangular die with spacers used to optimize the die width.

A wide variety of top dies is available to accommodate various applications. Goose neck dies are sometimes used to allow forming of narrow channels, for example.

While it is not the primary use of press brakes, a brake with sufficient clearance and capacity can be fitted with three-dimensional dies. These might be used for fluing of plates or for producing small heads. Crane access around the equipment might make a large C-press more efficient for these operations, but availability and other factors may make the brake a wise choice.

4.2.2 Brake work forming limits

Being restricted by material forming limits is more likely to occur in brake work than in rolling. This is because a brake is often used to form angles or boxes, such as in the construction of rectangular pressure vessels. The tight corner bends involve much higher strains than the large radii involved in rolling the typical cylinder.

For example, consider a cylindrical vessel of 100 in. (2540 mm) in diameter. Constructed of 1 in. (25 mm) plate, the extreme fiber deformation would be 1%.

Consider next a rectangular or square vessel constructed of the same 1 in. (25 mm) plate. If the vessel were fabricated with formed corners, a typical inside radius allowing for forming in a "single hit" would be the thickness of the material, or 1 in. (25 mm). The resulting extreme fiber strain in this case would be 33%.

Considering the forming limits described in Section 4.1.4 and depending on the material, this fabrication is much more likely to require heat treatment. In addition, however, the outside surface of the bend is more likely to suffer from orange peeling or cracking. Mild steel and many stainless steels can normally tolerate this degree of deformation regardless of grain direction. Higher strength steels and aluminum alloys are much more susceptible to cracking if formed across the grain.

Cracks often propagate from the edge of the plate due to roughness and material inconsistencies. Most such cracking can be avoided by locally grinding the edge of the plate smooth and rounding the corners prior to bending.

4.2.3 Crimping

Crimping, also referred to as pinching when performed using a set of forming rolls, is the process of pre- or post-forming the ends of a plate or sheet to be rolled. When a plate is to be rolled on a pyramid roll without pinch capability, a press brake can be used to reduce the flat that would otherwise occur at the ends of the plate. The ends of the plate will be laid out with equally spaced marks indicating where the top die of the press will contact the plate. A series of small bends are then made to approximate the required radius. Contour is checked using a template at both ends of the plate to verify accuracy.

It should be noted that even a plate that is crimped or pinched at the ends will have a flat section at the ends. Depending on the length of the flat, a cylinder may be subjected to further forming after welding. This can take place either in a set of forming rolls, using a brake, or with hydraulic rams and special fixturing.

4.2.4 Bending of pipes and tubes

Bending of pipes and tubes is usually accomplished by using pipe/tube benders. They come as manual benders, for light usage, or hydraulic/electric benders, Figure 4.10, for heavy usage.

Figure 4.10 Industrial bending machine (*Source:* Harris Thermal Transfer Products) and dies for pipes of various sizes (*Source:* Marks Brothers, Inc., ASME Fabricator)

Table 4.2 Brake forming loads

Thickness of metal		Width of Female Die Opening — *Approximate pressure in tons per linear foot for an air bending mild steel (60,000 psi tensile strength) to 90 degrees*																					
Gauge	Dec.	1/4"	5/16"	3/8"	1/2"	5/8"	3/4"	7/8"	1"	1-1/8"	1-1/4"	1-1/2"	2"	2-1/2"	3"	3-1/2"	4"	5"	6"	7"	8"	10"	12"
20	0.036	2.6	2.2	1.6	1.2	1.0																	
18	0.048		3.5	2.8	2.1	1.7	1.3																
16	0.060			5.3	3.7	2.8	2.2	1.7															
14	0.075				5.5	4.6	3.5	3.0	2.5	2.1													
13	0.090					6.4	5.5	4.3	3.6	3.2	2.8												
12	0.105					9.2	6.9	6.2	5.0	4.3	3.9	3.1											
11	0.120						10.1	8.0	7.0	6.1	5.3	4.3	2.9										
10	0.135							10.3	8.7	7.8	6.9	5.7	3.9										
9	0.150								11.9	9.8	8.8	7.0	5.0	3.7									
7	0.188									16.9	13.9	11.2	8.3	6.7	4.9								
1/4"	0.250										27.5	22.1	15.0	11.6	9.6	7.9	6.7						
5/16"	0.312											39.2	26.5	19.3	15.0	12.5	10.4	7.7					
3/8"	0.375												42.7	31.2	23.8	19.5	16.3	12.4	9.6				
7/16"	0.438													45.5	35.2	28.5	24.4	17.4	15.0	11.5			
1/2"	0.500														48.5	39.5	33.2	24.6	19.5	16.1	13.4		
5/8"	0.625															65.5	57.9	42.3	33.1	27.3	23.3	17.0	
3/4"	0.750																92.3	68.1	53.0	36.2	36.2	26.9	21.0
7/8"	0.875																	103	79.9	52.3	52.3	39.2	31.2
1"	1.000																		112	90.4	75.5	55.7	43.7

Heavy border or highlight indicates "ideal" material thickness/bottom die combinations (die width approximately ten times thickness).
Gray zone indicates danger of breaking either the part or the die.

Dies with various radii are usually furnished with benders. Since bending requires stretching of the outside surface of the tube/pipe, thinning of the material must be considered. The amount of thinning can be determined from theoretical and geometric considerations as explained in Appendix E.

4.2.5 Brake forming loads

Brake forming loads can be calculated, but working from a table is much more common. Brake manufacturers provide such charts for their equipment and fabricators use them with a factor for materials with different properties.

4.3 Roll Forming (Shells, Reinforcing Pads, Pipe/Tube)

Roll forming is the process of forming sheet or plate into cylindrical or conic sections using a set of forming rolls ("rolls"). Typical rolls involve three or four steel cylinders with bearings mounted to a heavy plate or casting at each end. The rolls have adjustment of relative location to accommodate various material thicknesses and forming radii. It is common for the top roll to have a release mechanism at one end to allow fully formed cylinders to be removed by sliding them off that end of the machine.

If provision is not made for "pinching" (also referred to as "crimping"), the ends of the plate as it enters and leaves the roll, a significant flat typically results.

Plate rolling machines ("rolls") are available in many configurations. In the past, most were "pyramid" rolls, but most plate rolls currently being produced are either single or double "pinch" rolls. Pinch rolls have been found to reduce labor as well as material waste.

4.3.1 Pyramid rolls

The industry-standard roll forming machine used to be a pyramid roll, Figure 4.6, comprised of two bottom rolls, usually at a fixed distance from each other, and an adjustable top roll located midway between and above them. Many of these rolls are still in service. In this arrangement, the bottom rolls were generally fixed and allowance for varying plate thickness and rolling radius was made by raising and lowering the upper roll.

In some cases, rolling equipment is arranged with the roller axes vertical. This is particularly advantageous when performing hot rolling because scale that forms on the metal surface can more easily fall out of the way. This prevents it getting embossed into the surface of the plate being rolled.

Construction of a pyramid roll is simpler and cheaper than that of a pinch roll because there are fewer moving parts to machine. The equipment is less flexible than more recent roll designs, however, and, as a result, ongoing labor costs tend to be often higher. A typical pyramid roll has two fixed bottom rolls at the same level, with a top roll centered above them and that can be moved vertically. Its use results in a significant flat section at each end of the plate after rolling. This is normally accommodated by either leaving the plate long and cutting off the ends, or forming (crimping) the ends using a press brake in advance of rolling. Even when pinched or crimped, some amount of flat will remain at the end of the plate. If not cut off prior to rolling, this is usually corrected after welding by returning the cylinder to the roll and re-rolling to eliminate the flats. Some pyramid rolls, referred to as pinch pyramid rolls, allow movement of one or both of the lower rolls to form the plate ends.

4.3.2 Pinch rolls

Pinch rolls are made in both three- and four-roll configurations.

In the three-roll configuration, Figure 4.11, two rolls (the pinch rolls) are usually located approximately vertically with respect to each other, with a third roll to one side. The roll at the side facilitates material handling, Figure 4.11(a), and enables bending the leading edge of the plate once the plate is captured by the pinch rolls. After the leading edge is bent, Figure 4.11(b), the plate is reversed and what was before the trailing edge becomes the new leading edge, Figure 4.11(c), and is bent, Figure 4.11(d). Rolling then proceeds to complete the forming of the cylinder, Figure 4.11(e) and (f).

The four-roll configuration of pinch rolls operates in the same manner as the three-roll configuration except that there is no need to reverse the plate as the trailing edge can be formed just before it exits the pinch rolls.

In either case, it may also be necessary to re-roll the cylinder after welding, both because of residual flats and because of welding distortion.

Pinch pyramid rolls operate in a similar manner. The top roll is centered between the two bottom rolls, however, and one or both of the bottom rolls can be moved inward or upward for pinching, Figure 4.12. One forms the leading edge of the plate, Figure 4.12(a), and rolling proceeds, Figure 4.12(b). The operation is completed, Figure 4.12(c), as the other bottom roll pinches the trailing edge, Figure 4.12(c).

4.3.3 Two-roll systems

Some forming rolls are manufactured with only two rolls. This system uses a urethane roll that forces the material against the top (steel) roll. The urethane roll is deflected, forcing the material to take the shape of the steel roll. To increase dimensional precision, it is usual to slip a tube of the required size over the steel roll such that the product is formed to its diameter. Two-roll systems are used for sheet gauge materials only and do not have a significant role in pressure vessel fabrication.

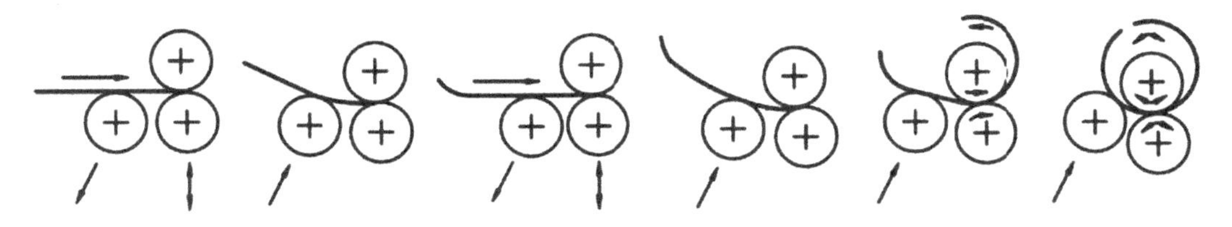

Figure 4.11 Pinch forming the ends of a plate

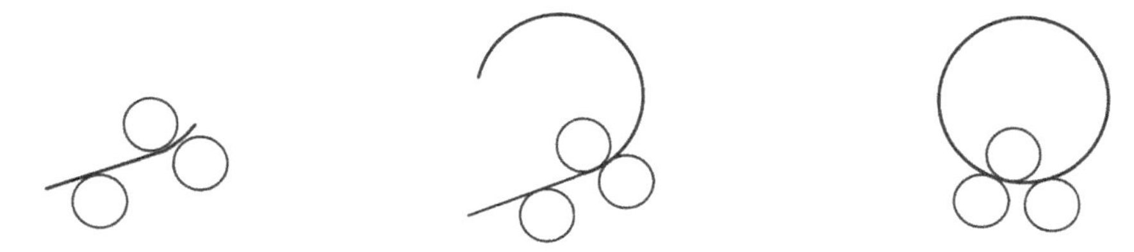

Figure 4.12 Pinch pyramid roll operation

4.3.4 Rolling radius variability compensation

Variability between plates is often such that plate rolls cannot just be set in advance, even if a plate of the same material and nominal thickness has just been rolled. Heavy plate rolls may have rolls with a larger diameter in the center to compensate for roll deflection. Sometimes, either because of roll deflection or plate thickness or property variation, the plate gets rolled to a different radius at the edges than in the center. If this effect is excessive, it can be compensated for. A piece of thin sheet is inserted between the top roll and the plate at the point on the cylinder that needs a smaller radius. The product is then run through the rolls again. This process can be repeated with successively thicker sheets until desired accuracy is achieved. Note that this must be planned for in advance. Flattening a plate that has been rolled to too small a radius is typically much harder than tightening a radius.

4.3.5 Heads and caps

Structurally speaking, the most efficient head is hemispherical (a "hemi"). Forming of a hemi head is much more difficult and time consuming than that of a shallower one. Thus, a hemi head costs more than the shallower head and is less often used. Next in structural efficiency is a semi-elliptical (SE) head, usually formed to the contour of a 2 : 1 ellipse. Finally comes the torispherical head, also referred to as flanged and dished. This head has a spherical dish radius for the crown and a smaller knuckle radius to transition to the flange. The inside dish radius (crown radius) and the inside knuckle radius of a torispherical head are usually specified as percentages of the outside diameter of the head. An 90% ICR–17% IKR head is a very close approximation to a 2 : 1 SE head. A 100% ICR-6% IKR head (often referred to simply as a "100% 6% head") is much less expensive to form, however, and still meets the requirements of the ASME BPVC. It is less efficient structurally, but the material is only a little thicker, and so for most applications this is the head of choice. It also has the advantage of a greater volume for the same crown to crown length, although it also requires an increase in the shell length.

For purposes of our discussion, the terms "head" and "cap" will be used interchangeably. The difference between them is usually only that caps are manufactured and stocked in pipe sizes and schedules (typically in accordance with ASME B16.9 Factory-Made Wrought Buttwelding Fittings), while heads are manufactured to order and in any size and thickness specified.

Most vessel manufacturers purchase heads from companies that specialize in their production. Several things contribute to this: The capital cost is high for much of the equipment used in head manufacturing. It is difficult to keep the equipment busy if your business is not manufacturing heads. And, the mechanics that operate the equipment develop their skills over years. It is hard to maintain these skills making heads only occasionally.

Heads and caps are produced using a number of technologies. Pressing, spinning, a combination of pressing and flanging (similar to a spinning operation), and forging are the most common.

4.3.5.1 Press forming

When material properties, size, and press capability coincide, the quickest way to form a head is to press a plate disk into a die. The operation is quick and requires relatively little skill on the part of the operator. While dies are expensive, if the cost can be amortized over thousands of manufactured parts, overall cost is minimized. It is common to design small pressure vessels around pipe sizes because caps are such an inexpensive way of closing the end of a vessel.

Figure 4.13 Forming of a large head (*Source:* Brighton Tru-Edge)

Caps often are made with a die set for each size. The male die gives the cap its contour while forcing the material into or through the female die, which sizes the outside diameter. Edges will be trimmed, usually while machining a weld prep, after forming.

A large press is used to form hemi heads and the inside crown radius of larger torispherical heads, Figure 4.13. The bottom die may have a spherical radius or it may be a torus (donut shape). A single die set is most efficient for its designed head size, but it can also be used for larger heads.

4.3.5.2 Spinning

Spinning is a progressive forming process. By moving a small amount of material at a time, relatively low-powered machinery can produce surprisingly large and deep heads. Sometimes material thicknesses that might buckle if pressed into a die are spun over a mandrel. The cost of a mandrel can be high, but it may be cost effective if other tooling is not available. If only a few heads are required, if the thickness is not great, and if no heating is required, then depending on the material and thickness to be formed, the mandrel might be made of wood or other inexpensive materials, reducing the cost.

4.3.5.3 Flanging

Many large heads are formed using a combination of press forming for the crown of the head and flanging for the knuckle and flange. While knuckles can be produced by working around the circumference of a head using special dies on a press, the flanger, Figure 4.14, is far more efficient. Similar to spin forming, the flanger uses a combination of inside and outside rollers to turn the flat perimeter of the plate into a knuckle and straight flange. The same dies may be used for several head sizes, reducing setup time, which is often a significant cost in fabrication. Likewise, the flanger is capable of handling a range of knuckle radii, perhaps only occasionally requiring changing of the rollers.

Figure 4.14 Flanging a large head (*Source:* Brighton Tru-Edge)

4.3.6 Hot forming

As noted elsewhere, hot forming can greatly reduce forming stresses. This is particularly useful when forming components from heavy plate.

A number of challenges present themselves when hot forming is performed. A heating furnace is required. Special handling equipment and fixtures are typically used. High levels of heat transfer from the product to the forming equipment can result in temperature drops that affect both formability of the component and the condition of the forming equipment. On the other hand, forming stresses are much reduced and spring back is essentially eliminated.

If the forming temperature is high enough, then post-forming heat treatment may not be needed. If this is the case, then the temperature of the component must be monitored and recorded during forming.

Note that hot forming of heads is sometimes referred to as forging.

4.4 Tolerances

In material forming, tolerances provide the permissible limits of physical dimensions. There is often a conflict between the dimensions needed and those that are readily produced. Actual dimensions of formed components, like all other components, are subject to variability.

Parts falling outside of the permissible tolerances are rejected and either reworked or disposed of, with attendant costs, but perfect dimensions are unobtainable.

Tolerance requirements on formed products typically stem from the need for strength, stability, fit, or a requirement for a particular size needed for a flow or a process.

Requirements for strength are generally addressed by specifying a material and a minimum wall thickness. Such requirements are found for the walls of pressure vessels or pipes, including shell and head sections. For a shell or pipe fabricated of plate, wall thickness will not change significantly as a result of the forming process. Thus, if the proper material and wall thickness are used for fabrication of a shell or a pipe, strength should not be an issue.

Section UG-80 of the ASME Section VIII, Div. 1 code specifies limits on out of roundness for cylindrical, conical, and spherical shells, both for internal and external pressure. For internal pressure, it specifies that the difference between maximum and minimum inside diameters at any cross section must not exceed 1% of the nominal diameter at the location under consideration. For external pressure, this same requirement must be met, but in addition there is a requirement for measuring and limiting deviation from the true circular form. This ensures that stability considerations are accounted for.

In the case of a head, thinning typically occurs in the crown region during forming, and must be accounted for. Similarly, if a pipe or tube is bent, the outer wall of the bend typically is subject to thinning that must be allowed for.

Tolerances for fit of pressure vessel components are often addressed at the fit up stage. A shell course may not fit well with either the head or with the next shell section. Whether this is because of out of tolerance rolling, deflection due to welding, or simply from elastic deflection, a good fitter up will jack or lever things into place, then tack them for welding. This ensures proper alignment, allows efficient welding of joints, and helps with overall roundness tolerance.

4.4.1 Brake forming tolerances

Tolerances achieved during brake forming are controlled to a large degree by the skill of the layer out and the brake operator(s), and are affected by the amount of care that goes into the forming process. Die motion can be made very repeatable, but variations in plate thickness and material properties still make plate forming something of an art.

4.4.1.1 Bump forming

Bump forming is the process of using many small bends to produce a radius or other bend in a part. It is particularly useful for forming cones and transitions, but can also be used to produce cylinders if a forming roll is not available.

Accurate layout and appropriate choice of bend spacing on the part go a long way toward producing an accurate product. With that basis, an accurate template and care on the part of the operator can produce parts of great accuracy. It is difficult to put a specific limit on the tolerances that can be achieved in brake forming because the limit is often determined by the time that a skilled operator puts into producing a quality part. Also, elastic deflection is often enough that a part can be pushed in or out of tolerance fairly easily. Careful fit up then becomes critical.

Meeting the 1% difference in measurements of minimum and maximum diameter mentioned in Section 4.4 is often determined by elastic deflection and by welding residual stresses from the longitudinal weld between halves of a cylinder. As a result, rounding up of the cylinder after welding is a critical part of producing a quality product.

4.4.1.2 Angle and channel forming tolerances

The same things apply to forming of angles as to bump forming of curves. Accurate layout, including consideration of forming radii and adjustment for the material in the bend, is the beginning of an accurate part. The biggest challenge is often to form an accurate pair of channels to be welded together to create a rectangular vessel. An angle can be trimmed on either leg, and the legs of a channel can be trimmed, but the inside dimension between the legs of a channel must be formed correctly. With sufficient reference lines and accurate punch marks for a centerline, an experienced brake operator is able to form a correct part. Careful measurement after forming the first leg of the channel will let the operator know what really happened with the forming allowance. This permits adjustment on the other end, if necessary. Later, if a sixteenth of an inch (1.6 mm) needs to be ground off the leg, it can be, but the space between the legs will be correct.

See also the discussion in Section 4.2 for a means of achieving accurate contours in the face of varying material thickness, properties, etc.

4.4.2 Roll forming tolerances

Like bump-formed cylinders, rolled cylinders are subject to elastic deflection and forming variability due to plate thickness and property variation. Section 4.3.4 provides a discussion of the means of compensating for radius variation. Meeting the Paragraph UG-80 tolerance requirements in ASME Section VIII, Div. 1 code is usually not difficult, but it typically requires re-rolling after welding to compensate for welding residual stresses as well as flats left at plate ends.

4.4.3 Press forming tolerances

When press forming is used to produce caps, dies are typically designed such that tolerances are met with the first hit. There is no reasonable correction for improperly designed tooling.

The dish portion of a large head is often formed on a C-press using a large radiused die set. Multiple strokes are used throughout the dish area to bring the radius to the required dimension. Because the process involves many strokes, repeated all over the head, it is a progressive forming process. Because of this it lends itself to achieving high levels of accuracy through minor local adjustments. As with other forming processes, the radius can easily be corrected to be smaller but not bigger.

4.4.4 Flanging tolerances

The typical flanging machine uses a pair of rollers to bend a flange at the outside of a plate. Often this is a dished plate for a flanged and dished head. Sometimes it will be a flanged only head. As with other progressive forming processes, careful work by the operator is critical. A flanged only head that is too large in diameter is usually more difficult to correct than a flanged and dished head. It is usual for head manufacturers to provide head "tapings" to fabricators in advance of shipping. A taping is an actual measurement of the circumference of the head. Having this in advance of arrival of the head allows the fabricator to produce shell sections that fit very closely to the head. While extra effort can be put into tightening head circumference tolerances, this is an effective way of compensating for any variability. It also allows compensation for any ovality of the head that may occur due to grain direction.

Reference

1 ASME. 2021. "*Boiler and Pressure Vessel Code, Section VIII, Division 1*". American Society of Mechanical Engineers, New York.

5

Fabrication

5.1 Introduction

Fabrication is a broad term used to cover many aspects of manufacturing. As applied in this chapter, the term fabrication encompasses those operations and procedures used to assemble the individual components of a vessel to create the vessel as a whole. Chapter 1 describes the process of producing a pressure vessel, from identification of need through receiving inspection at the user's facility. It is written at a fairly high level. Other chapters provide detailed discussions of specific operations and processes, such as the production of a rolled shell in Chapter 4 and welding processes in Chapter 7.

This chapter takes many processes described in detail in other places and ties them together for a more integrated look at the production of the vessel. It also addresses some of the issues that render fabrication more complex than it might seem when viewed from the outside. It generally follows the order in which fabrication operations occur in practice.

5.2 Layout

Layout, addressed in Chapter 3, is the process by which drawings and other information begin to be translated into a vessel. It combines drawing data with requirements from weld procedures, such as bevel angles and root gaps, and other sources to allow production of parts that will fit together as specified. It often involves geometry, trigonometry, and various triangulation techniques to transform angular and distance information into cut lengths, widths, and contours. Much contour development that used to happen using angular layout, projections, and a lot of work with dividers is now easily and quickly performed numerically. Of course, what is actually done when a computer performs layout work is simply the mathematical equivalent of what was previously done in the layout loft. In some cases, this layout work is transmitted directly to equipment that produces the required component, and in others it still requires a layer out to translate it into a form that is accessible to shop workers.

The layout function is discussed here to the extent that it must be integrated with other operations. Layout must often take place at intervals during the fabrication process, so planning for it is essential to efficient production of a pressure vessel.

Some layout functions must be performed in advance of other operations. An example of this is plate layout for rolling into a cylinder. Whether marked for crimping or not, the plate size, rolling

Fabrication of Metallic Pressure Vessels, First Edition. Owen R. Greulich and Maan H. Jawad.

direction, and rolling radius will be determined and marked. This includes making a decision regarding whether to leave extra material at the ends to allow for flats, which will be cut off later at either end of the plate. Weld preps may be indicated if they are to be produced prior to rolling. Locations of holes for nozzles, thermocouples, etc. may be marked, even if not cut in advance of rolling. And before all of this, to ensure a good fit, it is common to measure the vessel heads that will be welded to the rolled cylinder.

Some layout operations may be delayed until after completion of other portions of fabrication. These operations are typically things that are easier to do after fabrication is partially complete, or for which accuracy would be affected by prior operations. The layout of the hole contour for an offset nozzle could be accomplished in the flat. It is often delayed until after rolling and fit up of the shell to the head, however. This allows confirmation of the hole location with respect to other vessel features and components, ensuring an accurate layout.

5.3 Weld Preparation

Most welds require edge preparation prior to the weld process. Edge preparation is accomplished in a variety of ways depending on the parts to be welded, the material, the weld preparation (weld prep) configuration, material thickness, equipment availability, the weld process to be used, the need for cleanliness, etc. Equipment used for producing weld preps includes hand and automatic grinders, nibblers, routers, torches, mills, lathes, and various cutter arrangements that clamp to a pipe or shell and work their way around it, beveling as they go.

Applying the weld prep to a plate in the flat prior to rolling is often much quicker and easier than performing this operation after forming. While generally true, this is especially the case for the longitudinal weld of a single-piece shell course, since after rolling the abutting plate edges obstruct access to each other.

5.3.1 Hand and automatic grinders

Hand grinders, Figure 5.1, are low tech, easy to use, and versatile. They are typically used with an aggressive hard or soft wheel and are excellent for putting a weld prep on the material up to about 3/8 to 1/2 in. (10–13 mm). If used on aluminum or other soft metals, a special wheel must be used to avoid loading up the wheel with metal. For thick products, the amount of material to be removed makes hand grinders slow and difficult to use, and they become noncompetitive. They may be used, though, to smooth rough burned edges of thicker materials.

Automatic grinders are designed to increase speed and accuracy and to reduce the labor for producing weld preps. A common application uses various easily adjustable rollers to rotate a pipe against a spinning grinding wheel. Because the grinder is typically driven against the part by mechanical means, more powerful equipment is available than for hand grinders. Automatic grinders are therefore able to produce larger weld preps than can conveniently be accomplished by hand grinding. Such tools can reduce the time required to bevel a pipe end by an order of magnitude, at the same time producing a smooth, clean, and even surface for welding.

5.3.2 Nibblers

Nibblers, Figure 5.2, are handheld tools that drive an oscillating cutter at an angle along the edge of a plate. They produce a clean and consistent edge and land. Disadvantages are that they are

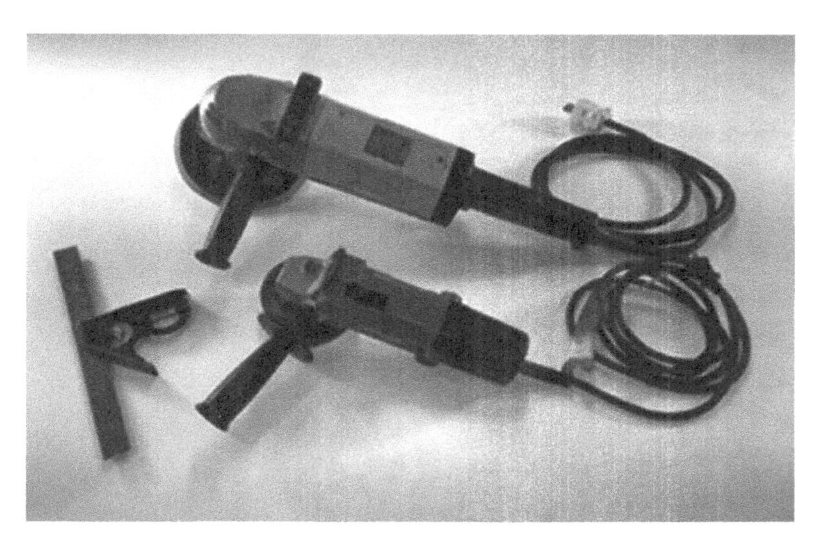

Figure 5.1 Angle grinders

relatively slow, are limited as to thickness, and may require a cutting-oil that must be removed before welding. The only setup they require, however, is setting them for the correct plate thickness, so they are often used for small amounts of weld prep when a grinder may be inconvenient.

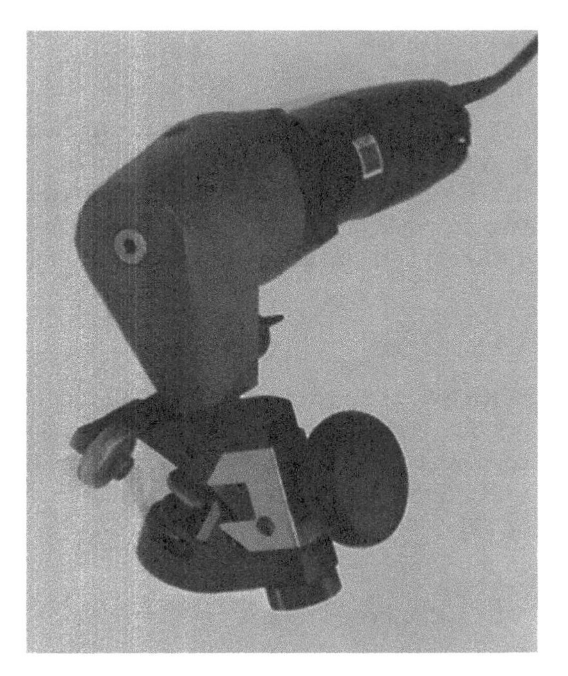

Figure 5.2 Nibbler (*Source:* Marks Brothers, Inc., ASME Fabricator)

5.3.3 Flame cutting

Various torch setups are available to prepare the edges of material, typically about 1/2 in. (13 mm) and above. Multi-torch setups can bevel both edges of heavy plate, pipe, or shell simultaneously, but cannot produce contoured bevels such as J- or U-grooves. As the material gets thicker, the volume of the weld required in a straight bevel configuration can be reduced by using a J-groove configuration. Therefore, the practical limit for flame cut weld preps is often considered to be about 3–3½ in. (75–89 mm) in total plate thickness, and many companies do not use them above about 2 in. (50 mm).

5.3.4 Boring mills

The use of horizontal boring mills, Figure 5.3, to produce weld preps is not uncommon for thick sections. This can be particularly cost effective in producing weld preps on plates that will be flat when installed. Numerically controlled mills can be used to produce the weld prep on contoured

Figure 5.3 Laser alignment verification on rotary table of horizontal boring mill (*Source:* Lawrence Livermore National Laboratory)

nozzle ends to fit on shells, whether concentric or eccentric. This allows J-grooves to be produced on heavy nozzle walls, for example. In the past, this would have been a complex and difficult operation. CNC boring mills can also be used for efficient programmed drilling of flange bolt holes. Certain materials may be processed using high-speed machining techniques, with cutting speeds measured in thousands of surface feet per minute (thousands of meters per minute).

5.3.5 Lathes

When weld preps are needed on the flat ends of cylinders, it is often desirable to produce them on a lathe, Figure 5.4. Practically, any desired weld configuration then becomes possible. This approach can be used for nozzles, ends of shell courses, heads, etc. Note that a vertical boring mill with a turret attachment for quick change of tool bits is often referred to as a vertical turret lathe, or VTL.

5.3.6 Routers

For soft metals such as aluminum, on which very high cutting speeds can be used, routers, Figure 5.5, are effective. High-speed machining techniques have been developed for these materials and the use of a router can be considered an offshoot of this. With carbide cutters, tool rotation rates of up to 25,000 rpm can be achieved, comparable to machining at about 5000 surface feet per minute (1500 m/min), depending on the cutter size. Smooth surfaces are often produced without lubricant because use of lubricant with a router is dangerous and can reduce tool life at high cutting speeds due to cold shocking of the tool. For weld preps up to about an inch (25 mm) in size, these inexpensive tools can provide very high productivity. Guide bearings or fixtures are used to control engagement and depth of cut.

Figure 5.4 (a) Lathe and (b) Lathe machining weld prep on pipe (*Source:* Harris Thermal Transfer Products)

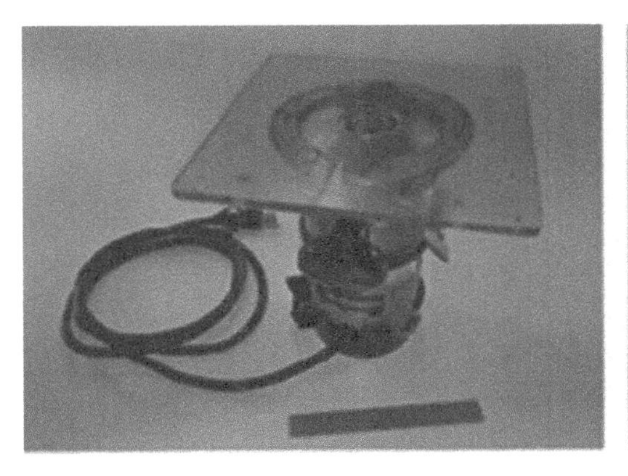

Figure 5.5 Router and detail of router bit for 45° bevel

5.3.7 Other cutter arrangements

Various manufacturers market assemblies that "walk around" a pipe or vessel to machine a weld prep on its end. One configuration uses a roller chain wrapped around the cylinder as a means of holding a cutting head. Another uses rings sized to fit various ranges of cylinder sizes. The roller chain design conforms naturally to any out of roundness of the cylinder, while the ring design ensures a round weld prep. The former ensures consistency of the weld prep with respect to the cylinder wall and the latter is particularly good for components with little out of roundness. If the pipe or shell end is flat, then the inner edge of the beveled surface will be round, ensuring that at least the outer portion of the lands on two shell courses can be counted on to align with each other.

Other designs include various mechanisms that clamp to the pipe and work from the end to produce a bevel, Figure 5.6.

Some of these tools can also be used for field cutting of pipe.

5.4 Forming

Chapter 4 addresses forming. Material that will be formed is typically formed before assembly, except in the case of plates that must be welded together to obtain a product of sufficient size. Large heads and similar components fall into this category. In such cases, care must be taken to ensure that the material properties in the weld area are reasonably matched with those of the parent material.

Material forming during fabrication is also practiced in a small way in the correction of welding distortion. This often occurs after the welding of the straight seam of a shell course or a large nozzle. Placement of the weld often results in either a crown or a dip in the shell at the weld. If it is excessive, this is typically corrected by either using a hydraulic cylinder to push the material back into

Figure 5.6 Split frame pipe beveler producing J-groove weld prep (*Source:* E.H. Wachs)

position, Figure 5.8, or by placing the cylinder back on the forming roll and rolling it again after welding.

In other cases, the material is typically moved only a small amount. Each case is as different from the next as the difference in weld distortion. This issue is addressed further in Section 5.7.

5.5 Vessel Fit Up and Assembly

Vessel fit up is the process by which the components of a vessel are assembled and made ready for welding. It includes the assembly and alignment of the straight seam of shell sections as well as the assembly of shell to shell, shell to head, nozzles to shell, supports, and any other joints that must be made up for welding. Thus, it is not unusual for vessel fit up to be somewhat scattered throughout the other fabrication processes.

Note that while the fitter may not have the same skill in welding as the welder, he/she will normally be qualified in the procedure used for tacking. Otherwise the tacks need to be removed, rather than melted in the weld itself.

The accuracy of construction of a pressure vessel as a whole depends on the fit up process. Properly rolled components that are improperly fit will not yield a properly constructed pressure vessel. During fit up, the following items typically occur:

- Alignment of plate edges in all directions and orientations to permit proper welds, Figure 5.7. Such alignment has a number of advantages for the welding process: The correct gap allows full penetration of the weld while requiring no more weld than necessary. Proper alignment of edges results in reduced discontinuity stresses. It ensures that the fit with the next shell course or the head can be accomplished easily. And it helps minimize undesirable weld distortion that could result in either a crown or dip in the vessel at the point of the weld.
- Shell courses may be rounded up. Shell courses must be sufficiently round to meet ASME BPVC roundness tolerances, to fit well with each other, and to fit with the vessel heads.
- "Spiders," Figure 5.9, will be installed to maintain roundness for shells that are slightly out of round or that are thin enough to flex under their own weight.

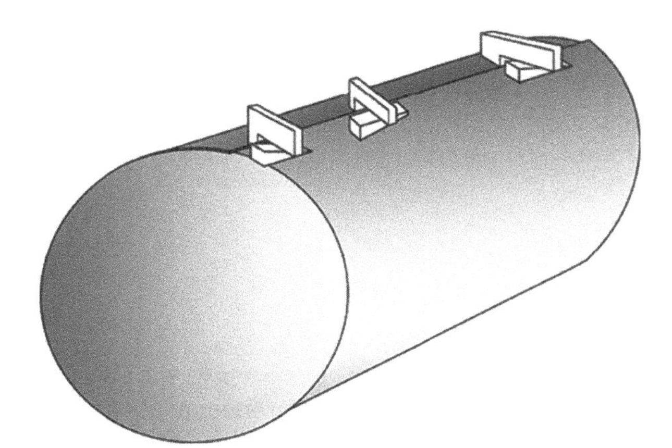

Figure 5.7 Edges of longitudinal seam being aligned using lugs and wedges (lugs and wedges enlarged for clarity)

- Other braces will be installed as needed to minimize vessel distortion due to welding.
- Heads and shells will be tacked together in their final configurations.
- Nozzle necks and flanges will be fit together.
- After welding of neck to flange, nozzles will be placed in their required locations and aligned.
- Couplings, thermowells, and any access fittings will be placed.
- At the appropriate time, reinforcing pads and wear pads will be installed.
- Vessel supports such as skirts, legs, or mounting brackets will be tacked in place, along with lugs and other appurtenances.

A look at this list makes it clear that some operations must take place prior to others, but to a large extent there is no "right" order. The right order is the one that allows the fabricator to produce the required product at the least cost.

5.5.1 The fitter

The role of the fitter may be filled by the welder, but often this responsibility is delegated to a specialist. The fitter is typically well qualified to perform general welding but has moved to a different role. This person has welder's qualifications for a number of the materials that the fabricator typically employs. He or she also has many additional skills, including the following:

- The ability to read and interpret drawings, work with the layer out, and to plan fabrication sequences if needed.
- Understanding of material handling as applicable to vessel assembly.
- Experience with welding of vessels so as to understand weld shrinkage and its effects.
- Understanding of the use of levers, wedges, hydraulic rams, and other tools to push components into place (including rounding up) prior to tack welding to hold them during the welding process.

5.5.2 Fit up tools

The following are the common tools of the fitter:

- Material handling equipment such as cranes, forklifts, turning rolls, and welding manipulators.
- Welding equipment.
- Measuring equipment, including tapes, calipers, levels, transits, etc.
- Marking tools such as soapstone marking sticks, which show up better than pencil on steel, markers, paint sticks, scribes, and hammer and center punch.
- Hydraulic jacks, "come-alongs" (hand winches), levers, lugs, wedges, and other tools for pushing or pulling material into place.
- Grinder and torch.

5.5.3 Persuasion and other fit up techniques

Assembling a vessel constructed of heavy metal components requires components to line up with each other. Misaligned joints are slower and more difficult to weld; they suffer from more extreme discontinuity stresses; they may not be ASME code compliant; and they may not fit with other mating parts. A rolled shell may come to the fitter somewhat out of round, or with edges misaligned either radially as shown in Figure 5.7 or axially, or at an angle to each other. Sometimes these

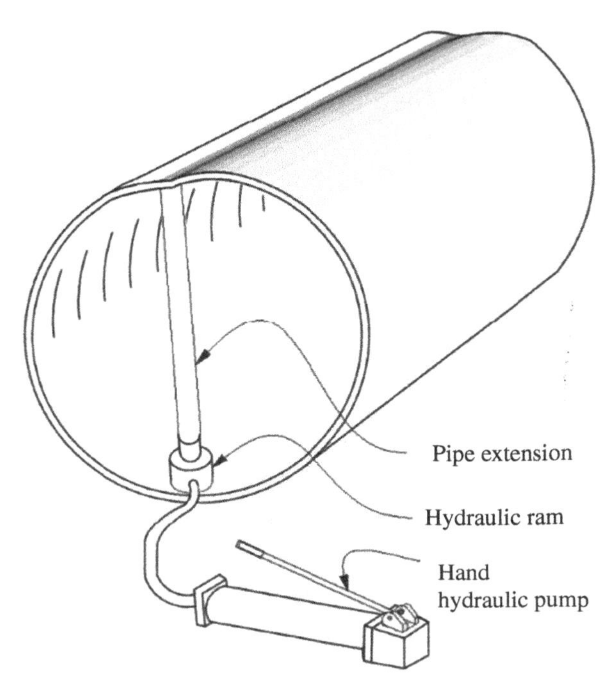

Figure 5.8 Cylinder end being rounded up using a pipe plus hydraulic cylinder

conditions are simply the result of deflection of a flexible component. Othertimes they may be caused by incorrect rolling.

The fitter will use the necessary tools to bring parts into tolerance and alignment.

For production of large quantities of similar vessels, the cost of specialized fixtures can be amortized. Special fixtures can make alignment of parts much quicker. Fixturing for a vessel shell might consist of a frame with hydraulic cylinders to align the straight seam, along with alignment pads for nozzles and other appurtenances.

For low volume or "one-off" construction, wedges and temporary lugs are often the tools of choice. Inward cusps at a prospective weld seam are often pushed out using a piece of pipe and a hydraulic jack, Figure 5.8. Outward cusps may require large lugs to bridge the joint and allow a means of pushing on it from the outside. The ways in which parts can be pushed into their required positions are limited only by the imagination of the fitter. The same tools can be applied in many different ways to address a wide range of fit up problems.

Welding procedures often call for a gap at the root of the weld. Maintaining the required gap during fit up is often accomplished by placing a piece of sheet of the proper thickness between the parts to be welded. Once the parts are tacked, the sheet is pulled or knocked out of the joint.

5.5.4 Fixturing

Even if a component seems to hold its position without support, it should be tack welded or otherwise held in the proper location during welding. As will be discussed in Section 5.5.6, the welding process often leads to distortions that can affect vessel configuration. A joint that starts out aligned may not remain so during welding if not held in place by tack welds or other means.

Figure 5.9 Adjustable spider assembly (*Source:* NASA)

Small components can usually be held in place sufficiently just by tacking them. Larger ones often require additional support such as temporary angles or gussets.

Large cylinders are rounded up, or held round, using spiders, Figure 5.9. A spider consists of a number of arms radiating from a central plate. Usually, one is placed at each end of a shell course and the outer ends of the arms are tack welded to the inside of the shell, holding it round for fit up and usually for a portion of the welding. The spiders also facilitate handling when a cylinder shows excessive flexibility. The number of arms depends on the size and flexibility of the shell. Four to eight are typical. Spiders may vary from simple angles or pipes to more complex assemblies with hydraulic cylinders or screw jacks used to push the shell into place.

Besides spiders, it is common to place supports across the interior of a vessel to keep a nozzle from sinking due to weld shrinkage. Pipes or structural steel members are commonly used.

As noted in Sections 5.5 and 5.5.7, from the point at which shell courses have been welded into cylinders, the assembly sequence may vary.

5.5.5 Welding fit up

Fit up of weld joints is generally done in two steps. The first step is to align the parts to be welded using mechanical fixtures such as jacks, straps, and wedges. The second step is to apply tack welds along the weld joint to hold the parts together. Sometimes the mechanical fixtures can then be removed, but they are often left in place to limit distortion during welding. The tack welds have substantial holding strength and are normally sufficient to keep the joint itself aligned while welding is in progress. They may not be effective, however, in keeping a nozzle or other component from sinking in the shell during welding.

5.5.5.1 Shell course fit up

Fit up of shell courses seems a straightforward operation. In practice, it is often more complex. First, the straight edges of the plate must be aligned with each other to facilitate welding of the longitudinal seam or seams in each shell course, Figure 5.7.

In the ideal situation, the shell cylinder will have been rolled such that the shell is truly round, the edges align with each other, and there is no excessive gap between them. Even in this case alignment and gapping will be verified and the edges tacked together. If the joint is not tacked, then once welding begins the unwelded portion of the joint will almost certainly move out of alignment due to thermally induced strain.

After these seams are welded, each shell course is rounded up. Then the shell courses are fit to each other and/or to the heads. In some cases, nozzles are fit and welded to the shells or heads before final closure of the vessel. Whether to fit heads and shells together or to install nozzles first is largely determined by the location and number of nozzles, their size and length, access to the vessel interior after assembly of the heads and shells, and alignment concerns. If a vessel has many nozzles in the shell sections, then easy access to the interior for welding is desirable. On the other hand, turning the vessel to perform circumferential welds later may be difficult if nozzles are protruding in all directions.

A useful compromise approach is to leave off one head. This approach is low risk if the missing head has no nozzles or only a single nozzle in the center. Fit up is then easy and only one circumferential weld remains.

5.5.5.2 Nozzle fit up

A typical pressure vessel nozzle consists of a pipe penetrating the vessel wall, with a flange on the outside end. Pipe to flange fit up, Figure 5.10, may be done either on a fit up table or a weld positioner, followed by welding on a positioner.

Nozzles, particularly large diameter ones that involve much welding, often are fitted with braces to keep them from sinking excessively due to weld shrinkage. Nozzles will be fit in place and may be supported with gussets or angles on the outside to maintain alignment. The gussets are usually removed after sufficient weld has been placed to stabilize the nozzles.

Figure 5.10 Flange nozzle neck fit up (*Source:* Marks Brothers, Inc., ASME Fabricator)

Figure 5.11 Head to shell and nozzle fit up on positioner and turning rolls (*Source:* Marks Brothers, Inc., ASME Fabricator)

Nozzle flanges must have the correct location in the shell, planar location and orientation and bolt hole orientation. The location on the vessel shell is marked by the layer out and the hole is made by flame cutting, laser cutting, machining, or some other process. This is sometimes done in the flat, but large holes in a shell affect the ability to roll cylinders accurately. Planar orientation is most easily achieved by positioning the shell such that the flange face is horizontal (facing directly upward). This allows easy use of a level to orient the flange face by measuring level in two perpendicular directions.

This is often done on a positioner or on turning rolls, which allow the shells to be rotated easily to the required orientation. If shell circumferential welds are performed on rolls, then the vessel is often left on the rolls for fit up and welding of the nozzles, Figure 5.11.

Standard bolt hole orientation in the industry is "bolt holes straddle centerline." For a flange whose face is vertical when the vessel is installed, the vertical centerline of the flange has bolt holes straddling rather than on it. For a flange with a horizontal face, the vessel centerline will be straddled, Figure 5.12.

Occasionally, when the relationship between nozzles is critical, an external fixture will be used to maintain that relationship. While producing such a fixture is not without cost, it can be particularly useful and cost effective if a number of vessels with identical nozzles and nozzle locations are being produced.

5.5.6 Weld shrinkage

Most welding processes involve molten weld metal plus additional energy that goes into heating the parent material. Once the metal – both parent and weld – has solidified, continued shrinkage toward ambient temperature results in stresses and strains.

Unbalanced tensile loading transverse to the weld results in strains that lead to bending.

Even for symmetric parts welding most often remains an asymmetric process. It proceeds from one end of a part to another, or around a component from one side to the other. If welding occurs

Figure 5.12 Bolt holes straddling centerline (*Source:* Marks Brothers, Inc., ASME Fabricator)

from both sides of the product, then it usually proceeds first on one side, then on the other. The asymmetric loads generated by this process have the potential to move components out of alignment. Also, the tensile loads generated by welding on the side of a curved part can cause that location to sink toward the inside of the curve. As a result, a circumferential weld and adjacent plate is often observed to have a slightly smaller diameter than the rest of the shell, and nozzles have a tendency to sink radially. Similar effects are sometimes observed with the welding of supports to the shell. Some ways of addressing these tendencies will be described in Section 5.7.

5.5.7 Order of assembly

While the order of operations may vary for a number of reasons, a common order would be the following:

1) Shell longitudinal weld fit up and welding.
2) Rounding up of shell courses, including re-rolling if required.
3) Nozzle fit up and welding (flange to pipe) if needed.
4) Fit up and welding of head nozzles, repads, couplings, and appurtenances (lugs, ladder clips, etc.).
5) Fit up of shell courses to each other and to heads.
6) Layout of shell nozzles, penetrations, lugs, and, for horizontal vessels, supports or wear pads, if not already done in the flat.
7) Fit up and welding of items in (6).
8) Layout, fit up, and welding in internals such as trays, supports, and studs for refractory liners. Some of these may precede head fit up to allow generally better access and especially for large components.
9) Fit up and welding of supports, legs or shell skirts, bolt rings, and bolt chairs.

Obviously, this order may vary by custom at a fabrication shop, because of component availability, or because of particular needs such as access to the vessel interior. In addition, operations that do not depend on each other will often be performed in parallel. Also, clearly the layer out will be involved at various points during these processes. Sometimes much of the shell layout is done in the flat, but this layout is also commonly performed after welding of shell courses to each other and heads. This order eliminates potential alignment errors due to fit up tolerances of those components.

5.6 Welding

5.6.1 Welding position

Ideally, welding will be performed in the flat position because working in this orientation generally allows the highest production rates with the least defects. With good planning and proper equipment, this can usually be achieved for most, but not all, welds on a pressure vessel. Because with the best of planning it is still common to have some out-of-position welds, it is usual to qualify welders for a range of positions as well as for a number of processes.

With this in mind, a shell course is most often oriented horizontally for welding, with longitudinal welds at the top when welding from the outside and at the bottom when welding from the inside. Similarly, circumferential joints will typically be welded at the top (outside) or bottom (inside).

Rolled shells, especially large diameter ones requiring two or more longitudinal welds, may require support to maintain roundness. See Section 5.5.4 regarding spiders and fixtures.

High volume weld processes for longitudinal welds are facilitated by using boom and column or other weld manipulators.

Cylinders of very large diameter are difficult to handle. For various reasons, including sometimes limited headroom beneath a crane, a decision may be made to assemble shell courses with their axes vertical. Longitudinal welds are then in the vertical position, and circumferential welds are horizontal. Welding vertically or horizontally is slower and requires more care than does welding in the flat position. Problems associated with flexing of a cylinder are much reduced, however. Also, gravity assists in keeping the ends planar, rather than pulling them out of alignment.

Once a shell is aligned to a head, even before welding, the three-dimensional structure of the head usually provides sufficient stiffness to hold the shell round and in place if properly tacked. To hold it round for tacking, however, often requires a "spider" such as that shown in Figure 5.9.

5.6.2 Welding residual stresses

Each weld pass that is applied results in a change in the residual stress distribution. Welding residual stress is a complex phenomenon affected by many factors. Weld configuration, process, speed, current, voltage, preheat or lack thereof, cooling rate, constraint on the weld, etc., all have an effect. Subsequent weld passes may anneal lower ones or put them into compression. Complex analytical techniques are used to estimate the stress state of a weld. Simplified models are used for rough approximations, while more sophisticated analyses provide somewhat greater accuracy. Validation of analytical models is challenging because of the difficulty in determining the actual triaxial stress state within a weld and the surrounding material. Figures 5.13 and 5.14 illustrate the importance of weld sequencing. The weld in Figure 5.13 was welded first on the outside, then on the inside, while for the weld in Figure 5.14, this was reversed. Note that the welding residual stresses at the interior in Figure 5.13 are far higher than those in Figure 5.14. This is an important factor in applications involving certain chemicals that make the metal susceptible to stress corrosion cracking.

Use of a double-welded joint requires less weld for thick sections, reducing both shrinkage and cost. Preheat, skipping around, one side to the other, controlled heat input, welding sides and then

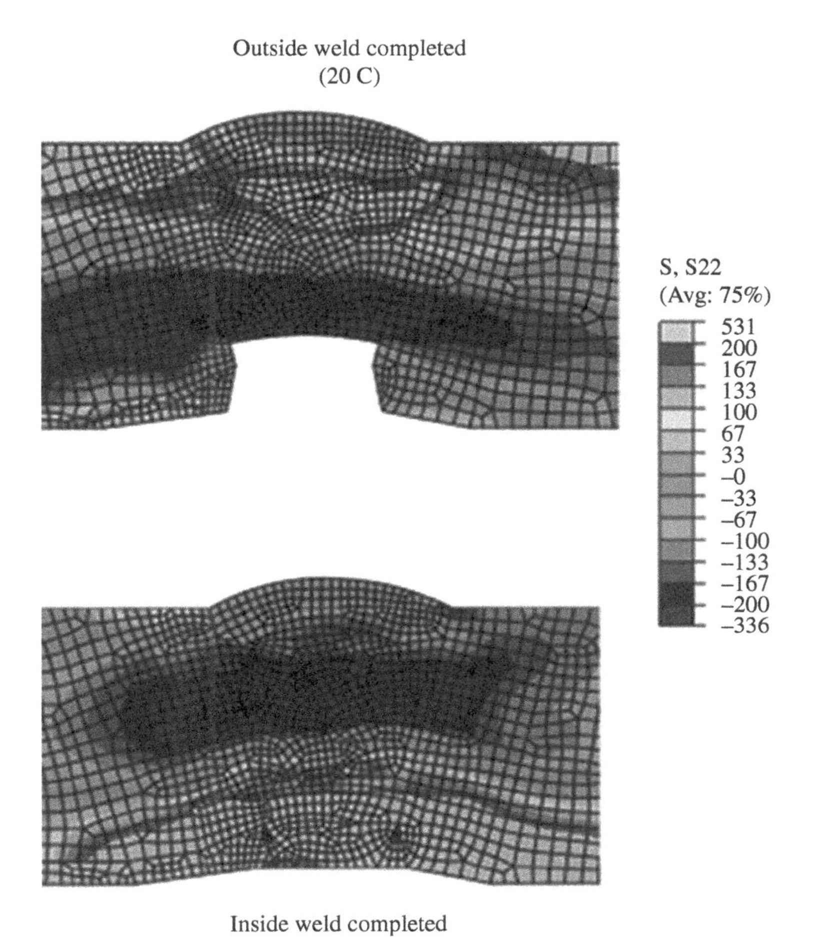

Figure 5.13 Welding axial residual stress, inside weld first, followed by inside weld (*Source:* Frederick W. Brust, 2020)

filling the center so volume pulling the sides together is minimized, and peening are all sometimes used. In addition, components are often braced so that yielding occurs in the weld metal while warm, rather than in the parent material.

5.6.3 Welding positioners, turning rolls, column and boom weld manipulators

Handling small individual components of a pressure vessel is easy. Handling larger components and their assemblies is more difficult. Once components have been tacked or welded together, weights can run from hundreds of pounds to hundreds of thousands of pounds. The use of cranes and forklifts for turning or positioning large assemblies is labor intensive, time consuming, and at times dangerous.

Also, while cranes can be good for moving and repositioning a vessel, they do not allow for smooth, controlled rotational motions as are desirable for facilitating welding. As mentioned in Section 5.6.1, welding is most efficient and less prone to defects if performed in the flat position. A variety of specialized handling equipment and fixtures have been developed to facilitate this.

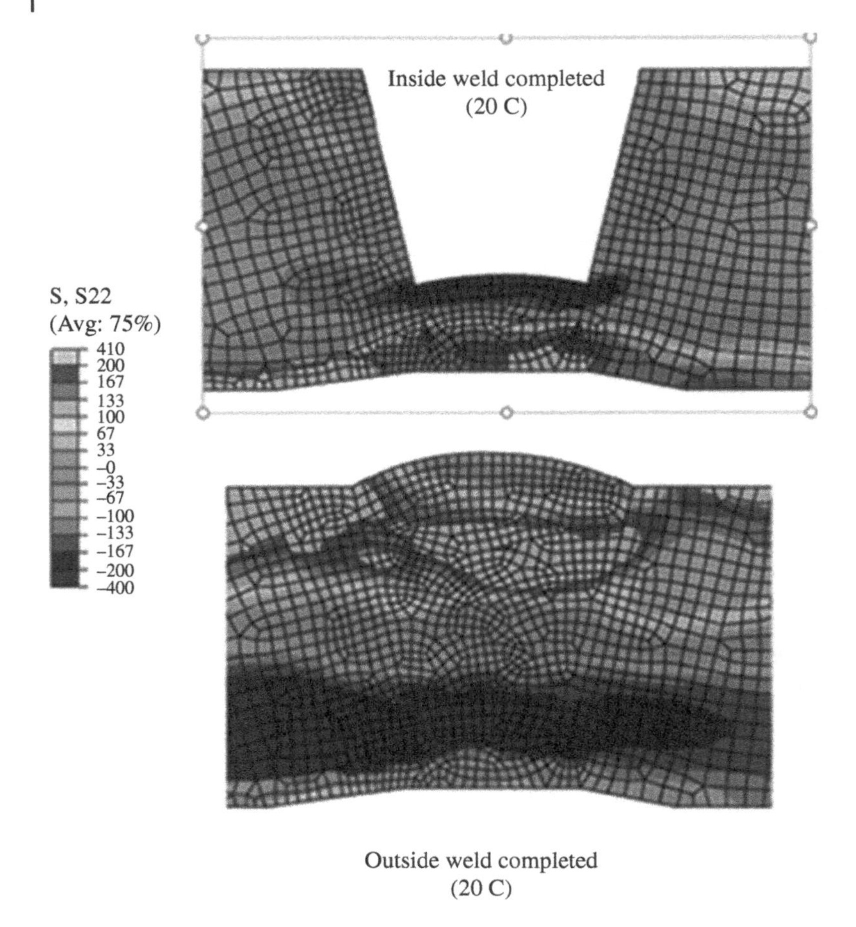

Figure 5.14 Welding axial residual stress, outside weld first, followed by outside weld (*Source:* Frederick W. Brust, 2020)

Fairly standard equipment available with a wide range of capacities include welding positioners, Figure 5.11, turning rolls, Figure 5.11, and weld manipulators, Figure 5.15.

5.6.3.1 Welding positioners

At one end of the range are benchtop welding positioners. At the other extreme, at least one manufacturer advertises a welding positioner capable of handling 175 tons (158 tonnes). Common features of such equipment are as follows:

- A base of sufficient size, mass, and stiffness that overturning is not a concern.
- Variable speed rotational ability supplemented by a rapid travel mode.
- Tilt capability of up to 135°.
- Tee-slots for mounting.
- Various types of controls such as manual, foot, and remote and programmable controllers.

Small to medium-sized welding positioners are commonly used, alone or in conjunction with turning rolls, to rotate pipe to flange and pipe to fitting joints to make the welder's job easier and more

Figure 5.15 Column and beam manipulator being prepared to weld vessel segment on positioner (*Source:* Lawrence Livermore National Laboratory)

controlled. If, for example, an elbow is to be welded to a pipe, the pipe may be centered on the positioner, with the elbow at the opposite end. If the length of the pipe is such that the overturning load is too much, turning rolls may provide support. This allows the use of idler rolls rather than powered ones, and it eliminates the possibility of the weldment slipping on the rolls due to the offset load of the elbow. See Figure 5.11.

Another use of a welding positioner is to rotate a head while nozzles are welded into it. This approach is often easier than trying to weld nozzles into the assembled vessel.

5.6.3.2 Turning rolls

Turning rolls are available in a variety of configurations. Some are powered, while others (idlers) are not. They typically have rubber or urethane wheels to minimize marring of the weldment, but are available with steel wheels for extremely heavy weldments. They come with various numbers of rollers to distribute the load of the part. Like welding positioners, the weight and size capacities of turning rolls are broad, and the ability to add more rolls, limited only by the available space, multiplies this further. Powered rolls typically are variable speed and numerous control options are available, Figure 5.15. This is similar to control options available for welding positioners, but with the addition of synchronization for applications when multiple powered rolls are needed.

5.6.3.3 Welding manipulators

Welding manipulators are used to place the welding head where it is needed for welding. A variety of manipulators are available, of which the most common is column and boom type. This tool, Figure 5.15, is available in a range of sizes, and it is most commonly used by a welding operator to produce longitudinal welds in cylinders or in the flat and to position the welding head for circumferential welds. Manipulators can also move torches for plate burning and perform other similar operations. The most common use of this type of equipment in pressure vessel fabrication is for submerged-arc welding, SAW. These capabilities have not changed, but additional ones have been added.

Current equipment can be purchased with all of the options previously available. In addition, the programming capabilities now available allow for autonomous welding, cutting, etc. Designs can be customized for special applications, but even a standard unit is, to some extent, customizable by how it is programmed.

Particular advantages of this type of equipment are high weld deposition rates and consistency of weld quality that stem from the ability to control travel of the welding head precisely.

Various other types of welding manipulators are available, including manipulators with sweep capability for welding curves and those capable of performing vertical welds with great consistency.

5.7 Correction of Distortion

Component distortion due to the heat of welding is inevitable. In some cases, the distortion is so limited that nothing need be done about it. In other cases, it must be corrected to ensure proper fit with other components.

The best way to deal with distortion due to welding is to avoid it by taking measures such as preheat, bracing during the design, or modification of the welding process. Failing that, it is usually addressed by one of three approaches as follows.

One is to allow sufficient material so that the component can be machined to tolerance after distortion occurs, which is often effective. The problems with this approach are: (1) knowing how much material to allow, (2) cost of the additional material and machining, and (3) the potential for the section to move further as material that is under stress is removed in the machining process.

A second approach is locally heating the weldment to relieve old stresses and develop new stresses to pull things back into tolerance. This approach is more an art than a science – results are sometimes unpredictable. Also, the additional heat sometimes affects material properties.

A third approach is to push or pull the material back into place.

While preheating is often used for reasons of both stress and control of material properties, pushing or pulling the material back into place is most often the approach of choice. In this approach, a load is applied to move the material, typically employing hydraulic cylinders or presses. It also requires the skill of a fitter with a knowledge of setting up equipment to move the part without causing damage anywhere else. It often does not require leaving additional material in place but is sometimes used in advance of a final machining operation. As with any forming process, correction of distortion by bending or stretching material involves strain and resulting residual stresses that may need to be considered.

5.8 Heat Treatment

Thermal issues related to fabrication are dealt with by heating prior to welding, control of temperature during the welding process itself, and post weld heat treatment. The requirements for preheat and post weld heat treatment are discussed at some length in Chapter 8, Welding procedures and post weld heat treatment. This section focuses on the application of heat as it is related directly to fabrication.

5.8.1 Welding preheat

Preheating for welding reduces temperature gradients in the area of the weld, reducing shrinkage stresses and thereby improving the residual stress distribution and reducing distortion. It also often improves ductility and reduces hardness of the weld heat-affected zone (HAZ).

Preheating does not normally require heating the whole weldment. That is often done for convenience, however, in the case of small weldments for which heating the whole assembly is the easiest approach. For large weldments, local heating is usually used. This is accomplished using a torch, electric resistance heaters, or sometimes induction heating. The large mass and heat capacity of weldments requiring preheating means that preheating is often a slow process.

Torches used for this process tend to be large and high capacity, sometimes referred to as "rosebuds." If preheat is required for a circumferential weld, then the torch will be placed to bear on the joint while the assembly is rotated on a set of turning rolls. This is often done for extended periods of time, depending on the material thickness and the diameter of the joint.

Electric resistance heaters may be strapped in place and covered with insulation. Their high energy consumption can require special power supplies. They, too, require a significant amount of time to heat a large mass of steel.

Induction heating can sometimes speed things up. It is most often used for standard products for which special fixtures and induction coil arrangements are produced. Induction heating generates heat within the metal of the pressure vessel by producing eddy currents. It is generally the most efficient of the means used for preheating, but the initial cost is the highest of the three approaches.

Whichever method of heating is used, insulation on the outside can help reduce losses to the environment and reduce the required time to temperature. Heating is sometimes done from the back side of the joint to ensure that the whole thickness has reached temperature.

5.8.2 Interpass temperature

The interpass temperature is the temperature range that must be maintained on a joint between weld passes. The minimum is often the same as the preheat temperature, though sometimes lower.

Maintaining a minimum interpass temperature serves several purposes. It helps to control hydrogen cracking for carbon, carbon–manganese, and ferritic alloy steels. For copper and copper alloys, it helps maintain good wetting of the molten pool onto the base metal. It also has the benefits attained by maintaining a larger portion of the weldment at a more uniform temperature during welding, particularly that of helping to reduce distortion.

A maximum interpass temperature is also sometimes specified. This is usually done when a minimum strength level is required, and for quenched and tempered steels. It is also done to reduce the risk of solidification cracking of various alloys, including austenitic stainless steels, nickel and nickel alloys, and aluminum and aluminum alloys. Further, such a limit helps control grain growth for fine grain steels and aluminum.

Often the heat from the weld process is sufficient to maintain the minimum interpass temperature. Sometimes the heat input from welding raises the material temperature too much, and it is necessary to wait for the weldment to cool between passes. This is particularly the case with automated welding processes because of their higher deposition rates and the resultant higher heat inputs.

5.8.3 Post weld heat treatment

Post weld heat treatment, PWHT, is most commonly performed to relieve stresses. In doing so, it reduces distortion that might otherwise occur during machining. It is also done to normalize a steel when welding has resulted in an extremely coarse grain structure (e.g., with electro-slag welding).

If the whole vessel assembly is heat treated in a furnace, all welds get PWHT at once. Because of labor costs associated with the setup of heating coils or induction heating, this is usually cost effective, but the requirement for PWHT actually applies only to the weld and surrounding area. ASME Section VIII, Division 1, UW-40 requires heating a minimum of the width of the weld plus $1t$ or 2 in. (50 mm), whichever is less, on each side or end of the weld, where t is the nominal thickness of the joint, though the need for protection from excessive thermal gradients may require heating a wider zone.

There are additional requirements regarding the rate of temperature rise, hold time, and cooling rate. Certain materials and certain processes have additional specific requirements. For example: Carbon steels designated by ASME as group P-1 require PWHT at 1100°F (595°C) when the thickness of the weld exceeds 1.50 in. (38 mm), with lower temperatures permitted for increased soak time. Low alloy steels require PWHT as high as 1250°F (675°C) and the requirement takes effect for thinner materials than for P-1.

In most cases, the material can be allowed to cool after welding and prior to heat treatment. However, ASME Section VIII, Division 1, Part UHT addresses certain cases in which heat treatment must occur before the weldment cools below the minimum preheat temperature.

5.9 Post-fabrication Machining

It is often necessary to machine portions of a fabrication after completion of welding, and if applicable, PWHT. Typical components requiring such machining are large fabricated flanges, which are likely to distort during welding. Figure 5.16 illustrates such a case.

5.10 Field Fabrication – Special Issues

Most pressure vessel fabrication takes place in a shop. In general, the equipment and amenities available in the shop make fabrication there more cost effective than field fabrication. Sometimes shop fabrication is infeasible, however. Among the things that prevent shop fabrication are the following:

- The vessel is too large to ship.
- Inadequate access is available for moving the entire vessel into place, e.g., replacement of some vessels in a refinery.
- The number and configuration of field connections make field fabrication necessary.

A number of aspects make work in the field different from shop fabrication. All must be recognized and accounted for to make field fabrication safe, productive, successful, and profitable.

As noted in the following sections, complexity and resource requirements increase as the project size increases.

Figure 5.16 Vertical boring mill facing large vessel flange (*Source:* Lawrence Livermore National Laboratory)

5.10.1 Exposure to the elements

One of the most obvious differences between shop and field fabrication is exposure to the elements. A large shop may have open ends for ventilation and be subject to outside temperatures, but it usually has a roof and two walls on the long sides. The air temperature is perhaps the same as that of outdoors, but rain and snow do not fall and there is protection from the sun. Work in the field has only the protection that has been planned and arranged.

When the weather is good, fabrication proceeds fairly easily, though the sun may be hot or the air may be cold. Precipitation poses a great hazard to the quality of welds. Damp welding flux results in porosity and potential hydrogen embrittlement. Wet surfaces are slippery. Wind can blow shielding gas away from a weld, resulting in porosity, surface oxidation, and other problems. Cold weather requires heavier clothing, makes movement more difficult, and reduces dexterity.

In some cases, it is possible to schedule around bad weather, simply waiting for the skies to clear. For small and short jobs, the use of tarps to break the wind and keep off the water may be enough. Such tarps can be stretched using whatever structure is available in the vicinity, or poles that are installed if nothing else is available. Large tents may be set up if work must continue through an extensive period of bad weather. Tents can, to some extent, mimic shop facilities, but they do not come equipped with bridge cranes.

5.10.2 Staging area

A significant benefit of working in a shop is often the presence of a large, wide, flat floor. A large floor area is often used for layout, for staging, and for assembly. Fabrication of pressure vessels often depends on having a flat and solid surface to support the work and to permit layout and assembly.

There are two obvious approaches if the site lacks such a surface. One is to level up pedestals or footings, or to place structural members and/or cribbing, etc. beneath the fabrication or beneath turning rolls or positioners that support the work. The other, at times a better choice, is to pour a concrete slab to support the work just as if it were in a shop. Such a slab is relatively inexpensive if the fabrication is large and work will continue for some time. Sometimes it can be left in place, though in other cases demolition costs must be included in the job cost.

5.10.3 Tool and equipment availability

Field fabrication is able to do without some of the largest and heaviest machines in a shop provided advance work is performed on various components. Rolls, presses, and the like need not be available if components are rolled and otherwise formed in the shop. Machine tools are generally not required if components can be machined in the shop and properly aligned in the field. For large sealing surfaces and the like, however, setup of field capability may be necessary.

Organizations that engage in field operations only occasionally prepare for it on a job-by-job basis. In such a case, the required tools, welding machines, etc. will be gathered for each job, and many of them may be rented.

Companies whose businesses support a large amount of field work will maintain tools and equipment for just such work. Generators, compressors, and welders will be stocked. Shipping containers may be kept available and stocked with miscellaneous equipment and supplies. Likely, equipment will include tools such as torches, grinders, drills, levels, etc. Common field supplies would include gas bottles, grinding wheels, marking paint, couplings, flanges, common sizes and schedules of pipe, and the like. The cost of equipment and supplies that are stocked must be built into the job cost, but ensuring their ready availability facilitates smooth flow of the job. Without it, unnecessary time is spent on trips to acquire supplies and there is a risk of idle hands in the field and delays to the job.

5.10.4 Staffing

Companies doing large amounts of, or mostly, field work often use a mixed staff for that work, comprising a number of regular company employees along with welders and other labor hired in the local area. Project and field management and other core staff are provided by regular company employees. There also may be regular hires who move from job to job with the company. If the job is located far from sources of skilled labor, then all labor may need to be shipped in and boarded on-site. If this is the case, it must be considered at the time of the bid, and these expenses must be built into the project cost.

5.10.5 Material handling

Fabrication shops typically are equipped with overhead cranes for lifting, moving, and loading and unloading of major elements. There are often jib cranes for handling smaller components and assemblies in local areas such as where flanges are fit and welded to pipes, and in the area of machine tools. Fork lifts and pallet jacks are also available.

How material is handled in the field depends on the weight and size of components and the magnitude of the job. Small jobs may be accomplished with the use of levers and winches, but most projects will require forklifts, mobile cranes, or even large construction cranes installed on-site. Construction cranes will typically require significant foundations and/or large amounts of counterweights, but can be essential to large fabrications.

Figure 5.17 Large spheroidal segment on tilt table (*Source:* CB&I Storage Solutions)

A particular aspect of material handling that is often an issue in the field is the positioning of extremely large assemblies for fit up and welding. Figure 5.17 shows a portion of a large spherical vessel mounted on a large tilt table that allows changing the position of the component to facilitate welding. The effect is similar to the use of rolls, except that for an assembly either too large to be maneuvered on rolls or which requires welding to provide structural integrity prior to placement on rolls, the use of turn tables or turn fixtures is desirable.

5.10.6 Energy sources

Energy tends to be taken for granted in the shop. The shop will have been constructed or upgraded to include sufficient electrical capacity to run the necessary fabrication, heating, and welding equipment. Natural gas, propane, and other gas supplies will be in place for cutting, heating, and heat treatment.

Field fabrication requires that all of these energy sources be verified and, in some cases, planned and arranged for. A site without easy access to electrical power for welding will require bringing in power, either as temporary power drops or in the form of generators, or else gas-powered welders will be needed. If gas welders are used, then fuel must be supplied.

Gas or electrical supplies for PWHT must be planned for.

5.10.7 PWHT

Large fabrications often require PWHT. If this is required, means of heating must be supplied. Gas-fired furnaces may be built on-site, requiring a significant supply of heating gas, typically natural gas or propane. Electric resistance heaters or induction heating systems require large amounts of electrical power.

Shop heat treatment has the benefit of calibrated furnaces, ports for recording thermocouple readings, etc. Field heat treatment installations must be designed to provide for accurate temperature control and use calibrated thermocouples as well. Even when a field furnace is constructed, it may only be large enough to heat a portion of the pressure vessel. If this is the case, then local heating of final closure welds will be needed. This will normally be performed using electrical resistance or induction heating systems.

5.10.8 Layout

Much layout work for field fabrication is likely to be done "back at the shop." Most likely shell courses and gores for large heads will be laid out, cut, and formed in the shop, then moved to the field for assembly. Field layout will include work to facilitate the work of the fitters as well as layout, or at least verification, of nozzle locations after shell fit up.

5.10.9 Fit up

Field fit up proceeds in a manner very similar to shop fit up, except that often a mobile rather than an overhead crane is used, sometimes necessitating an extra person on the job. The assembly may be pre-fit in the shop so that edges can be trimmed and welds prepped. Generally better construction is achieved when the work done in the field is minimized. Also, for large fabrications, scaffolding is often required.

5.10.10 Welding

Field welding in some cases mimics shop welding. If the magnitude of the project allows for placement of turning rolls for positioning of shell courses, then use of automatic processes may be very similar. If the fabrication is too large for this, then either special fixturing, Figure 5.18, or manual or semiautomatic processes are likely to be used. Much more out-of-position welding is likely to occur

Figure 5.18 Special fixturing for field welding of large fabrication (*Source:* CB&I Storage Solutions)

due to issues of handling. Preheating of the weld area may be more challenging due to wind. As noted in Section 5.10.6, gas-powered welders may be used rather than electric ones. This requires familiarity of the welders with another type of equipment.

Shop facilities will have bake-out ovens for weld rod and flux. These must also be provided in the field.

5.11 Machining

Designs for field installation avoid as much as possible the need for field machining. For example, a large sealing surface that must be welded in the field will be designed, if possible, so that joints can be aligned sufficiently that field grinding or polishing are enough to make the seal. If this is not feasible, the surface may be field machined after assembly and welding. If this is unavoidable, then special tools are designed for use in the field. As with shop equipment, accuracy and stiffness of the equipment are paramount. The requirements for flatness and surface finish normally remain the same, so significant care is required.

5.12 Cold Springing

Cold springing to compensate for inaccurate fit up is not good practice and is generally not permitted by the codes. If a system operates at elevated or reduced temperatures, however, it may be desirable to cold spring the associated piping to average out the thermal stresses. Such cold springing must be specified as a part of the design and must identify the most desirable neutral position. Fit up and welding then proceed, allowing either for a gap or an interference, as needed to achieve the desired stress state at ambient temperature.

A typical case might involve allowing for operation of a system at several hundred degrees Fahrenheit. In this case, a specified gap may be allowed for between flanges, to be pulled together at assembly. As the system is heated up, thermal expansion increases the length of the piping, bringing the system to a neutral position, zero springing stress, at approximately half the expected temperature rise required to close the gap.

6

Cutting and Machining

6.1 Introduction

Cutting and machining involve closely related operations, sometimes with ambiguity as to what constitutes cutting and what constitutes machining.

Cutting is a process used when one workpiece is to be made into two or more. The part of the product not used, unless too small, generally remains available for use in producing some other component. Thus, for example, when a shape is cut from a full mill plate, the balance of the plate remains available for cutting other shapes.

Machining involves bringing a roughly sized or shaped part to its finished configuration, usually in three dimensions. As the result, the material removed is most often in the form of chips or swarf and is of no further use except as scrap.

As a result, cutting is often used as a process preliminary to machining, creating a part that will be placed on the equipment used for machining and that is close enough to the final shape that machining can proceed efficiently. Machining is then used to refine the shape of a component to meet the specified shape, tolerance, and finish requirements. These requirements usually involve tolerance and finish requirements much more stringent than those for cutting, typically running in the thousandths of inches while cutting tolerances are often measured in fractions of inches.

Cutting processes are discussed first, followed by machining.

6.2 Common Cutting Operations for Pressure Vessels

Cutting is used for the following sorts of operations:

1) Cutting plate into flat rectangles or shapes suitable for incorporation into weldments as is, or after forming, or for machining or other processing to produce finished components.
2) Cutting to length of bar, pipe, tube, or structural shapes, as well as sometimes mitering and/or beveling.
3) Removal of braces, fixtures, and other temporary attachments that have been welded or otherwise affixed to components to facilitate the fabrication process, but that are not needed in the final product.
4) Cutting holes in pipes or shells for installation of nozzles.
5) Contouring pipe ends where the nozzle fits into the vessel shell.

Fabrication of Metallic Pressure Vessels, First Edition. Owen R. Greulich and Maan H. Jawad.

6.3 Cutting Processes

The following tools and methods are commonly used for producing rectangles and shapes in plate:

- Shear (straight cuts only).
- Flame (torch).
- Plasma cutter.
- Laser.
- Water jet.

These tools are typically used for cutting bar, structural shapes, pipe, and tube to length:

- Abrasive saw.
- Cold saw.
- Band saw.
- Hack saw.
- Torch.
- Laser.

Plate cutting is addressed first, followed by cutting of bars, pipes, tubes, and structural members.

6.3.1 Plate cutting

6.3.1.1 Shearing

As used in a fabrication shop, a shear may be likened to a large pair of scissors for cutting metals up to about 3/4 in. (19 mm) in thickness. Most shop metal shears are for straight cuts, but circle shears efficiently cut circles of metal, typically under 3/16 in. (5 mm), and handheld shears with an oscillating blade allow cutting of shapes in mild steel up to 10 gauge (0.135 in., 3.4 mm). Cutting of rectangles, and sometimes pieces such as gussets, in metal plate not greater than about 3/4 in. (19 mm) is often most efficiently performed using a shear. Beyond that thickness, the equipment becomes very heavy and expensive, and it is common to have more rollover on the edge of the plate than can be tolerated.

Shears, Figure 6.1, used for sheets and thin plates are often electrically powered with a flywheel, but those meant for thicker materials are powered by hydraulic cylinders. Shearing is a very cost effective way of cutting materials, requiring only a second or two to cut from one edge of a plate to the other.

The process has the following limitations: Shearing is limited to straight cuts of non-brittle metals. Because as one side of the plate is held flat and the other is pushed down, distorting the plate, cuts must proceed across the full width of the plate in a single cut. Thus, an 8 ft (2.4 m) shear, for example, is actually limited to the 8 ft length of its blades, and it does not work well to try to cut a piece out of a plate, rather than off the end of it as illustrated in Figure 6.2.

For some operations, shearing is efficient, but if a portion of the finished plate outline involves curves, other means of cutting will normally be used.

The typical shear consists of a large base made up of castings or plate, with two heavy end plates. The base is surmounted by a table, typically with rollers for easy movement of plates, the back edge of which has a heavy blade made of tool steel. Above, and mounted to the end plates, is a ram, provided with another blade. The ram is mounted on ways that provide for its motion in the vertical direction, while providing very close alignment between the opposing blade faces. The gap between the blades is a function of the material being cut and its thickness and it is critical to good cutting.

Figure 6.1 Plate shear (*Source:* Marks Brothers, Inc., ASME Fabricator)

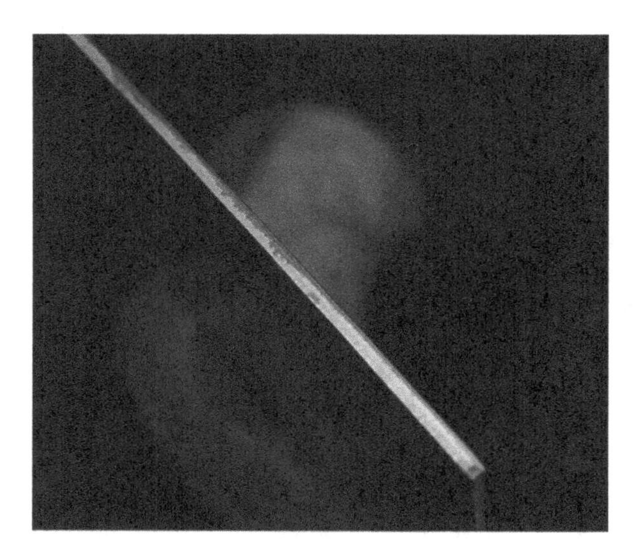

Figure 6.2 Edge of sheared plate (*Source:* Harris Thermal Transfer Products)

Mechanical or hydraulically operated hold-downs are located above the table, near the edge, to hold the plate in position while the ram descends, shearing the plate from one edge to the other.

An adjustable backstop that can be used to help position the plate accurately for cutting is usually mounted on the back of the ram.

Advantages of the shear are its quick and easy operation, straight cuts, and short learning curve for operators. Disadvantages are limited thickness, a sheared, often granular appearing, edge that is often quite sharp, and rollover of the edge of the cut. Also, shearing is limited to full-width cuts, so use of material is sometimes not as efficient as that of other cutting processes.

Shearing is often used to make components of saddle supports, gussets, etc. It is less often used for pressure parts because the sheared surface often requires grinding or other treatment prior to welding, due to the contour and the granular surface.

6.3.1.2 Flame cutting

Flame cutting uses a fuel gas supplemented with oxygen to increase heat and a jet of cutting oxygen that actually burns the metal. Flame cutting of plate varies from hand cutting, practiced on a small scale, to the use of large burning machines with multiple torches. Because of the fine speed control that is achieved in an automated process, burning machines can cut through plate as thick as 16 in. (400 mm). The finish of the cut edge can be quite smooth, Figure 6.3, on plates up to several inches thick, but tends to have flow lines and occasional gouges in thicker plates. Flame cutting machines are programmable and often are equipped with a number of torches, allowing the production of multiple parts at once. Cutting speeds range from 1 or 2 in./min (in./min), 2.5 to 5 cm/min, for very thick materials, to upward of 30 in./min (0.75 m/min) for 1/4 in. (6 mm) plate.

Large water tables support the plates being cut. Such tables are made of steel, filled with water to a level close to the bottom of the plates being cut, and are usually equipped with steel slats on edge every few inches to support the plate being cut without excessively disrupting the gas flow. The water below the plate cools the combustion gases and the slag, reduces noise, and decreases significantly the quantity of undesirable byproducts going into the air. The slats are removed periodically to allow removal of built-up slag at the bottom of the table. They are replaced when the cuts in the top edge compromise the support of the plate above.

Flame cutting is a flexible and efficient process for cutting carbon steels, burning the steel with the cutting oxygen. A brittle foamy slag results, but the cut edge will normally be quite smooth, about 63–125 μin. (1.6–3.2 μm), if the torch is adjusted properly. For materials such as stainless steel or aluminum, however, the inability to burn the metal means that these metals can only be melted through with a torch. For these materials, process is slow and the edge quality is poor, so other cutting processes such as plasma, laser, and water jet are used instead.

The workhorse of cutting for carbon steel pressure vessel plates is flame cutting. Pressure vessel fabrication often makes use of it for shell plates, head plates prior to and after forming, plate flanges, large nozzle roll-ups, reinforcing and wear pads, saddle plates, and more. It is also frequently used

Figure 6.3 Typical 3/4 in. (19 mm) flame cut components (*Source:* Marks Brothers, Inc., ASME Fabricator)

to produce weld preps for v-groove and double v-groove welds. While shearing is quick and relatively easy, flame cutting allows nesting of gussets and other shapes to save material that might otherwise go in the scrap bin.

This situation is beginning to change, however, at least for relatively thin materials. In this arena, the speed and high tolerances of laser cutting have taken it from a costly anomaly to a fairly common process. Multi-torch machines, however, continue to be effective for large quantities of parts of a size that can be cut 8 or 12 at a time across the width or length of a plate.

6.3.1.3 Plasma cutting

A plasma torch, Figure 6.4, uses an electric arc to create an arc and a plasma that may be between 20,000°F (11,100°C) and 45,000°F (25,000°C). A shielding gas guides the plasma and the arc that follows it to the work piece, which must be conductive and grounded, and the combined gas and plasma flow ejects molten metal from the kerf of the cut. The process can be used to cut essentially any conductive material. While the process is very hot, the heated metal is ejected so quickly and the torch moves so quickly (typically 20–50 in./min (0.5–1.25 m/min)) that the heat transfer to the surrounding metal is usually less than for conventional torch cutting. This results in a relatively small heat-affected zone.

It is common to mount a plasma torch on the same burning machine as is used for torch cutting, taking advantage of the same system programmability, water tables, etc. Plasma cutting sometimes results in a tapered kerf, which is reduced in newer machines by better control of gas flow and swirl, and in some cases by using automated capability to change the angle of the torch slightly to compensate for the taper.

Because plasma cutting does not depend on burning of the material, it can be used for any conductive material, including stainless steel, high nickel alloys, and aluminum. Advantages of the system are the ability to cut materials that cannot be cut with conventional torches, fast cutting,

Figure 6.4 Plasma torch (*Source:* Harris Thermal Transfer Products)

Figure 6.5 Plasma cut edge (*Source:* Marks Brothers, Inc., ASME Fabricator)

small heat-affected zone, and lower total heat input than that of a conventional torch. Disadvantages are the need for a large electrical power supply, potential taper on the edges of parts, the need for a supply of gas, and a limit on thickness of about 3–6 in. (75–150 mm) depending on the material, torch, and power supply.

Figure 6.5 shows a piece of plasma cut stainless steel. The near edge was machine plasma cut, while the rough-edged cutout section was done manually.

Plasma cutting is used for all of the applications identified in Section 6.3.1.2 for flame cutting, but is used both for carbon steels and for metals that are not readily flame cut.

6.3.1.4 Laser cutting

Laser cutting machines work by focusing a laser to a very fine point to produce a point of very high intensity energy on the workpiece. The material then burns, melts, vaporizes or is blown away by a jet of gas. The result is an edge with a good-quality surface finish as shown in Figure 6.6. The nature

Figure 6.6 Laser-cut parts ready for processing (*Source:* Marks Brothers, Inc., ASME Fabricator)

of the finely focused laser beam is to produce a very small kerf. Combined with a properly constructed computer numerical control (CNC) machine, this allows production of very high tolerance work. Thus, while laser cutting is generally classified as a cutting operation, its tolerances rival those of machine tools. In some cases, depending on material thickness, part geometry, and the design and condition of the machine, tolerances within 0.001 in. (0.025 mm) are achieved. In addition, the process is very fast, especially for thin materials. Carbon steel plate of 3/8 in. (10 mm) can be cut at up to 500 in./min (13 m/min).

Advantages of laser cutting are good tolerances and finish, small heat-affected zone, and fast work on thin plates. Disadvantages include high power requirements, limited table and part size due to optical limitations, and cost.

In the past, laser cutting was used only for unique applications due to its high cost. With current equipment and cutting speeds, it is a highly competitive process if the workload is sufficient to keep the machine operating continuously. A typical high production machine will have multiple tables to allow loading to proceed at the same time as cutting, eliminating downtime for loading and unloading of material.

6.3.1.5 Water jet cutting

Water jet cutting of metals is accomplished with an extremely high pressure jet of abrasives (often garnet) containing water. Very small jets that are used, focused through a jewel orifice, provide small kerfs and allow for tight tolerance work. Originally, water jet cutters had only two, or at most, three axes of motion. Currently available machines include five axes or more, allowing tilting and rotation of the head. While generally not essential for applications related to fabrication of pressure vessels, these capabilities allow the cutting of complex shapes. They can also be used to perform as simple an operation as compensating for the angle of kerf. Water jet cutters capable of cutting up to 6 in. (150 mm) of metals are available.

Advantages of water jet cutting are the absence of a heat-affected zone, tight tolerances and good finish, lack of part distortion due to heat, fast cutting for thin plates, and a relatively clean operation. Disadvantages include relatively slow cutting speeds. Depending on the required finish, the abrasive used, and the pressure of water, 1 in. (25 mm) plate is cut at rates varying from about 1 to 7 in./min (25–175 mm/min).

A water jet cut edge will always be acceptable for pressure vessel applications, but the slower cutting speed for thick sections may make the process undesirable.

6.3.2 Pipe, bar, and structural shape cutting

6.3.2.1 Abrasive sawing

An abrasive saw is a circular saw using an abrasive disk rather than a blade with teeth. It is also referred to as a cut-off saw or a chop saw. The radius of the abrasive disk, minus that of the arbor and support plate, determines the size of the product that can be cut. Many shops maintain saws of 12–16 in. (300–400 mm) in diameter, allowing cutting stock of up to about 6 in. (150 mm) in diameter, but saws and wheels of up to about 3–1/2 ft (1100 mm) in diameter are available. Allowance must be made for a decrease in diameter of the wheel as it wears, but setups that allow turning the workpiece as the wheel comes against it can permit cutting through the wall of almost any diameter pipe.

Advantages of abrasive sawing are its relatively low cost and its ability to cut a variety of metals. Disadvantages include a tendency for the blade to wander at times, requiring a large amount of grinding for cleanup, and the process is typically quite hot. Blade wander is minimized if the pipe is rotated against the wheel.

6.3.2.2 Cold saw

In configuration, a cold saw is similar to an abrasive saw, but it operates at lower revolutions per minute and uses a blade with teeth designed to put most of the heat generated in the cutting process into the chips, leaving the part cool. In addition, many cold saws include a coolant pump and reservoir to reduce blade wear. The typical cold saw uses a carbide blade and is capable of cutting various types of metals, depending on the blade characteristics. Saws are available with a variety of options to permit semiautomatic or automatic operations, and in sizes that permit cutting pipe or tubing up to about 6 in. (150 mm) outside diameter. Miter cuts are available on some saws. With proper fixturing, tight tolerances may be achieved.

Advantages of a cold saw are clean cuts, often burr-free, no heat-affected zone, good tolerances, and the ability to cut miters. The major disadvantage is the limitation on size.

6.3.2.3 Band saw

Band saws use a continuous blade running over two or more wheels to cut pipe, tube, bar, and structural sections in a wide range of sizes. Size capacity ranges widely and selection is typically based on the type of work done in a particular shop. Saws are available in both vertical and horizontal configurations, and with a number of options for automatic operation. Coolant is used to improve blade life. Tolerance and finishes are not as fine as with cold saws, but are generally acceptable for welding, fit up, and other operations performed during pressure vessel fabrication.

Advantages include no heat-affected zone, clean cuts, and reasonable tolerances. Disadvantages may include slower cutting than some other processes.

6.3.2.4 Hack saw

Hack saws used for cutting metal sections are typically arranged to cut material that is oriented horizontally. They use straight blades up to about 20 in. (50 cm) in length, but most common is about 14–16 in. (350–400 mm), allowing cutting of round bar up to about 8–9 in. (200–230 mm). Coolant is necessary for blade life. Automatic options are available.

Advantages of hack saws include straight cuts, a finish of approximately 125 μin. (3.2 μm), and the ability to cut most metal alloys.

A hack saw finds frequent use in fabrication of pressure vessels for cutting pipe for nozzle necks and occasionally for cutting bar to make special fittings.

6.3.2.5 Torch

The traditional means of cutting large structural or tubular members is the flame cutting torch. Such torches are usually eclipsed by various types of saws when a bar needs to be cut. When structural members must be cut, however, a torch will often do the job more quickly than any such saws. The operation requires careful layout and work by the operator, but with care, and a little cleanup grinding at the end, this is a very viable approach. Tolerances typically cannot match those of saws, but the work is accomplished quickly and much structural work does not require tight tolerances.

Torch cutting is also used for cutting nozzle contours on the ends of pipes. While not a high technology solution, it is readily available in the smallest of shops and is highly versatile. For small

Figure 6.7 Laser-cut nozzle contours (*Source:* Marks Brothers, Inc., ASME Fabricator)

quantities of parts, torch cutting remains competitive because it is inexpensive and eliminates the need to ship components elsewhere to be laser cut.

6.3.2.6 Laser

Laser cutting is used both for plate and tubular products. While plate cutting occurs on tables similar to those for flame cutting, tubular products are held in a chuck or collet and rotated to expose to the laser the portion to be cut. At the same time, either the laser or the part can be moved axially to produce required contours at the end of the part as shown in Figure 6.7. The laser optics are also automatically moved to the optimal distance from the part for focusing during part rotation. Thus, even rectangular tubes for vessel supports or internals can be cut with great efficiency. Cutting speeds of up to 500 in./min (13 m/min) may be attained on parts of about 0.25 in. (6 mm) thick. This speed is vastly greater than those for either torch or plasma torch cutting. As a result, some laser cutting machines are equipped with multiple tables to allow loading of plates while cutting is taking place. This helps to use machine capacity fully, aiding in amortization of the relatively high cost of these machines.

6.4 Common Machining Functions and Processes

This section first considers the functions that need to be done using machining processes, followed by some issues generally associated with machining processes, and finally discusses machining processes and equipment used to perform them.

The machining most commonly performed in the production of pressure vessels consists of the following functions:

1) Preparing of weld surfaces ("weld prep")
2) Machining of flanges, including facing, boring, drilling, and in some cases fully machining open die forgings
3) Preparing tubesheets, including facing, drilling, reaming, grooving, and putting weld preps on edges
4) Facing mating surfaces of heat exchanger plenums
5) Drilling heat exchanger baffles

The following machining and related processes are commonly used to perform the aforementioned functions on pressure vessel components:

- Milling
- Turning and boring
- Drilling
- Tapping
- Water jet cutting
- Laser cutting

Other machining processes less commonly used on pressure vessel components and that will be covered in less detail are as follows:

- Reaming.
- Electrical discharge machining (EDM), plunge and wire.
- Electrochemical machining.
- Electron beam machining.
- Photochemical machining.
- Ultrasonic machining.
- Planing and shaping (no longer in common usage).
- Broaching.

Each of these operations, and the required equipment, will be described in its own section.

6.5 Common Machining Functions for Pressure Vessels

6.5.1 Weld preparation

Weld preparation is the process of preparing edges and surfaces for welding.

For fillet welds, typically the only required weld prep is removal of excess rust and scale from the surface. In cases involving close tolerances, the abutting edge of a plate or pipe may be machined for a close fit, but no machining of the actual surface to be welded is usually needed.

Groove welds in most cases require some preparation of the weld groove, though for thin materials, straight or angled edges on one or both sides of the plate are often sufficient. For cost reasons, weld preps for v-grooves are usually produced using flame cutting, grinding, or similar processes; see Section 5.3. For thicker materials, a U-groove or J-groove is desirable to reduce both weld quantity and welding residual stresses.

Groove weld preps are often applied to flat shell plates using a milling machine, especially for large-diameter shells. In the past, when planers and shapers were more prevalent, these tools were used as well.

A weld prep to be produced on a circular section such as a flat head, a pipe or already rolled shell, or a flange or tubesheet will typically be done either on a lathe or a vertical boring mill or vertical turret lathe (VTL), depending on the diameter and length.

6.5.2 Machining of flanges

A number of operations are involved in production of a flange. Standard flanges will usually be procured whenever possible because companies that produce them have tooled up for efficient production. If standard flanges are available, then individual production will not normally be competitive. For simple plate flanges, special designs of flanges, flanges of materials that are not usually stocked, or large flanges that are rolled out of plate, however, in-house or contract machining of the component is often justified. Also, special categories of components such as quick opening locking rings and their associated flat covers are often not readily available and require fabrication and machining.

Simple plate flanges may require only facing and drilling, if burned inside and outside diameters (IDs and ODs) are acceptable and if the raw plate surface is acceptable on the back of the flange. For circular flanges, the facing will be done on a lathe or vertical boring mill, as will the finishing of IDs and ODs, if needed. Rectangular and other shapes of flanges will likely be faced and machined on a milling machine, often a computer numerically controlled (CNC) mill. Drilling is performed using a radial arm drill, a large drill press, or a milling machine. Depending on the quantity of parts to be produced, use of drill fixtures with hardened drill bushings may be justified; see Section 6.6.

Specially designed flanges, flanges made of exotic materials, and flanges rolled out of plate are typically all subject to the same machine operations. In these cases, it is typical to machine all surfaces of the flange, and welding neck flanges are the most common configuration. Facing, boring, back surface machining and contouring of the welding neck, and weld preps will all be performed on a lathe or vertical boring mill. As before, drilling involves a radial arm drill, a large drill press, or a milling machine.

6.5.3 Tubesheets

Preparation of tubesheets involves facing, drilling, reaming, and grooving. In some cases, in addition a hub must be machined for welding. For round tubesheets, the facing, outside machining, and hub will normally be done on a vertical boring mill or VTL, Figure 6.8. If a large surface

Figure 6.8 Vertical boring mill (VTL) (*Source:* Harris Thermal Transfer Products)

Figure 6.9 Machining a typical tubesheet (*Source:* Harris Thermal Transfer Products)

grinder is available and no hub is required, however, the surface grinder provides a quick and inexpensive means of facing the two sides of a plate. Surface grinders easily produce flat and parallel surfaces to tight tolerances, but sometimes the surface is too smooth to grip the surface of a gasket effectively.

Depending on the size of the tubesheet, drilling, reaming, and grooving for rolling of tubes may be accomplished using a drill press or radial arm drill. Accurate repetitive spacing of holes, however, is more easily achieved using a CNC mill, Figure 6.9.

6.5.4 Heat exchanger channels

Heat exchanger channels and similar components requiring a flat surface for mating can be produced in various ways. If they are round, a lathe or vertical boring mill is usually most efficient. For rectangular components, a milling machine will usually be used. A surface grinder with proper setup or tooling can also be used, but the surface may have less "bite" on the gasket than desired, risking gasket extrusion.

6.5.5 Heat exchanger baffles

Heat exchanger baffles are usually made of sheet or thin plates. Historically, they have been drilled, and this is still a feasible way of producing them. A drill or a mill, particularly a CNC mill, provides a valid approach. It is possible to stack multiple baffles in a fixture for drilling, expediting the process. Flame cutting was usually not an acceptable solution because the high heat input caused distortion of the plates. However, laser and water jet cutting are now available and can be programmed to cut all the holes. Laser cutting occurs quickly and most of the heat is removed as the molten metal is blown out of the kerf, so distortion is minimal. Water jet cutting is inherently a cool process, but is comparatively slow. Both of these processes are capable of holding tolerances of better than ± 0.005 in. (0.125 mm) on plates up to about 1/2 in. (13 mm) thick, though accuracy of large dimensions will be affected by machine wear, part geometry, and temperature.

6.6 Setup Issues

Setup issues for machining involve maintaining the proper relationship between the workpiece and the cutting tool. Because machining almost always involves tight tolerances and requires forces high enough to cut the metal, the stability of the workpiece relative to the machine tool is important.

One aspect of this involves holding the part in place. With extremely large and heavy workpieces placed in front of a horizontal boring mill, it may sometimes be possible to machine without restraining the part, but this is a rare exception. In general, the object being machined is always held in place. Bolts, clamps, tee-nuts, chucks, and fixtures are the usual means.

The other half of the problem is to place the tool in the correct location. Many machine tools have the capability of indexing the cutting tool very accurately with respect to the workpiece. In some cases, fixtures are used to align either the part or the tool, or both.

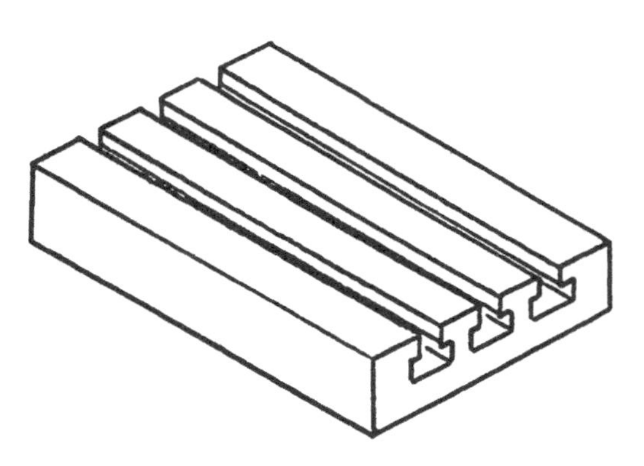

Figure 6.10 Typical tee-slots for clamping parts

On the tables of mills, on rotary tables, on some chucks, and on various other tools, tee-slots are provided to assist in clamping parts in place, Figure 6.10. Many companies produce a wide range of devices that use tee-slots and tee-nuts that can be used to clamp parts to a table or other machining base, as shown in Figure 6.11. In some cases, the part is just bolted down. In other cases, "over-center" or "toggle" clamps, available in a multitude of configurations, are used for quick installation of parts on a small mill.

For pieces being turned on a lathe or a small to medium-sized VTL, the most common way of holding the part is with a chuck. A three-jaw

Figure 6.11 Clamping lugs using tee-slots to hold part for machining (*Source:* Marks Brothers, Inc., ASME Fabricator)

chuck works well for round or almost round components, holding them either from the outside or the inside. Most three-jaw chuck designs move all three jaws radially at the same time in order to center a part. This works well for round components and when a concentric turning or boring operation is needed. A four-jaw chuck is used to hold parts that are not round and parts that need to be held off-center. The jaws are moved individually. Thus, a four-jaw chuck requires more work and care to mount a workpiece, but provides greater flexibility for out-of-round and irregular parts.

It is also common for parts on a large boring mill or VTL to be bolted down, either directly or using tee-slots, tee-nuts, and hold-down bars of one sort or another.

The setup arrangements described above are used when the machine itself provides the capability of moving the tool and the part accurately with respect to each other. This capability allows tight tolerances to be held.

When multiple identical parts must be produced, a fixture is often used to hold them during machining.

A drill press or a radial arm drill has little capability for accurate positioning. When using these tools to produce multiple parts, it is common to use fixtures to hold the parts in place and jigs to provide the required locational accuracy of the drill. Drill jigs can be made to locate the holes on a bolt circle, for example. The jig will be produced either using an extremely careful layout followed by careful hole drilling on the drill or by indexing accurately to position on a mill. For production of large numbers of parts, the holes on the jig, Figure 6.12, will be fitted with hardened drill bushings to minimize wear. The use of this accurate jig allows precise drilling of the required bolt pattern on a

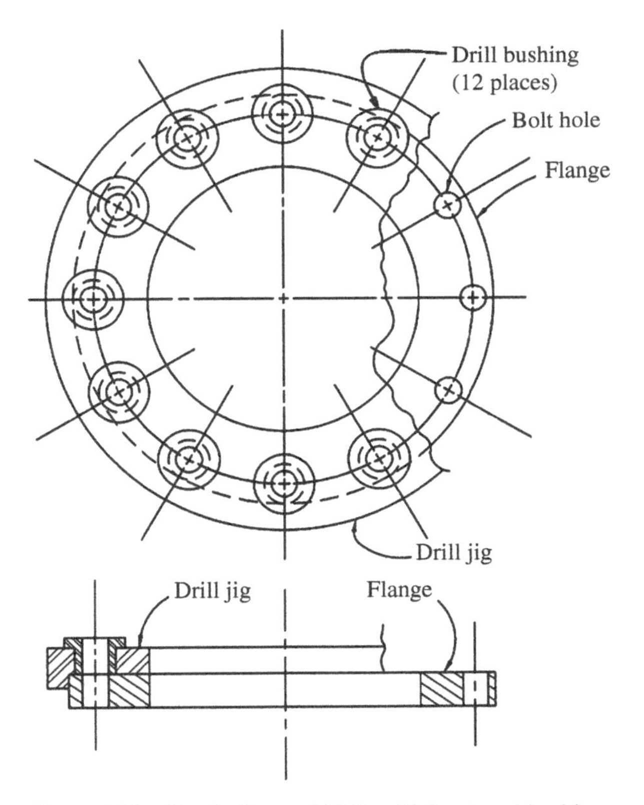

Figure 6.12 Simple flange drill jig with hardened bushings

drilling machine that otherwise is limited in its accuracy. Each time that a part is required with that particular drill pattern, the jig will be placed on the part, the assembly will be clamped down, and drilling can proceed. Proceeding in this manner allows an inexperienced operator to use a relatively inexpensive machine to produce a very precise part.

6.7 Material Removal Rates

While tooling, setup, and other factors also affect the cost of machining, a major component of cost is the time required to perform the actual machining operation. This is a function of the material removal rate and the amount of material that must be removed. Material removal rates in turning and boring vary with the feed, speed, and depth of cut. When milling, the feed, speed, depth of cut, the OD of the tool or part, and the number of cutters govern material removal. And all of these are affected by size, rigidity, and power of the equipment, strength, stiffness, and machining characteristics of the part material, and the capability of the tool and cutter material.

6.7.1 Feed

Feed refers to the rate of advance of the tool into the work. In turning or boring, it is typically expressed in inches per revolution (in./rev) or millimeters per revolution (mm/rev), whether fed across the face or along the cylindrical surface of the part. In milling, it may be expressed either as in./rev (mm/rev) or in inches per minute (in./min) or millimeters per minute (mm/min). Either way, the number of flutes or teeth on the cutter is included in determining the appropriate value.

6.7.2 Speed

Speed, or cutting speed, refers to the relative speed between the cutting tool and the material. It is measured in surface feet per minute (SFM or sfpm), also referred to as feet per minute, or meters per minute (MPM). For consistency with other units, this book will use ft/min and m/min, though in the machining industry SFM is more common. On a milling cutter, it is the speed in ft/min of the outermost cutting edge. In turning, it is the rev/min (rpm) times the circumference at the point of the cut.

6.7.3 Depth of cut

Depth of cut is the thickness of material that is removed per pass as a workpiece is machined. It measures the distance that the tool bit cuts into the part in a direction perpendicular to the tool or part travel.

In turning, whether on a lathe or a vertical boring mill, depth of cut is the thickness of material being removed, measured in the direction perpendicular to the feed of the tool.

Thus, in a facing cut on a lathe or vertical boring mill, the depth of cut is measured axially with respect to the machine, while the tool moves radially.

When cutting a cylindrical surface on the lathe, depth is measured radially and the tool is fed axially. Note that when cutting a cylindrical surface, the reduction in diameter per cut is twice the depth of cut, once on each side of the diameter.

It is common practice to use a depth of cut approximately five times the feed. Thus, for a feed of 0.010 in./rev (0.25 mm/rev), the depth of cut would typically be set at about 0.050 in. (1.25 mm).

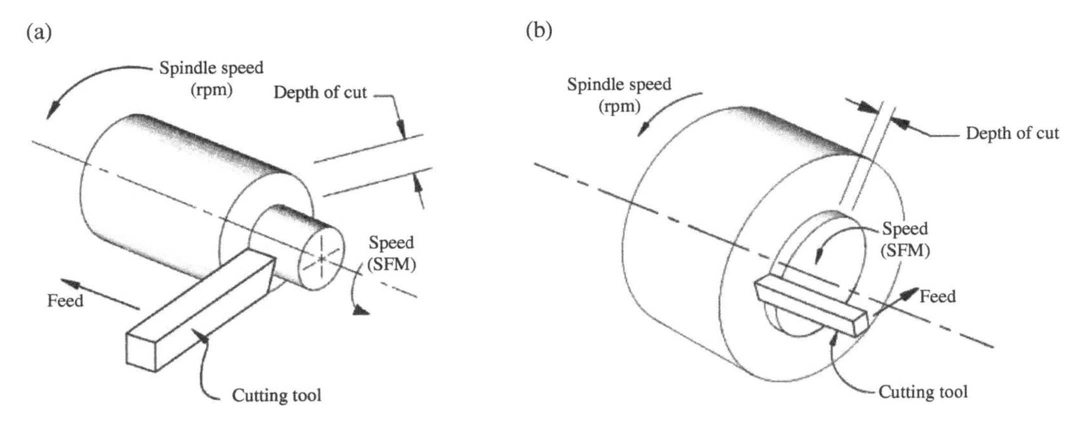

Figure 6.13 Feeds and speeds of lathes. (a) Axial cut on lathe. (b) Face cut on lathe

Different metals are machined at different feeds and speeds. If mild steel is taken as the standard, stainless steels are generally machined at significantly lower rates, while aluminum alloys are typically machined much faster. Also, different cutting tool configurations are used, depending on material properties. Metalworking books [1] provide information for determining these values, and exactness is not a requirement. Approximations are often used, and higher or lower values may be used, based on experience, variations in materials, and whether chatter or vibration occurs. Figure 6.13(a) and (b) illustrate machining parameters for axial and face cutting on a lathe. All parameters are comparable for a vertical boring mill or VTL.

6.8 Milling

Milling removes material using rotary cutters. Machines that perform this operation are usually referred to as mills, vertical mills, or sometimes vertical boring mills or horizontal boring mills, depending on the orientation of the cutter axis. Note the use of the term "boring mill" to refer both to a mill with a rotating vertical cutter and to a vertically oriented lathe.

Small machines typically move the part into the cutter for most operations. See Figure 6.14(a) for a small vertical boring mill of the type typically referred to as a "Bridgeport" after the company that popularized this type of machine. Figure 6.14(b) illustrates a larger gantry-style vertical drill/mill drilling a tubesheet. Such machines often make use of the bridge structure shown in this figure. The bridge typically moves the headstock horizontally in both directions. On some machines, the head can also tilt in one or both directions.

Figure 5.3 shows a large part mounted in front of a large horizontal boring mill. On the machine shown, the horizontal travel is achieved by moving the column from side to side. Horizontal travel in the other direction is done by moving the spindle axially.

How relative motion between the cutting tool and the part is achieved varies on large horizontal boring mills. Vertical motion is normally accomplished by moving the headstock vertically on the column of the machine, and the spindle moves in and out. The table may move crosswise relative to the spindle, crosswise and in and out, or not at all. On extremely large mills it is not unusual for the table to be stationary and all motion be accommodated by moving the column or bridge, headstock,

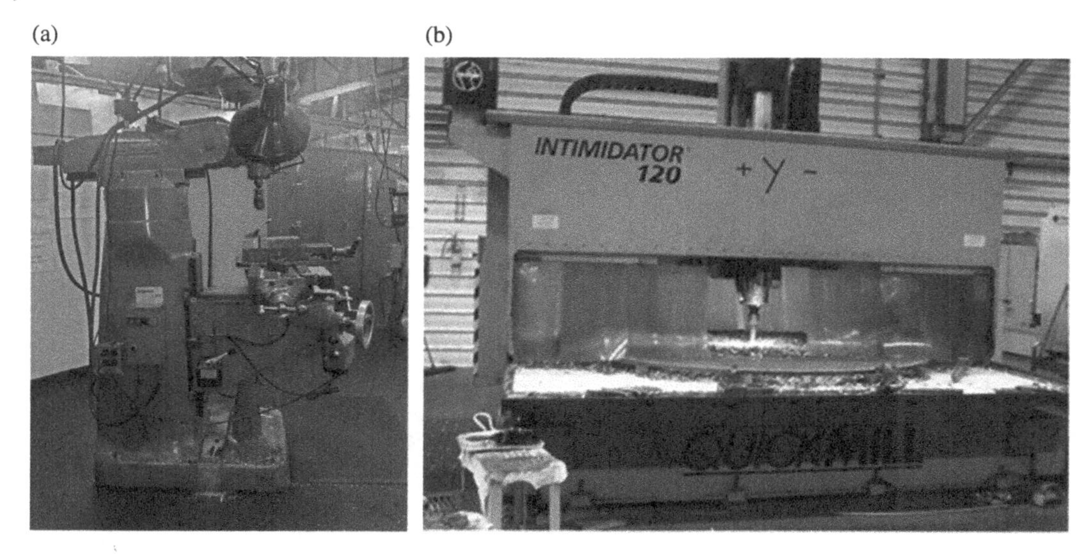

Figure 6.14 Boring mills. (a) Bridgeport-type vertical mill. (b) Large gantry-style drill/mill (*Source:* Harris Thermal Transfer Products)

and spindle. As with smaller machines, a rotary table may be used to provide an additional axis of motion.

Milling provides high levels of dimensional accuracy. Mill dials and readouts are typically graduated in thousandths of an inch, or even in ten-thousandths. Actual tolerances achieved are typically wider than these, however. A number of factors make a ±0.001 in. (0.025 mm) tolerance difficult to achieve in many applications – tool deflection, part deflection, and wear of leadscrews, etc. all have an effect. That said, with reasonable care a tolerance of ±0.005 in. (±0.12 mm) or ±0.010 in. (±0.25 mm) is often fairly easily achieved on small to medium-sized parts.

The most common milling operations performed on pressure vessel components are preparing plate or forging ends or edges for welds, facing sealing surfaces on flanges and other components, drilling flanges, and drilling tubesheets.

Facing operations are usually performed using the largest cutter available and within the capacity of the machine, balanced with part size, to reduce cutting time. Preparation for welding may be done with specially ground cutters so that a single pass, or a few passes, of the tool will complete the contour. Small endmills will be used to produce o-ring grooves, and if a dovetail groove is required, then a pass with a straight endmill will be followed by a pass with a dovetail bit, followed by polishing of the bottom of the groove to provide a suitable sealing surface.

Bolt patterns and tubesheet drilling were tedious in the past, requiring indexing in both *x*- and *y*-directions to locate each bolt hole. CNC mills have reduced this process to a fairly simple programming exercise. The mill will index from hole to hole much more quickly than a machine operator can, and without the chance of placing a hole 0.100 in. (2.5 mm) off its intended location through misreading of a scale.

Relative motion between the tool axis and the workpiece has always been available in orthogonal *x*-, *y*-, and *z*-directions. Addition of a rotary table adds a rotary axis, and if the table can be tilted, an additional axis. Large indexable "fly-cutters" allow facing of flanges with spiral or concentric cuts. This provides better gasket sealing capability than is normally achieved with the toolmarks produced by a rotating tool being moved back and forth across a surface.

While not common, it is also possible to produce a curved bore using a horizontal boring mill and specialized fixtures. Such an operation requires a specialized program for a CNC mill, or a setup involving a linkage designed to produce the particular curve in the bore of the part. Because of the design and tooling costs, it would only be done to produce a part that was not otherwise available. An example might be a sweep wye forged out of a high alloy. Such a setup in the past would only be accomplished by a skilled machinist working with a highly qualified manufacturing engineer. Today, using a CNC mill or machining center, production of such a component is relatively straightforward, though not trivial.

6.9 Turning and Boring

Turning of ODs, boring of IDs, cutting the tapered necks of welding neck flanges, and facing are most often performed on lathes or vertical boring mills. Threading can also be performed, but only concentric with the axis of rotation, and for small-diameter threads, the use of taps and dies is nearly always more productive.

Vertical boring mills are sometimes referred to as vertical lathes, or, if equipped with a tool changing turret, vertical turret lathes (VTLs). Note that there is overlap of names between the Bridgeport-type vertical boring mill and the vertically oriented lathe.

Turning operations take place by spinning the workpiece and feeding a cutting tool across the part to remove material. Radial, axial, or angled tool feeds are standard on conventional machines. On current CNC machines, the feed can also involve a contoured cut. In some cases, two cutting tools are used at the same time, on opposite sides of the part. This is usually only on a vertical boring mill, and it is not common.

Lathes, Figure 6.15, are generally designed and constructed with the capability to machine components that are longer in the axial direction than their diameters, though a lathe is able to turn short workpieces within its diametral capacity.

Figure 6.15 24 in. (60 cm)-diameter lathe (*Source:* Harris Thermal Transfer Products)

Vertical boring mills, Figure 6.8, and VTLs normally turn a diameter that is equal to or larger than the axial length of the part. They are typically for diameters greater than those of conventional lathes, since they would otherwise duplicate lathe capacity in what is for many applications a difficult orientation.

Carbide and other ceramic cutting tools are most often used. They offer a much longer life than comparable cutting tools of conventional tool steels. However, for some applications, high-speed steel or other alloys are used because they are easily modified in the toolroom. Ceramic cutting inserts do not lend themselves to modification for specialized applications and custom inserts must be purchased, sometimes with a substantial lead time.

6.10 Machining Centers

The majority of turning and milling in fabrication shops is performed on conventional lathes, vertical boring mills, and mills. The trend, however, is toward machining centers. A machining center combines the capabilities of a mill and a lathe in one unit. A typical machining center can often produce a finished part on a single machine. A machining center may be equipped with a turret, allowing any one of a variety of tools to come into play. If more tools are needed, an automatic tool changer, or ATC, is the usual solution. Drum-type tool changers can accommodate up to about 30 tools, and many more can be stored on a chain-type changer.

One common configuration includes a lathe with the addition of drilling and milling capability, often in multiple dimensions. This would allow a flange to be turned, bored, and drilled, all on the same machine. Such machines are particularly good for high production rates.

Extremely large machining centers are not common. The mass, space requirements, and costs of such machines are typically cost prohibitive unless they can be kept busy all the time. Thus, the use of conventional machine tools is likely to continue in the fabrication industry, though the trend toward computer numerical control (CNC) will also continue.

6.11 Drilling

Drilling is accomplished using a drill press, radial arm drill, a vertical or horizontal milling machine, or a numerical control drill. The differences among these various pieces of equipment are discussed in the following:

Drills are designed to accommodate large axial loads, but essentially no lateral loads. It is therefore unable to cut sideways, as a milling machine can. Each type of drill has its advantages.

A drill press is a fairly small piece of machinery. It is relatively inexpensive and does not require much space in a shop. A bench or floor standing drill press can typically drive a twist drill bit of 1/2–1 in. (13–25 mm) in diameter in steel. Locating the hole is done by moving the workpiece beneath the quill, often on a cast iron table. This works well enough for relatively small and light pieces, but as the part being drilled increases in size, handling it and moving it back and forth so that the drill bit is accurately aligned with a laid-out punch mark can be difficult.

A radial arm drill is designed to overcome the weaknesses of a drill press. It does so at greater cost and a need for more floor space, but in return it allows for easy positioning of the drill bit above the part and allows drilling larger holes. It typically has a more powerful motor than a drill press and it is gear driven, rather than belt driven. It has a large column supporting a radial arm that can be rotated, and the drilling head can be moved in or out along the arm. The speed range of a radial arm drill may go lower than that of a drill press to accommodate a larger bit, and the drill will have a power feed. Thus, the radial arm drill allows drilling multiple holes without repositioning the part, and depending on its size, might drive a 3–4 in. (75–100 mm) diameter drill bit. If multiple parts with the same drilling pattern are needed, a radial arm drill and a drilling fixture with hardened drill bushings can be used. The drill bushings help with quick and accurate location of the bit above the part and they eliminate wear on the fixture.

Mills are capable of sustaining the lateral loads developed in the milling process, but this adds little value if the only required operation is drilling. On the other hand, mills are constructed to be very rigid, and often have powerful motors and low speeds and feeds. They allow easy positioning of the drill with respect to the part, and if they are CNC, this positioning can be programmed and performed quickly. Otherwise, it will be easier and quicker to position a radial arm drill. Thus, while a mill is very capable of drilling holes, it is often overkill. Mills are far more expensive than comparably sized drills, require a significant commitment of space, and may not be as easy to position.

CNC drills have locating capability of CNC mills, but typically lack capacity for lateral cutting. For drilling alone, this is not a disadvantage and there is a cost saving that comes from the lighter construction of this machine. Such a machine is most useful for repetitive operations, for example the drilling of multiple flanges. In such a case, a fixture would be developed to locate the workpiece accurately and quickly, helping to minimize setup labor. The CNC drill would then automatically locate and drill the holes. Sometimes the CNC drill is actually a multifunction machining center that can mill and/or turn as well.

6.12 Tapping

Tapping is the cutting of internal threads using a single rotating tool with multiple cutting edges. Tapping can be used for either straight or tapered threads. Most pressure vessels do not involve threaded holes other than for couplings, thermowells, and similar applications, and the sizes are usually not excessive. Taps are readily available up to about 2 in. (50 mm) in diameter. Specialty taps are available for a variety of applications. The number of flutes varies from 2 or 3 for small taps, up to about 6 or 8 in the larger sizes. Having less flutes is a benefit in the smaller sizes because it allows for both more space for chip clearance and stronger flutes. In the larger sizes, a higher number of flutes helps with centering the tap and reduces loading on the individual cutting edges.

Straight taps are usually classified as taper, plug, or bottoming. A taper tap has a long tapered section, usually covering 7–10 threads, to facilitate alignment and beginning of cutting. A plug tap is similar, but with a shorter taper. A bottoming tap is designed to allow threading to the bottom of a blind hole and has no taper.

A taper tap should not be confused with a tap for cutting a tapered thread such as a pipe thread. A tap for a pipe thread will have full threads along the taper to create the required full tapered thread in the part.

Most taps have straight flutes, but spiral fluted taps are sometimes used. Spiral taps facilitate chip removal in power tapping. Their use is especially common when power tapping in materials in which the chips do not readily break off without backing up the tap.

Small holes may be hand or power tapped, Figure 6.16, while larger ones are typically tapped with power because of the required loads and because the mill or drill used helps with alignment. Tapping speeds are much less than the typical cutting speed for a particular material.

Figure 6.16 Hand cutting threads in aluminum using a taper tap

6.13 Water Jet Cutting

Water jet cutting has not typically been considered to be a machining process because it does not involve a traditional solid tool with a cutting edge. It has evolved, however, to a process that can cut metal, stone, and other materials, hard or soft, to tolerances rivaling those of many traditional machining processes.

While not new as a concept – the idea was used in hydraulic mining in the mid-1800s and in cutting paper in the 1930s – water jet cutting has clearly advanced significantly. Early systems were limited by the lack of reliable high pressure pumps, nozzle wear, and the lack of affordable control systems.

The current process uses a high pressure pump with a nozzle with an abrasive inlet and a jewel orifice, a controller, and a mechanism for manipulating the jet, either in x–y coordinates for cutting sheets or plates of material, or in a more complex fashion for cutting three-dimensional shapes.

Characteristics of water jet cutting are as follows:

- No heat input and no heat-affected zone.
- No distortion due to heat.
- Narrow kerf.
- Tight tolerances.
- Slight taper of the cut, which can be adjusted out by angling the nozzle.
- No flash or burrs.
- Cuts a wide range of materials.
- Cutting range up to about 6 in. (150 mm) in metals.
- Speed can be traded for a better finish.
- Relatively slow cutting.

6.14 Laser Machining

Laser cutting, like water jet cutting, was not historically considered a machining process, but can also produce parts to high tolerances. As a result, the process is now sometimes referred to as laser machining, particularly when used to produce three-dimensional parts. While not avoiding heat input altogether, and therefore still leaving a heat-affected zone, this is a very fast process and most of the heat is removed as the molten material is blown from the melt zone.

Advantages:

- Small heat-affected zone.
- Minimal distortion due to heat.
- Narrow kerf.
- Tight tolerances.
- Cuts a wide range of materials.
- Cutting range from foil up to about 1 in. (25 mm) of steel.
- High cutting speed.

Disadvantages:

- Cutting table sizes limited by optical needs.

See Section 6.3.1.4 for a further discussion of laser cutting.

6.15 Reaming

A multi-fluted cutting tool used to slightly enlarge a drilled hole, leaving a precisely sized and smoothly finished hole is called a reamer. Typical reaming tolerances are 0.001 in. (0.025 mm) or less. The operation is performed either by hand or on a drill or mill. As a general machining operation, reaming is done when the tolerance or finish of a drilled hole is not sufficient, or sometimes to bring two closely aligned holes in mating parts into perfect alignment.

The amount of material removed in the reaming process is small, usually between 0.005 in. (0.125 mm) and 0.010 in. (0.254 mm). The reaming allowance varies depending on the material and application.

Reamers are available in straight and spiral fluted varieties. Both types have tapered lead-ins to facilitate starting the cut. Hand reamers usually have a longer taper than machine reamers, to help with alignment. Reamers for tapered holes are also available, permitting precise production of holes for tapered pins, or to produce tapered sleeves to accept Morse Taper Shank Drill Bits and other such tools.

In pressure vessel fabrication, the most common application of reaming is the sizing and finishing of the holes in a tubesheet prior to grooving and inserting and rolling the tubes.

6.16 Electrical Discharge Machining, Plunge and Wire

Electrical discharge machining (EDM) uses erosion of the workpiece by sparks to remove material. It can be used for any conductive material and is therefore particularly useful for already hardened materials or brittle metals that would otherwise be difficult to machine. It can also machine shapes that would be extremely difficult to produce using conventional machining techniques, since electrodes can be produced in the shape of a required hole.

The workpiece is submerged or filled with a dielectric fluid and a spark is developed between the tool (electrode) and the workpiece by exceeding the dielectric breakdown voltage of the fluid. This sparking occurs many times per second, while the dielectric fluid flushes out the material that has been removed. Because of the arcing between the tool and the workpiece, both the tool and the workpiece may be subject to erosion, though techniques have been developed to minimize tool wear.

The electrode is advanced slowly. If the control system functions as intended, the electrode will never actually touch the workpiece. The system therefore operates essentially without contact forces.

Both solid tools and wires are used for EDM. Wire EDM performs an operation somewhat similar to sawing. To avoid wire breakage due to erosion, supply and take-up spools move the wire continuously through the part.

Surface finish is controlled by optimizing the arc and by controlling speed. Speed can be optimized, however, by using one electrode for rough cutting then moving another electrode through the part at a lower speed.

Because extremely hard materials are not often used in pressure vessel construction, and because of the relatively high cost of EDM compared to conventional machining, it is not often used in construction of pressure vessels. Probably, the most common use is removal of broken taps from holes in the process of being threaded.

6.17 Electrochemical Machining

Electrochemical machining removes material using a process that might be described as reverse electroplating. It has similarities to EDM in that it is a noncontact process; it can be used on vary hard materials; and it uses an electrode, but involves no arcing. Instead of a dielectric fluid, it uses an electrolyte (conductive fluid) and it dissolves the workpiece with no wear of the electrode.

This process has found applications in construction of aerospace pressure vessels because of the importance of minimizing weight, but is not generally used in construction of ground-based pressure vessels.

6.18 Electron Beam Machining

Electron beam machining (EBM) uses a concentrated pulsed beam of high-velocity electrons to heat the workpiece locally and vaporize it. EBM is normally performed in a vacuum to minimize collisions of the electrons with air molecules and to minimize contamination. It is not normally used on pressure vessels.

6.19 Photochemical Machining

Photochemical machining (PCM), also called photochemical milling or photo etching, uses etchant chemicals to selectively corrode away a portion of a material. The "photo" portion of the process is actually in the development of the tool, or mask, that is placed over the material to protect the portions of the workpiece that will *not* be machined.

An offshoot of the production of circuit boards, PCM can produce very fine detail and complex parts. Its use is limited to metal foil or sheet in the range of 0.0005–0.080 in. (0.013–2 mm). It does not have common applications in the production of pressure vessels.

6.20 Ultrasonic Machining

Ultrasonic machining is another process generally used for very hard materials. Its use consists of vibrating a tool at frequencies typically in the range of 18–40 kHz, against a workpiece, in a slurry of abrasive particles. Ultrasonic machining is most commonly used in machining of ceramics and glasses. It is not typically used in production of pressure vessels.

6.21 Planing and Shaping

Planing and shaping are machining processes performed on machine tools referred to as planers and shapers. They are two closely related processes. In planing, a cutting tool is held stationary while the workpiece is moved past it, scraping off a layer of material. In shaping, the workpiece is held stationary and the tool is moved. While used to produce many other shapes as well, the linear motion of the two processes particularly lent itself to production of weld preps on straight edges of plate materials. Both processes have been largely supplanted by the use of milling machines, which are generally much more versatile.

6.22 Broaching

Broaching is a process most commonly used for the production of axially oriented features, such as splines and keyways, in bores of components that must turn, or be turned by, shafts. The tool and typically the machine used to perform the operation are generally both referred to by the term "broach."

In broaching, the shape of the tool provides the configuration of the finished feature, with individual teeth of the broach increasing in height so as to deepen the cut.

While used to produce keyways and splines for components such as valves, broaching is not generally used in production of pressure vessels.

6.23 3D Printing

3D printing, also referred to as additive manufacturing (AM), is the process of producing components by successively adding material. As such, it is not a machining process, and it might be considered the opposite of machining. It is briefly addressed in this section, however, because the types, configurations, and tolerances of components it produces are typically more similar to machined parts than to those assembled by other means.

AM components are produced in various ways. Some AM machines add material by extruding it onto the surface of the previous layer. Others first lay down a layer of a fusible powdered material, typically a metal, then use a laser or an electron beam (EB) to melt, or sinter, the portion of the powder where the part will be. Another layer of powder is laid down, and the process is repeated.

Either way, the process is relatively slow. A strong advantage, however, is that the material stock is the same regardless of part shape and size. In addition, the laser or EB sintering process allows production of parts within parts, something that conventional machining is unable to achieve. Production of parts with internal passages is also sometimes easier using this approach.

The field of AM, while expanding rapidly, is still in its infancy. Engineers and scientists are working to better understand material properties and tolerance capabilities for the process. Development of standards for production, process control, and quality are continuing.

The use of AM parts in pressure vessels is limited, if happening at all. The cost, the lack of understanding of material characteristics, and the availability of needed components in other forms have limited its use. Small and specialized AM parts for valves and other components that may be difficult to produce or require stocking of many small items are likely to become more prevalent. Flanges, vessel shell plates, and the like have less need for a new process because the use and availability of these components are high and the cost of producing them using AM is relatively high.

6.24 Summary

The discussion of machining introduces some of the most common, and a few less common, machining processes and machine tools. It is necessarily only a brief introduction. Much more detailed information may be found in references such as Machinery's Handbook [1]. This reference provides much more thorough introduction to many aspects of machining. It has been published for many years, is updated on a regular basis, and continues to be useful. Other books provide more detailed insight into individual types of machine tools, processes, applications, tooling and tool design, and myriad related topics.

Reference

1 Erik Oberg, Franklin D. Jones, Holbrook Horton, and Henry Ryffel. *Machinery's Handbook*, 31st Ed, 2020. Industrial Press, South Norwalk, Connecticut.

7

Welding

7.1 Introduction

Welding is the process of fusing materials together by locally melting the base material as well as any added filler material, with or without the application of pressure, or by the application of pressure alone and with or without the use of filler material. The definition of welding is not commonly used to refer to joining techniques such as brazing, which is described in Section 7.10, and soldering. These are processes in which a molten metal with a melting point below the solidification temperature of the parent material is used to join parent materials that are not melted.

In pressure vessel fabrication, welding is primarily used for joining components together; however, it also has been used for cladding (applying a layer of material with properties more compatible with the process fluids or environment) and for repairing gaps or defects in the parent material. While the process is also used for plastics, only welding of metals has been addressed in this book.

While welding is most often used to join materials of the same alloy, it can also be used for assembly of dissimilar metals. When this is done, it is common to use a filler metal of a somewhat higher alloy than the higher of the two alloys being joined to relieve concerns about dilution of alloying elements.

The molten metal involved in the welding process tends to be highly reactive with oxygen in the air. Therefore, reaction with the surrounding air is prevented by providing either an inert slag or inert gas layer as a shield or by preventing contact with the reactive gases by performing the weld in a vacuum. Using a gas shield or a vacuum can result in cleaner welds, but these processes also tend to be more expensive when large volumes of weld are required.

7.2 Weld Details and Symbols

Fabrication drawings follow drawing conventions to ensure clear and accurate communication of the configuration of the product being produced. A part of this communication is that of weld configuration. While detailed drawings of special weld cross sections are not uncommon, especially for large or unique cross sections, standard weld configurations are generally communicated using standardized weld symbols. These symbols describe weld characteristics such as whether the weld is a groove or a fillet, weld dimensions, root openings, and whether the weld is continuous or intermittent. Some of the weld joints described by weld symbols are described in the following sections:

Fabrication of Metallic Pressure Vessels, First Edition. Owen R. Greulich and Maan H. Jawad.

7.2.1 Single fillet welds

Single fillet welds, Figure 7.1(a), are used in many applications:

a) Socket weld flanges.
b) Nozzle welds, often on top of groove welds for larger sizes where reinforcement is needed, and sometimes sized to add reinforcement area.
c) Certain small fittings and bolting pads.
d) Flat covers.
e) Jacket closures.
f) Head-to-shell lap joints.
g) Reinforcing and wear pads.
h) Tube-to-tubesheet joints.
i) Structural attachments, support rings, and lifting and other lugs.

7.2.2 Double fillet welds

Figure 7.1(b) provides an example of a double fillet weld, which is used for the following applications:

a) Through nozzle penetrations, either alone or with groove welds.
b) Slip on flanges.
c) Certain small fittings and bolting pads.
d) Structural attachments, including stiffening rings and lifting and other lugs.

7.2.3 Intermittent fillet welds

Figure 7.1(c) illustrates the use of intermittent fillet welds, which are used for certain specific applications in process equipment. Their use has two common purposes, depending on the application:

a) Their most common use in an ASME code vessel is to reduce the total weld volume, and, thereby, cost. This is done when the reduced amount of weld is sufficient and the discontinuous weld is adequate for the design conditions. The most common case is for attaching reinforcing rings used for stability under vacuum.

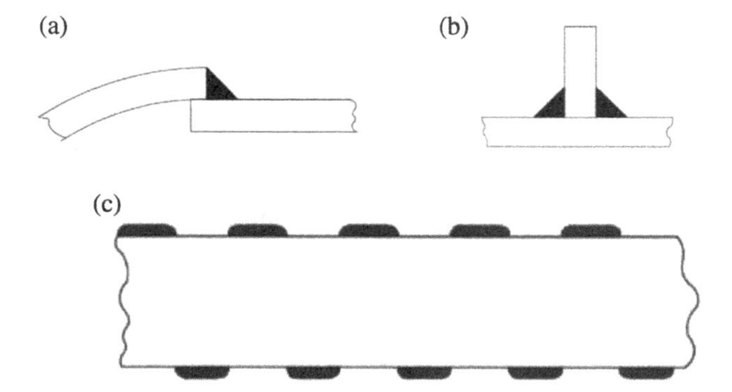

Figure 7.1 Some types of fillet welds: (a) single fillet weld, (b) double fillet weld, and (c) intermittent weld

b) In a non-code vessel built for high vacuum applications, it is common to use an intermittent weld where a continuous fillet weld could result in a virtual leak. A virtual leak occurs when a very small leak from an enclosed volume results in a continuous bleed of gas to the inside of the vessel, preventing pump-down to the required level. This is not an actual leak in the vessel, since there is no connection to the outside, but in high vacuum applications, the bleed from the space under the repad can continue for extended periods of time, preventing effective pump-down.

   It is usual to search for leaks in a high vacuum vessel or system using a helium mass spectrometer connected to the vacuum pump. A small jet of helium is directed to the outside of the vessel in the area of the suspected leak, and the mass spectrometer will indicate when the helium is pulled into the vessel. By this means, the area requiring repair can be localized.

If the gas comes from a weld pinhole connected to porosity, or to the space under a sealed reinforcing pad, then there is no connection to the outside of the vessel and the mass spectrometer is of no use. If the reinforcing pad is welded intermittently, then there is a path for the helium to be drawn into the vessel when the jet is in the neighborhood of the leak. Otherwise, the whole internal weld will be suspect because the helium has no way of entering the space under the reinforcing pad.

7.2.4 Single-bevel butt welds

Single-bevel butt welds, Figure 7.2(a), are usually for fairly thin sections, typically about 1 in. (25 mm) or under:

a) Shell-to-shell or head-to-shell joints.
b) Piping joints.
c) Pipe-to-welding neck flange joints.
d) Inset contour fittings.
e) Contoured insert nozzle-to-shell joints.

7.2.5 Double-bevel butt welds

Double-bevel butt welds, Figure 7.2(b), are generally for sections above about 3/4 in. (19 mm), or for special applications:

a) Shell-to-shell or head-to-shell joints.
b) Segmental flat head or formed head welds.
c) Piping joints.
d) Pipe-to-welding neck flange joint.
e) Inset contour fittings.

7.2.6 J-groove or double J-groove welds

Figure 7.3, as well as U-groove or double U-groove welds, Figure 7.4, are often used instead of simple V-groove welds to reduce total weld volume and welding distortion. The use of the J or U configuration allows access of the welding head to the root of the weld even with steep weld side walls, while with a narrow V-groove, weld placement of the weld root would be more difficult.

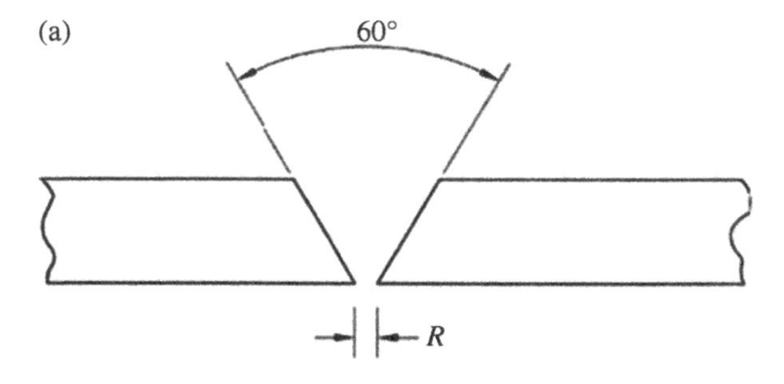

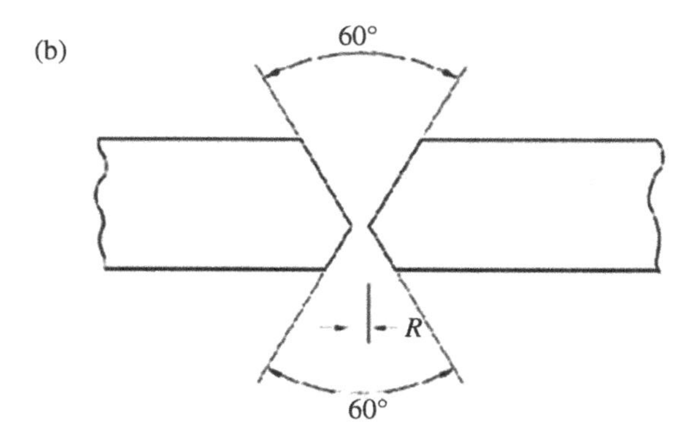

Figure 7.2 Bevel joints: (a) single bevel joint and (b) double bevel joint

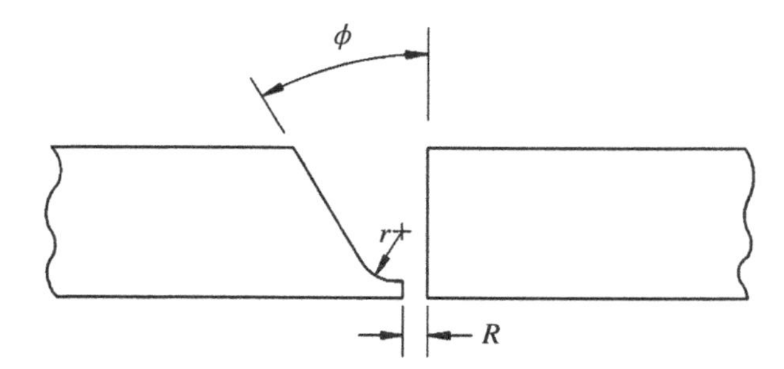

r = 1/4 inch
ϕ = 30° for flat and overhead only
ϕ = 45° for all other positions

Figure 7.3 Typical J bevel joint

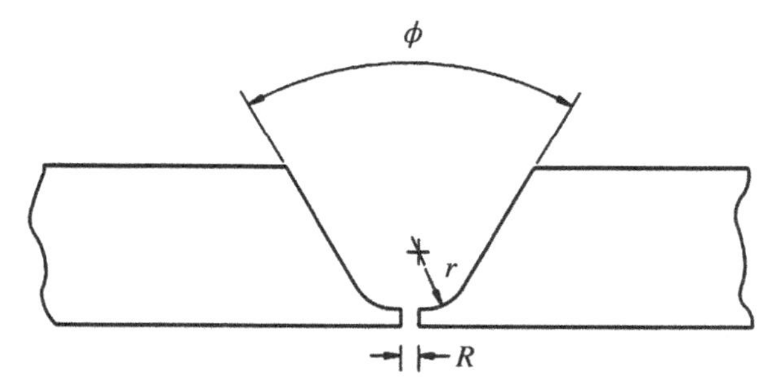

r = 1/4 inch
ϕ = 20° for flat and overhead only
ϕ = 45° for all other positions

Figure 7.4 Typical U bevel joint

7.2.7 Backing strips

Backing strips, also referred to as backup, or backing bars provide a means of containing the weld in a single-bevel groove while the metal is still molten, preventing blow-through and minimizing back-side oxidation during the welding process. They are also generally used for electroslag and electrogas welding. Backing strips are usually just tacked in place, after which the weld can proceed normally. Then the more efficient processes such as SAW can be used for the full volume of the weld. Installation and use of a backing bar are typically less expensive than either beginning a weld with a GTAW root pass and then switching to SAW or, for thin sections, welding from both sides.

A backing bar results in crevices, potential corrosion locations, and stress intensifications, so it is not appropriate for all applications. Sometimes a backing bar is used during welding, and is then removed after completion of the weld but before any nondestructive evaluation. When this is done, some back-side repair may be needed. If the backing strip is not removed after welding, then the ASME Boiler and Pressure Vessel Code limits efficiency of the joint to 0.90, usually leading to a thicker wall.

7.2.8 Consumables

Consumable inserts fill a role similar to backing bars but are fully consumed and become part of the weld. They are typically fused in place using GTAW. This provides a means of easily and consistently producing a good-quality weld root, capable of passing inspection and with no crevices to increase the chance of corrosion or cracking.

7.2.9 Tube-to-tubesheet welds

Several configurations are used for tube-to-tubesheet welds. The tube end is sometimes flush with the face of the tubesheet, sometimes inset, and sometimes extends beyond the face of the tubesheet. In the past, such welds were performed manually, but it is now common to weld any of these joint types using an orbital welder.

7.2.10 Weld symbols

Standard weld symbols used in the United States to identify weld configuration and details are based on the terminology of the American Welding Society (AWS) as shown in Appendix B. The AWS, ISO, and British Standards documents defining the use of welding symbols are AWS A2.4 Standard Symbols for Welding, Brazing, and Nondestructive Examination; EN ISO 2553 Welding and allied processes – Symbolic representation on drawings – Welded joints; and BS EN 22553 Welded, brazed and soldered joints – Symbolic representation on drawings. The terminology in accordance with AWS consists of a Reference Line, Figure 7.5, which points to the weld joint, with the associated symbols and dimensions. Weld symbols such as those shown in Figure 7.6 are used to indicate the type of welding desired.

The weld symbols are inserted on the top or bottom of the Reference Line to indicate the type and location of the weld as shown in Figure 7.6.

Supplementary information can be added to the Reference Line to indicate further welding instructions as illustrated in Appendix B.

The weld symbols in Figure 7.7 show a combination of the terminologies of Figures 7.5 and 7.6.

When a large amount of information needs to be shown on the Reference Line and welds are performed successively, such information may be shown on two or more lines as illustrated in

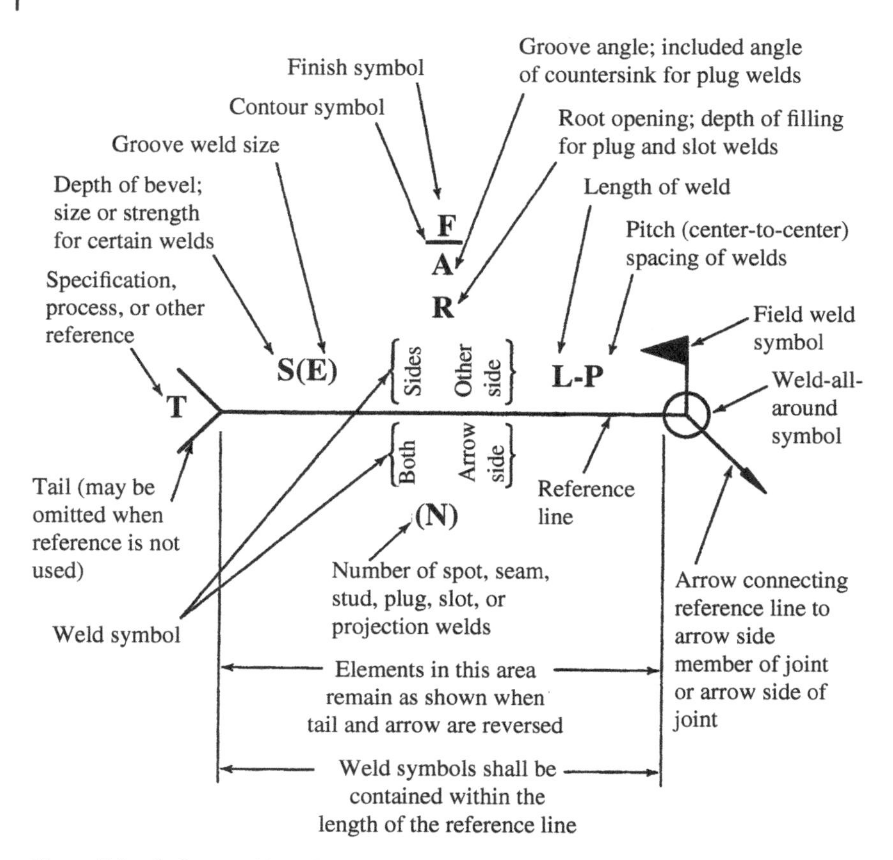

Figure 7.5 Reference Line (*Source:* American Welding Society, 2007)

Figure 7.8. The first welds to be placed are shown on the line nearest the arrow, later welds on other lines, in order of deposit.

Figure 7.9 shows a nozzle neck-to-shell attachment. The weld symbol of the nozzle neck-to-shell attachment shows a full-bevel weld on the outside of the shell that is welded all the way around the nozzle and is ground flush with the shell, all in the field. The weld attaching the pad to the nozzle neck is a 1/2 in. (13 mm) bevel of the pad plus a 3/8 in. (10 mm) fillet weld on top of the bevel. The weld is around the nozzle and is performed in the field. The weld attaching the pad to the shell is 1/2 in. (13 mm) fillet welded all the way around the pad in the field.

7.3 Weld Processes

Pressure vessel fabricators have access to numerous welding processes approved by the respective codes and standards for use in construction. Each code typically specifies which processes are acceptable within its scope. The designer and welding engineer must be aware of these limitations when specifying welds.

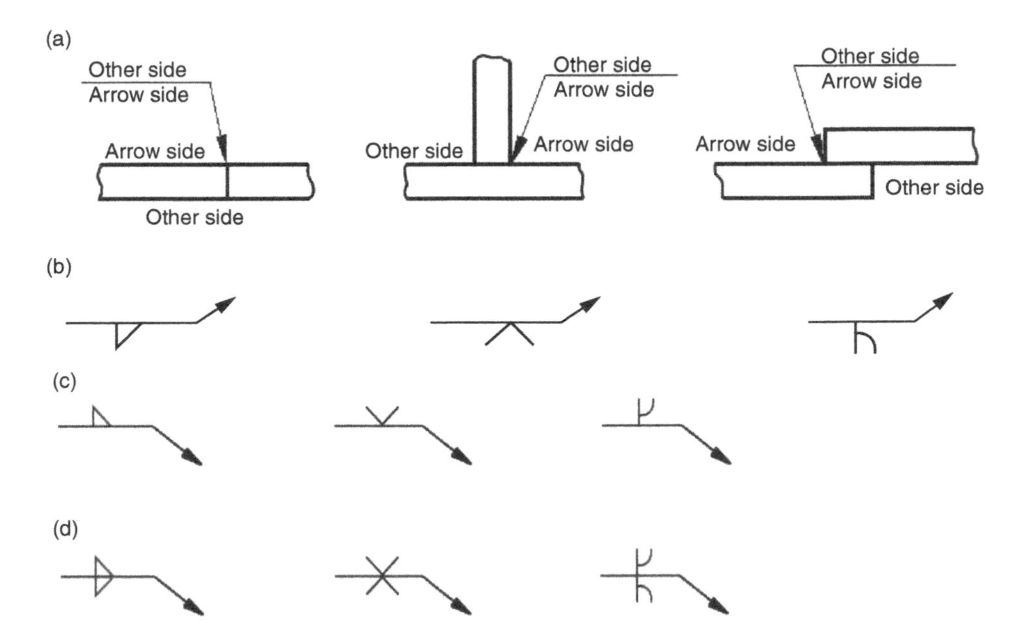

Figure 7.6 Weld symbols inserted on Reference Line: (a) arrow side and other side of the weld, (b) weld on the arrow side, (c) weld on the other side, and (d) weld on both sides

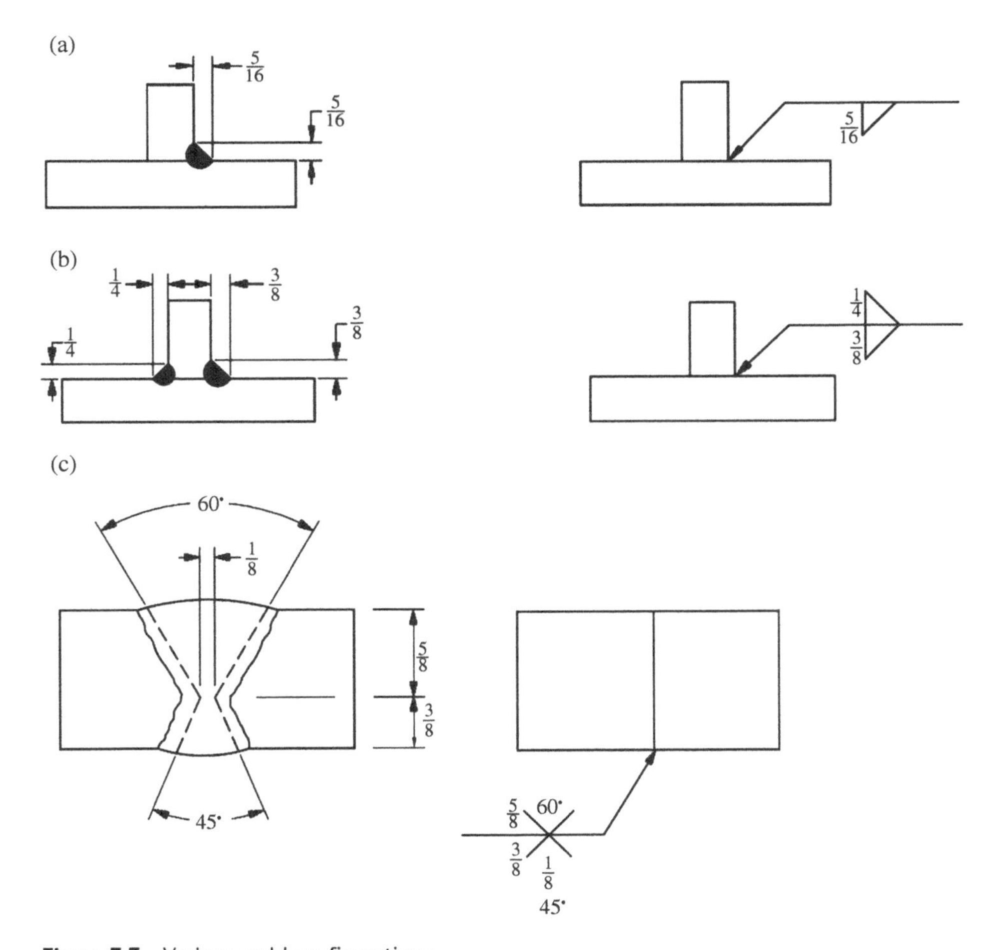

Figure 7.7 Various weld configurations

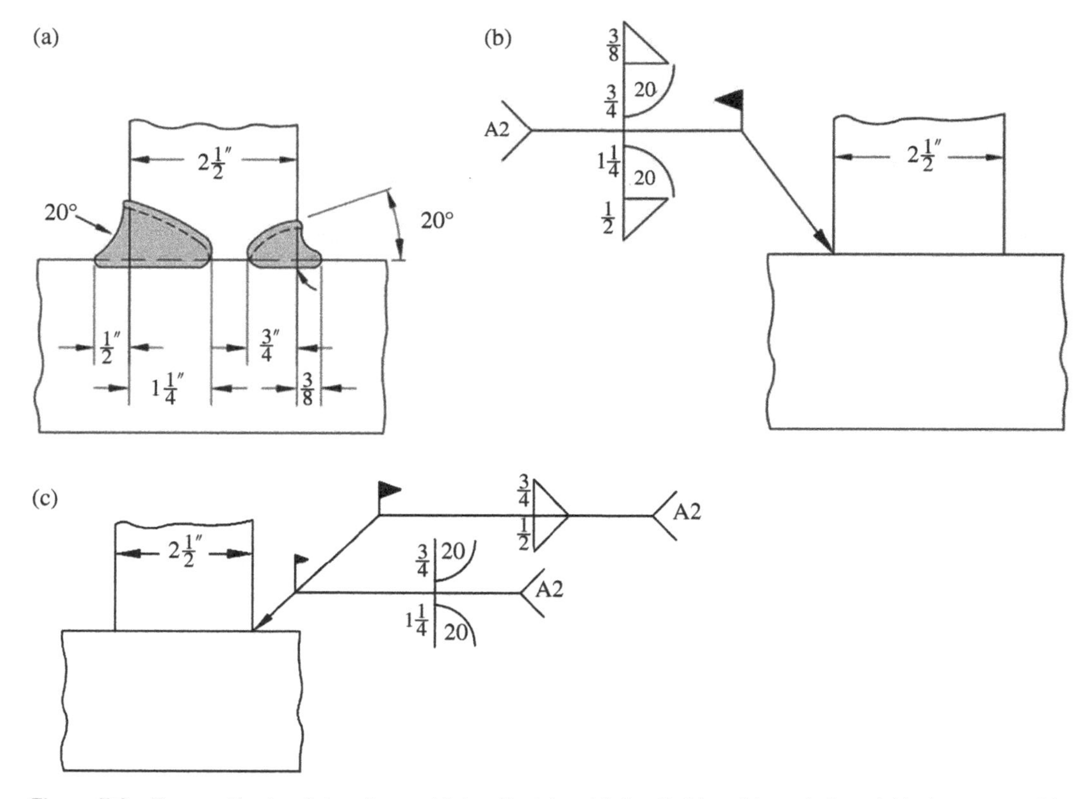

Figure 7.8 Two methods of showing weld details: (a) weld detail, (b) weld symbol, and (c) alternate weld symbol

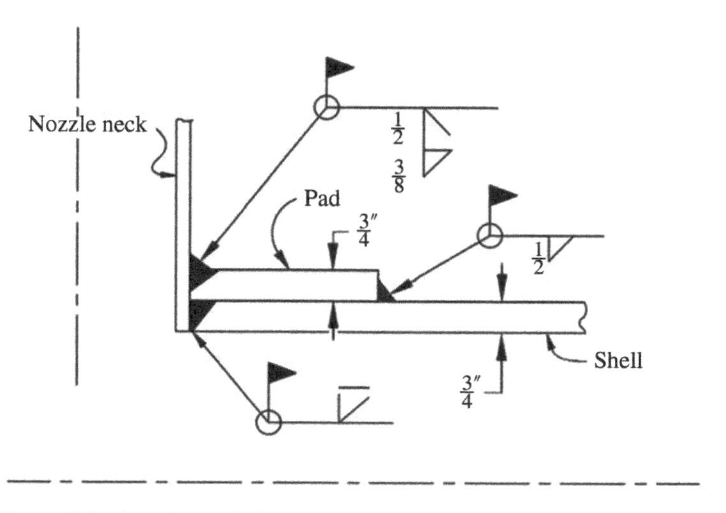

Figure 7.9 Nozzle-to-shell attachment

Any of a number of processes can produce suitable welds for most applications [1]. This is convenient for fabricators who may have limited types of welding equipment or who may not be able to access certain equipment because it is tied up with another job, but in many cases the choice of weld process can have a significant effect on both weld quality and cost. Some of these processes are suited for mass production, some for production of a high volume of weld metal, and others are for limited applications.

Issues common to most or all processes include the following:

1) Most molten metals are highly reactive and means of preventing oxidation must be employed such as a shielding slag, a shielding gas, or removal of oxygen by welding in a vacuum.
2) The high amounts of energy involved in the welding process often require means of cooling portions of the welding equipment, and sometimes the component being welded. Coolant may be channeled around or through equipment as illustrated in Figures 7.11 and 7.15.

Some weld processes approved by ASME are described next.

7.3.1 Diffusion welding (DFW)

7.3.1.1 Process

Diffusion welding, also referred to as diffusion bonding, is a solid-state bonding process that occurs due to solid-state diffusion when two metal pieces are held in intimate contact at a temperature near their melting points. At such temperatures, though the metals remain solid, there is sufficient atomic activity that the atoms of the two surfaces intermix and the surfaces bond to each other.

The process requires a firm consistent pressure along a clean bond line, so the surfaces to be bonded are generally first machined and polished. The plates are then placed in a vacuum furnace or a furnace with a slightly reducing atmosphere and subjected to a high temperature, typically 50–80% of the melting temperature of the plates while pressure is applied by a hydraulic press, dead weight, or differential pressure. After a period of time determined by the material(s) being bonded and the applied temperature, the two machined and polished surfaces fuse together to form a homogeneous bond.

7.3.1.2 Applications

1) Welding titanium and zirconium alloys.
2) Bond dissimilar materials such as copper-to-zirconium and copper-to-titanium.
3) Cladding.

7.3.2 Electron beam welding (EBW)

7.3.2.1 Process

In this process, electrons are generated by a cathode, Figure 7.10, using a high voltage, typically in the range of 150,000 V. The resulting high-energy beam melts a narrow zone through the parts to be joined, allowing the parts to fuse together. The process is conducted in a vacuum chamber to allow for optimal heat control and focus of the beam, as well as to preclude oxidation and other unfavorable chemical reactions.

7.3.2.2 Applications

1) Welding of reactive alloys such as titanium and zirconium.
2) Fuel housings, impellers, valves.

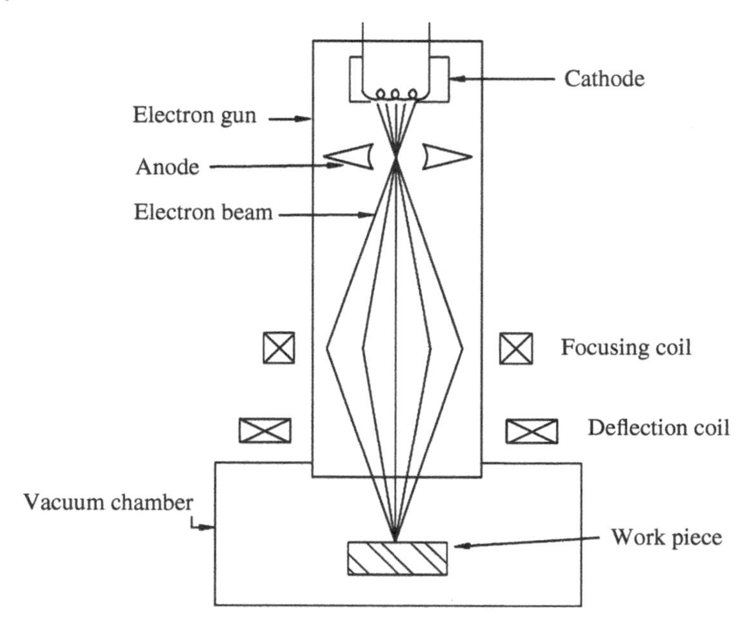

Figure 7.10 Basic components of electron beam welding

3) Welding thin materials to thicker materials.

7.3.3 Electrogas welding (EGW)

7.3.3.1 Process

Electrogas welding is the same as electroslag welding except in that an inert gas shield is used instead of flux.

7.3.3.2 Applications

1) Used in the pressure vessel and shipping industries for welding thick parts.
2) Used for plates up to 14 in. (355 mm) thick.

7.3.4 Electroslag welding (ESW)

7.3.4.1 Process

Electroslag welding is essentially a submerged-arc welding process intended to weld thicknesses up to 14 in. (355 mm). Butt welding is usually done in the vertical position, Figure 7.11, with very little joint preparation. Weld metal is added at such a rate that individual weld passes do not exist. Rather, the weld is put in place largely as a single mass of molten metal. For this reason, the weld is, even more than for more conventional weld processes, effectively a casting.

7.3.4.2 Applications

1) Used in the pressure vessel and shipping industries for welding thick parts.
2) Can weld plates up to 14 in. (355 mm) thick.

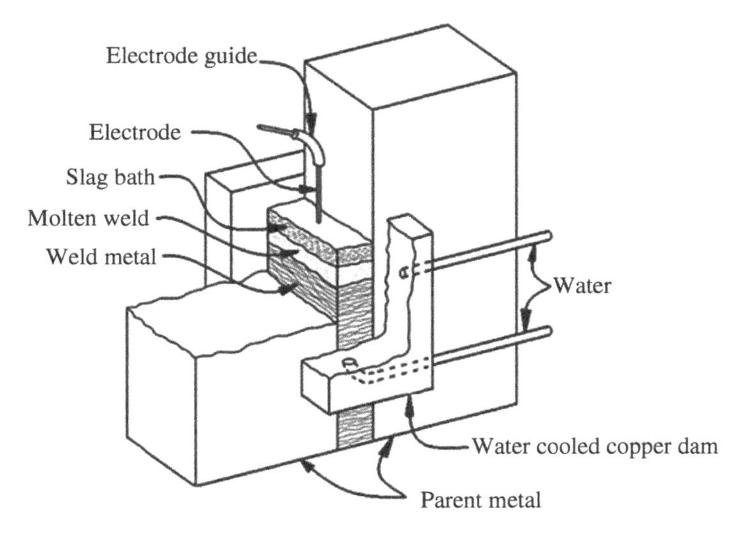

Figure 7.11 Electroslag welding

7.3.5 Flux-cored arc welding (FCAW)

7.3.5.1 Process

Flux-cored arc welding is similar to GMAW. It uses a continuously fed, flux-filled hollow wire. When molten, the flux floats to the top of the weld and provides the weld shield. Sometimes FCAW is performed using an inert gas shield in addition to the flux.

7.3.5.2 Applications

1) Desirable for small jobs or high production operations.
2) Suitable for a wide range of thicknesses.

7.3.6 Flash welding

7.3.6.1 Process

FW is a resistance welding process in which two pieces are spaced apart at a distance based on material properties, thickness of the pieces, and the desired weld properties. A voltage differential is then applied to the pieces to generate an arc between them, melting the ends. Force is then applied at the ends of the pieces to forge them together.

7.3.6.2 Applications

1) Joining ends of long tubes and pipes.
2) Thick structural members.

7.3.7 Friction stir welding (FSW)

7.3.7.1 Process

FSW uses a spinning rod driven into a tight butt weld joint as a means of heating the joint sufficiently to bond the mating pieces. While heat is involved, the parent material is not actually melted and this is a solid-state bonding process with similarities to both diffusion welding and forge welding.

7.3.7.2 Applications

1) Welding together large plates or shell courses.
2) Useful for a variety of materials.

7.3.8 Gas metal-arc welding (GMAW)

7.3.8.1 Process

Often referred to as MIG (metallic inert gas), GMAW is a very common welding process. It uses a continuously fed wire, Figure 7.12, with inert gas to shield the welding. It can be used in most positions.

7.3.8.2 Applications and characteristics

1) Desirable for small jobs and certain types of high production operations.
2) Suitable for a wide range of thicknesses.
3) Not as portable as some weld methods due to gas cylinders.
4) Can result in large amounts of weld "spatter."

7.3.9 Gas tungsten-arc welding (GTAW)

7.3.9.1 Process

Commonly referred to as TIG (tungsten inert gas) or heli-arc, GTAW is a very common welding process. It uses a nonconsumable tungsten electrode with an inert gas, Figure 7.13, to protect the hot surfaces from excessive oxidation. A consumable filler metal is typically added, particularly

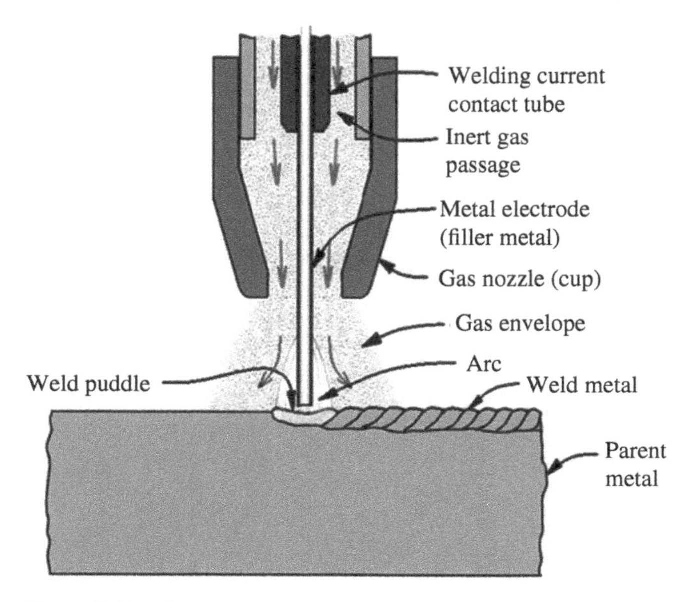

Figure 7.12 Gas metal-arc welding

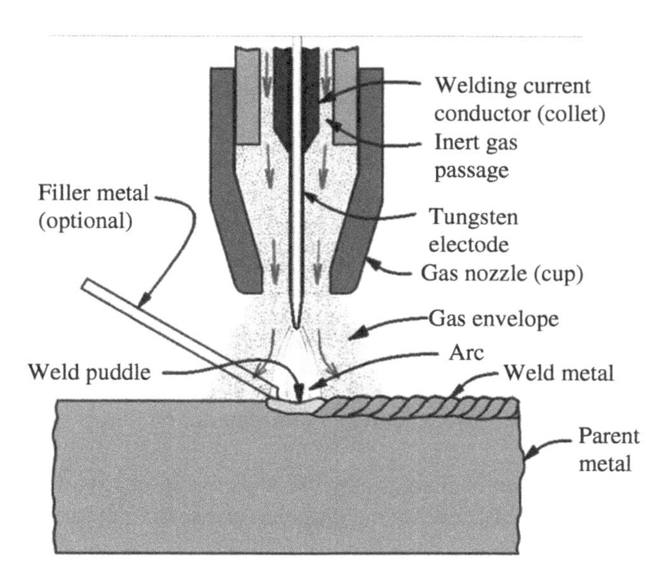

Figure 7.13 Gas tungsten-arc welding

for field welding when joints may not be optimal, but for thin-walled tubes, the process is frequently used in the absence of filler metal with a tight square butt joint. Sometimes it is used with a continuous wire feed (sometimes referred to as a cold wire feed because the wire is not the electrode and has no current flowing through it) for improved production. The high temperature of the arc between the tungsten electrode and the joint melts the metal to form the weld.

Use of GTAW as an automatic process in which the electrode rotates around the outside of a tubing joint is referred to as orbital welding and is common in applications where large numbers of the same size tube must be joined together. With good process control, orbital welding produces high-quality welds with great consistency. Deposition rates are relatively low.

7.3.9.2 Applications

1) Desirable for small jobs or high production operations.
2) Suitable for a wide range of thicknesses.
3) High purity applications.

7.3.10 Laser beam welding (LBW)

7.3.10.1 Process

A concentrated high-energy laser beam is pointed at the joint, causing the metal to melt and fuse together.

7.3.10.2 Applications

1) Complex geometries.
2) Welds of different thicknesses of material (e.g., convoluted bellows or hose-to-end connectors).
3) Welds of very thin materials.
4) Especially useful in high-production situations.

7.3.11 Orbital welding

Orbital welding is not a distinct process itself, but rather the automated use of other processes, typically GTAW. It uses an automated welding head that is clamped in place around the joint to be welded, or in the case of tube-to-tubesheet welds, clamped so that most of the unit is outside the tubesheet. The head rotates around the tube to be welded, following a preprogrammed sequence that controls arc ramp-up, speed, pulse magnitude and frequency, number of "orbits" (typically one for thin tubes, sometimes two or more for thicker tubes), and ramp-down. The technique and its associated hardware were developed to remove welder variability and have been demonstrated to produce high-quality welds consistently if the weld parameters have been properly specified, the process is controlled, and equipment is maintained.

Because orbital welding is an automated process, the person performing it is a *welding operator* rather than a *welder*. This places somewhat different qualification requirements on the person responsible for performing the weld.

The ASME Boiler and Pressure Vessel Code, Section IX, Paragraph QW-304 specifies the requirements for welders, while Paragraph QW-305 specifies those for welding operators, and Paragraph QW-360 addresses Welding Variables for Welding Operators.

7.3.11.1 Applications

1) Tubing butt welds, especially when performed in high volumes.
2) Tubesheet welds.

7.3.12 Oxyfuel gas welding (OFW)

7.3.12.1 Process

Oxyfuel gas welding (OFW; also commonly called oxyacetylene welding (OAW), oxy welding, or gas welding in the United States) is a manual process and is the oldest welding method used in ASME code pressure vessel fabrication. It combines oxygen with other gases, Figure 7.14, to

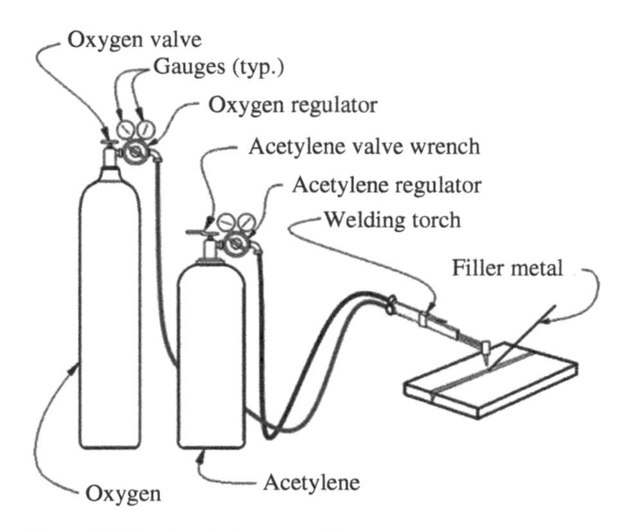

Figure 7.14 Oxy-fuel gas welding

produce a hot flame for depositing stick electrodes into a molten weld surface. The fuel is commonly acetylene although other gases such as Mapp, propane, or natural gas may be used.

7.3.12.2 Applications

1) Primarily used in carbon and low alloy steels.
2) Easy to use for repairing or welding small sections.
3) Used where access to electricity and water is limited.

7.3.13 Plasma-arc welding (PAW)

7.3.13.1 Process

The process is similar to that of GTAW except that the nonconsumable tungsten tip is inside the torch and the arc is thus restricted, Figure 7.15.

7.3.13.2 Applications

1) Thick sections.
2) Long welds.

7.3.14 Resistance spot welding (RSW)

7.3.14.1 Process

RSW uses two electrodes, typically mounted on manually or automatically actuated arms that press two or more layers of the parent metal between them, at the same time applying a modulated current to heat and fuse the metal. It results in locally fused zones between the pieces of the parent metal. It is sometimes referred to simply as "spot welding."

7.3.14.2 Application

Used for dimpled heat exchanger plates and strap-on jackets.

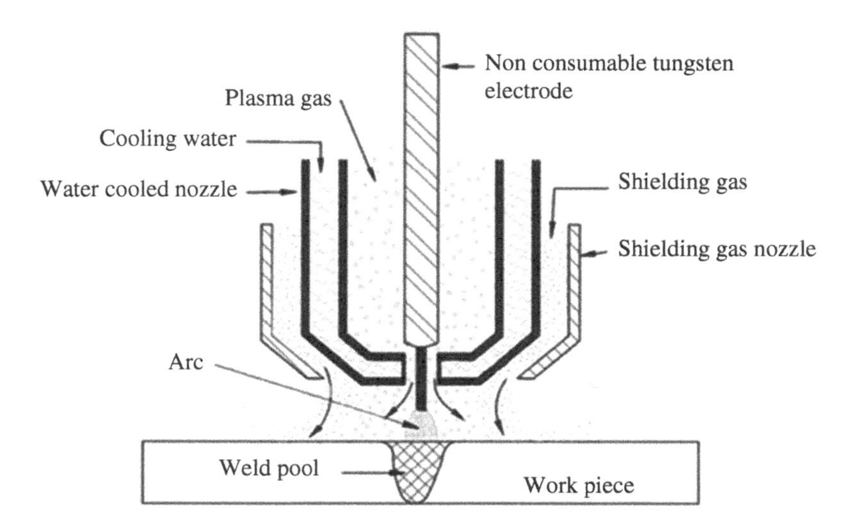

Figure 7.15 Plasma-arc welding

7.3.15 Resistance seam welding (RSEW)

7.3.15.1 Process

RSEW is similar to resistance spot welding except that it is performed using a pair of conductive rollers as electrodes, enabling its use as a continuous process.

7.3.15.2 Applications

1) Braid-retaining collars on flexible metal hoses.
2) Heat exchanger plates.

7.3.16 Submerged-arc welding (SAW)

7.3.16.1 Process

Submerged-arc welding is primarily an automatic process. Welding is done using an arc between a bare wire or ribbon and the parent metal beneath a blanket of granular flux, Figure 7.16. Sometimes the flux includes alloying elements. SAW can also be performed using a strip of material (ribbon) off a roll, for overlay purposes. Figure 7.16 shows a configuration in which the flux is fed around the weld wire. Other setups use a separate feed involving one or more wires introduced behind the feed of slag from a hopper.

7.3.16.2 Applications

1) Extensively used for carbon, chrome–moly, and stainless steels.
2) Used in welding both longitudinal and circumferential joints, especially for large weld volumes.
3) Used for weld overlay or cladding.

7.3.17 Shielded metal-arc welding (SMAW)

7.3.17.1 Process

Commonly referred to as "stick welding," this is the manual welding process that the most people are probably familiar with. It uses a welding electrode ("rod" or "stick") covered with a flux except at one end which is clamped in an electrode holder, a "stinger." The stinger provides the current

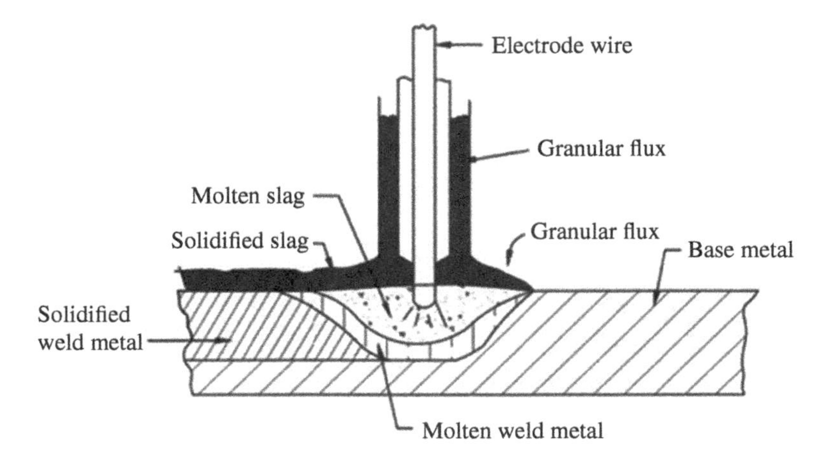

Figure 7.16 One type of submerged-arc welding

through the rod to the parent material, the parent material being connected to the other pole of the welding machine. This results in an arc between the electrode and the parent material. If the current and voltage are properly balanced, the electrode and the parent material are melted in a proper proportion to create a good weld. The process can be performed using either positive or reverse polarity, depending on the application. Sometimes alloying elements are contained in the flux.

7.3.17.2 Applications

1) Welding of joints in all positions.
2) Applicable to all materials.

7.3.18 Stud welding

7.3.18.1 Process

It is a high-speed resistance welding process in which a stud is welded to a base metal, often for attaching insulation or refractory material. A high current is directed through the stud to the base metal, heating the base of the stud and fusing it to the base metal. The fastener can be slotted, threaded, unthreaded, or tapped.

7.3.18.2 Applications

1) Used extensively to attach studs to structural members or vessel components, or jacket bars to heads and shells.
2) Can be used for various materials such as carbon steel, stainless steel, and nickel alloys.

A comparison of the advantages and drawbacks of some of the welding processes is provided in Table 7.1

7.4 Weld Preheat and Interpass Temperature

A minimum parent material temperature required before welding begins is referred to as preheat. The parent metal temperature that must not be exceeded during welding is referred to as the interpass temperature. Information on how these requirements are met is provided in Chapter 5, while how they are determined is addressed in Chapter 8.

7.5 Post Weld Heat Treating

Many carbon steel welds and some stainless steel welds require post weld heat treating (PWHT). Details on determining PWHT requirements are provided in Chapter 8. Information on accomplishing PWHT is given in Chapter 5.

7.6 Welding Procedures

Use of qualified welding procedures and welders or welding operators is required by applicable codes and helps to ensure that welds will be of adequate quality and meet specified material

Table 7.1 Some welding comparisons

Weld process	Typical weld deposition rate lb/hr (g/s)	Benefits	Drawbacks
SMAW (submerged manual arc welding, "stick")	3–6 (0.4–0.8)	Very low initial cost for both material and equipment, many welders available, low project setup costs; often used for fit-up	Limited deposition rates, greater chance of slag, porosity, and other defects, rough weld surface depending on welding rod selected and skill of welder, some weld spatter
SAW (submerged-arc welding)	6–20 (0.8–2.5)	High-quality welds, high deposition rates, consistent quality with qualified operators, good control of weld surface and contour, no spatter	High initial cost for larger weld power source and positioning equipment, higher fixturing costs
FCAW (Flux-cored arc welding, "flux core")	3–8 (0.4–1.0)	Relatively low initial cost, good deposition rates, good-quality product with competent welders	Somewhat greater chance of defects than SAW, and greater tendency toward weld spatter
GMAW (Gas metal-arc welding)	3–8 (0.4–1.0)	Relatively low initial cost, good deposition rates, good-quality product with competent welders	Somewhat greater chance of defects than SAW, and greater tendency toward weld spatter
GTAW (gas tungsten arc-welding, "TIG," "heli-arc"	0.5–2 (0.06–0.3)	Extremely clean, high-quality welds, can be used to produce clean and even weld root passes when other processes might result in burn-through	Very low deposition rates, high heat input per unit weld
EGW or ESW (Electroslag or Electrogas)	35–50 (4.4–6.3)	Extremely high deposition rates, low cost of joint preparation, thick sections may be welded in a single-pass. Slow cooling prevents formation of martensite and associated cold cracking in the heat-affected zone	High heat levels make cooling arrangements desirable, low cooling rates result in a coarse grain weld, high heat input may result in a low toughness weld, may produce welds that have hot cracking and heat-affected zones with low toughness and high transition temperatures, flux must be stored in an oven to prevent absorption of moisture, good process control is required to avoid slag inclusions, and can only be used in the flat position

property requirements. Details on actual qualification requirements are provided in Chapter 8. All standards for pressure vessels and pressurized components include requirements for qualification of welding procedures.

7.7 Control of Residual Stress and Distortion

High temperatures and thermal gradients associated with welding can result in significant distortion and residual stress upon cooling. Much effort goes into designing welding procedures that are as efficient as possible from a fabrication standpoint and which reduce heat inputs so as to control distortion upon cooling. More information is provided in Chapters 5 and 8.

7.8 Material Handling to Facilitate Welding

Proper positioning and process control improve productivity and weld quality immensely. Chapter 5 addresses the use of equipment for positioning the product to facilitate welding, both with respect to position and control of other aspects of the process.

7.9 Weld Repair

When weld repairs are allowed and when they are needed, as well as impact and heat treatment requirements for them, are discussed in Chapter 13.

7.10 Brazing

Brazing is the process of joining two or more metal components using a filler metal with a melting point (liquidus) above 840°F (450°C) and below the solidification temperature (solidus) of the base materials. The process differs from welding in that parent materials are not melted. It is sometimes referred to as silver soldering because silver is used in a number of brazing alloys, but many other such alloys contain little or no silver, and the temperature range at which brazing occurs is above that for soldering.

7.10.1 Applications

Brazing is often used to join dissimilar metals, in cases in which welding would result in unacceptable distortion, and when component tolerance is better achieved with snug fitting parts. Brazed pressure vessels are not common because of the tight tolerances required to facilitate flow and full coverage of the filler metal.

7.10.2 Filler metal

In brazing, the filler metal is most often flowed into the joint using capillary action. It is also sometimes plated on one or both of the surfaces to be joined, or combined with the flux and applied at the same time. Premanufactured filler metal inserts can also be used. Cleanliness of the parts to be joined is important, as the filler will not flow well on a contaminated surface.

7.10.3 Heating

Heating is accomplished using a gas torch, in a furnace, using a dip process, by induction, and very occasionally using a heli-arc (GTAW) torch. An advantage of furnace brazing over torch brazing is that the whole assembly is heated together and there is less tendency toward residual stress and distortion. It typically requires a special furnace and often must use fixturing to hold parts in the correct relative orientations and to prevent sagging.

7.10.4 Flux

A flux is typically used to prevent formation of oxides during heating and serves as a cleaning agent for small amounts of contamination. It is also possible to braze without flux in some cases. Furnace brazing with an inert or reducing atmosphere, or in a vacuum, can permit brazing without flux if parts are thoroughly cleaned prior to insertion in the furnace, and a few braze alloys are self-fluxing with certain metals. When flux is used, removal of residual flux after brazing is important to avoid corrosion.

There are a number of ways of applying flux to the joint. The most familiar is probably simply painting the joint surfaces with a brush, using a paste or liquid flux. Sometimes it is combined in a paste with the filler metal, or applied along with the filler metal either using coated rods or flux-cored rods.

7.10.5 Brazing procedures

As with welding procedures (see Section 8.4), the use of brazing procedures is critical to the quality of the braze and the final brazed product. The production and use of brazing procedures are entirely parallel to those for welding procedures, so reference is made to that section.

Reference

1 R. Sacks and E. Bohnart. 2005. "*Welding Principles and Practices*". McGraw Hill, New York, NY.

8

Welding Procedures and Post Weld Heat Treatment

8.1 Introduction

All of the weld processes described in Chapter 7 are capable of producing high quality welds. As further described in that chapter, some processes are limited either by the process itself, or for practical reasons, to the production of certain types of welds. Regardless of the weld process or type, numerous variables affect weld quality. Examples of these are the process itself, base metal and filler metal, groove design, filler metal diameter, weld preheat, electrical characteristics, and post weld heat treatment (PWHT). The ASME BPVC and other codes for pressure vessels and components have long recognized the need to control these variables in order to ensure the quality of welds. this chapter describes how this is accomplished.

8.2 Welding Procedures

Quality of welds in pressure vessels and other process equipment is achieved by use of welding procedures that have been properly qualified and documented and by welders who have demonstrated their ability to produce good welds using those procedures.

The process of ensuring quality welds begins with the development of a welding procedure specification, WPS. This process is detailed in the ASME Section IX code. A procedure specification is a written document providing direction to the person applying the material joining process. Details for the preparation and qualification of weld procedure specifications (WPS) are given in the ASME Section IX code, and section references in this chapter labeled QG or QW refer to that code. Table 8.1 shows a typical WPS for submerged-arc welding (SAW). It shows essential and nonessential variables pertaining to such items as joints, base metals, filler metals, positions, preheat, PWHT, electrical characteristics and technique.

Procedure specifications used by a fabricator must be qualified by that fabricator, or shall be a standard procedure specification acceptable under the rules of the applicable code of construction. Procedure specifications address the conditions, including ranges, if any, under which the material joining process must be performed as shown in Table 8.1. These conditions are referred to in this section as "variables." When a procedure specification is prepared by the organization, it shall address, as a minimum, the specific essential and nonessential variables that are applicable to the material joining process to be used in production. When the referencing code, standard, or specification requires toughness qualification of the material joining procedure, the applicable

Fabrication of Metallic Pressure Vessels, First Edition. Owen R. Greulich and Maan H. Jawad.

Table 8.1 Listing of essential, supplementary essential, and nonessential variables for submerged-arc welding [1]

Paragraph		Brief of Variables	Essential	Supplementary Essential	Nonessential
QW-402 Joints	.1	ϕ Groove design			X
	.4	– Backing			X
	.10	ϕ Root spacing			X
	.11	± Retainers			X
QW-403 Base Metals	.5	ϕ Group Number		X	
	.6	T Limits		X	
	.8	ϕ T Qualified	X		
	.9	t Pass $^1/_2$ in. (13 mm)	X		
	.11	ϕ P-No. qualified	X		
QW-404 Filler Metals	.4	ϕ F-Number	X		
	.5	ϕ A-Number	X		
	.6	ϕ Diameter			X
	.9	ϕ Flux-wire class.	X		
	.10	ϕ Alloy flux	X		
	.24	± or ϕ Supplemental	X		
	.27	ϕ Alloy elements	X		
	.29	ϕ Flux designation			X
	.30	ϕ t	X		
	.33	ϕ Classification			X
	.34	ϕ Flux type	X		
	.35	ϕ Flux-wire class.		X	X
	.36	Recrushed slag	X		
QW-405 Positions	.1	+ Position			X

Table 8.1 (Continued)

Paragraph		Brief of Variables	Essential	Supplementary Essential	Nonessential
QW-406 Preheat	.1	Decrease > 100°F (55°C)	X		
	.2	ϕ Preheat mainL			X
	.3	Increase > 100°F (55°C) (IP)		X	
QW-407 PWHT	.1	ϕ PWHT	X		
	.2	ϕ PWHT (T & T range)		X	
QW-409 Electrical Characteristics	.1	> Heat input		X	
	.4	ϕ Current or polarity		X	X
	.8	ϕ I & E range			X
QW-410 Technique	.1	ϕ String or weave			X
	.5	ϕ Method cleaning			X
	.6	ϕ Method back gouge			X
	.7	ϕ Oscillation			X
	.8	ϕ Tube-work distance			X
	.9	ϕ Multi to single pass per side		X	X
	.10	ϕ Single to multi electrodes		X	X
	.15	ϕ Electrode spacing			X
	.25	ϕ Manual or automatic			X
	.26	± Peening			X
	.64	Use of thermal processes	X		

Legend:
+ Addition > Increase or greater than ↑ Uphill ← Forehand ϕ Change
- Deletion < Decrease or less than ↓ Downhill → Backhand.

Table 8.2 Suggested form for welding procedure specifications (front) [1]

FORM QW-482 SUGGESTED FORMAT FOR WELDING PROCEDURE SPECIFICATIONS (WPS)
(See QW-200.1, Section IX, ASME Boiler and Pressure Vessel Code)

Organization Name ____________ By ____________
Welding Procedure Specification No. ____________ Date ____________ Supporting PQR No.(s) ____________
Revision No. ____________ Date ____________

Welding Process(es) ____________ Type(s) ____________
(Automatic, Manual, Machine, or Semi-Automatic)

JOINTS (QW-402) — Details

Joint Design ____________
Root Spacing ____________
Backing: Yes ____________ No ____________
Backing Material (Type) ____________
(Refer to both backing and retainers)

☐ Metal ☐ Nonfusing Metal
☐ Nonmetallic ☐ Other

Sketches, Production Drawings, Weld Symbols, or Written Description should show the general arrangement of the parts to be welded. Where applicable, the details of weld groove may be specified.

Sketches may be attached to illustrate joint design, weld layers, and bead sequence (e.g., for toughness procedures, for multiple process procedures, etc.)

*BASE METALS (QW-403)

P-No. ____________ Group No. ____________ to P-No. ____________ Group No. ____________
OR
Specification and type, grade, or UNS Number ____________
to Specification and type, grade, or UNS Number ____________
OR
Chem. Analysis and Mech. Prop. ____________
to Chem. Analysis and Mech. Prop. ____________
Thickness Range:
Base Metal: Groove ____________ Fillet ____________
Maximum Pass Thickness ≤ 1/2 in. (13 mm) (Yes) ________ (No) ________
Other ____________

*FILLER METALS (QW-404)	1	2
Spec. No. (SFA)		
AWS No. (Class)		
F-No.		
A-No.		
Size of Filler Metals		
Filler Metal Product Form		
Supplemental Filler Metal		
Weld Metal		
Deposited Thickness:		
Groove		
Fillet		
Electrode-Flux (Class)		
Flux Type		
Flux Trade Name		
Consumable Insert		
Other		

*Each base metal-filler metal combination should be specified individually.

(07/17)

Table 8.2 (Continued)

FORM QW-482 (Back)

WPS No. ____________ Rev. ____________

POSITIONS (QW-405)
Position(s) of Groove ____________
Welding Progression: Up ____________ Down ____________
Position(s) of Fillet ____________
Other ____________

POSTWELD HEAT TREATMENT (QW-407)
Temperature Range ____________
Time Range ____________
Other ____________

PREHEAT (QW-406)
Preheat Temperature, Minimum ____________
Interpass Temperature, Maximum ____________
Preheat Maintenance ____________
Other ____________
(Continuous or special heating, where applicable, should be specified)

GAS (QW-408)

	Gas(es)	Percent Composition (Mixture)	Flow Rate
Shielding			
Trailing			
Backing			
Other			

ELECTRICAL CHARACTERISTICS (QW-409)

Weld Pass(es)	Process	Filler Metal		Current Type and Polarity	Amps (Range)	Wire Feed Speed (Range)	Energy or Power (Range)	Volts (Range)	Travel Speed (Range)	Other (e.g., Remarks, Comments, Hot Wire Addition, Technique, Torch Angle, etc.)
		Classification	Diameter							

Amps and volts, or power or energy range, should be specified for each electrode size, position, and thickness, etc.

Pulsing Current ____________ Heat Input (max.) ____________

Tungsten Electrode Size and Type ____________
(Pure Tungsten, 2% Thoriated, etc.)

Mode of Metal Transfer for GMAW (FCAW) ____________
(Spray Arc, Short-Circuiting Arc, etc.)

Other ____________

TECHNIQUE (QW-410)
String or Weave Bead ____________
Orifice, Nozzle, or Gas Cup Size ____________
Initial and Interpass Cleaning (Brushing, Grinding, etc.) ____________
Method of Back Gouging ____________
Oscillation ____________
Contact Tube to Work Distance ____________
Multiple or Single Pass (Per Side) ____________
Multiple or Single Electrodes ____________
Electrode Spacing ____________
Peening ____________
Other ____________

(07/17)

supplementary essential variables become essential variables and shall also be addressed in the procedure specification as well.

This document provides written direction to the person performing the welding, addressing specific code defined *essential variables* that affect the quality of the weld, and if there are toughness requirements, then it addresses *supplementary essential variables* as well. Procedure specifications are addressed in section QG-101 of Section IX of the ASME BPVC.

Table 8.3 Groove-Weld Tension Tests and Transverse-Bend Tests [1]

Thickness *T* of test coupon, welded, in. (mm)	Range of thickness *T* of base metal, qualified, in. (mm) [Note (1)] and [Note (2)]		Maximum thickness t of deposited weld metal. qualified, in. (mm) [Note (1)] and [Note (2)]	Type and number of tests required (tension and guided-bend tests) [Note (2)]			
	Min.	Max.		Tension, QW-150	Side bend, QW-160	Face bend, QW-160	Root bend, QW-160
Less than 1/16 (1.5)	*T*	2*T*	2*t*	2	...	2	2
1/16 to 3/8 (1.5–10), incl.	1/16 (1.5)	2*T*	2*t*	2	[Note (5)]	2	2
Over 3/8 (10), but less than 3/4 (19)	3/16 (5)	2*T*	2*t*	2	[Note (5)]	2	2
3/4 (19) to less than 1½ (38)	3/16 (5)	2*T*	2*t* when t < 3/4 (19)	2 [Note (4)]	4	...	...
3/4 (19) to less than 1½ (38)	3/16 (5)	2*T*	2*T* when t < 3/4 (19)	2 [Note (4)]	4	...	...
1½ (38) to 6 (150), incl.	3/16 (5)	8 (200) [Note (3)]	2*t* when t < 3/4 (19)	2 [Note (4)]	4	...	...
1½ (38) to 6 (150), incl.	3/16 (5)	8 (200) [Note (3)]	8 (200) [Note (3)] when *t* ≥ 3/4 (19)	2 [Note (4)]	4	...	...
Over 6 (150) [Note (6)]	3/16 (5)	1.33T	2*t* when *t* < 3/4 (19)	2 [Note (4)]	4	...	...
Over 6 (150) [Note (6)]	3/16 (5)	1.33T	1.33*T* when t ≥ 3/4 (19)	2 [Note (4)]	4	...	...

NOTES:
(1) The following variables further restrict the limits shown in this table when they are referenced in QW-250 for the process under consideration: QW-403.9, QW-403.10, and QW-404.32. Also, QW-202.2, QW-202.3, and QW-202.4 provide exemptions that supersede the limits of this table.
(2) For combination of welding procedures, see QW-200.4.
(3) For the SMAW, SAW, GMAW, PAW, LLBW, and GTAW welding processes only; otherwise per Note (1) or 27, or 2*t*, whichever is applicable.
(4) see QW-151.1, QW-151.2, and QW-151.3 for details on multiple specimens when coupon thicknesses are over 1 in. (25 mm).
(5) Four side-bend tests may be substituted for the required face- and root-bend tests, when thickness *T* is 3/8 in. (10 mm) and over.
(6) For test coupons over 6 in. (150 mm) thick, the full thickness of the test coupon shall be welded.

Once developed, the procedure must be qualified to demonstrate that the proposed process is capable of producing joints with the required properties for the intended application. The production of a weld using the procedure is documented in a procedure qualification record (PQR), including both the specified essential and supplementary essential variables and the results of the required tests. The PQR is certified by signature or other means and must be accessible to the Authorized Inspector.

A procedure specification may be supported by one or more PQRs, and one PQR may support one or more WPSs. These situations occur if, for example, a procedure specification includes more than one process, such as a submerged-arc weld (SAW) with a GTAW root pass. The GTAW and SAW may be qualified separately, but the GTAW procedure may also be used as the root pass for a weld completed using, for example, SMAW. Procedure Qualification Records are addressed in section QG-102 of the ASME BPVC. The type and number of tests required, along with the range of thickness qualified by a given test, are found in QW-451. See Table 8.3 for an example of some of the test requirements.

The purpose of qualifying the procedure specification is to demonstrate that the joining process proposed for construction is capable of producing joints having the required mechanical properties for the intended application. Qualification of the procedure specification demonstrates the mechanical properties of the joint made using a joining process, and not the skill of the person using the joining process. However, the welder or welding operator who prepares the WPS qualification test coupons meeting the requirements of QW-200 is also qualified within the limits of the performance qualifications listed in QW-304 or QW-305, respectively.

The procedure qualification record (PQR) documents what occurred during the production of a procedure qualification test coupon and the results of testing that coupon. As a minimum, the PQR shall document the essential procedure qualification test variables applied during production of the test joint, and the results of the required tests. When toughness testing is required for qualification of the procedure specification, the applicable supplementary essential variables shall also be recorded for each process. The organization shall certify the PQR by a signature or other means as described in the organization's Quality Control System. The PQR shall be accessible to the Authorized Inspector.

A welding performance qualification (WPQ), brazing performance qualification (BPQ), or fusing performance qualification (FPQ) demonstrates the ability of the person to produce a sound joint using a procedure specification, and this is documented in the performance qualification record. The performance qualification record documents the essential and supplementary essential variables used for production of the test coupon, the ranges of variables qualified, and the results of required testing and/or nondestructive examinations. The performance qualification and the PQR are addressed in sections QG-103 and QG-104 of ASME Section IX, respectively. Table 8.4 summarizes essential variables for welder performance for various welding processes, from ASME IX [1]. Table QW-416.

QW-451 provides the testing requirements for procedure qualifications and welder qualifications. Production weld testing must meet the requirements specified in Section VIII of the ASME BPVC.

The requirements are very much like those for the procedure qualification.

8.3 Weld Preparation Special Requirements

For the standard grades of carbon and stainless steel, preparation for welding can usually be accomplished by standard techniques such as flame cutting, plasma cutting, grinding, or machining. Some materials, however, require special treatment.

Table 8.4 Essential welding variables for welder performance [1]

Paragraph [Note (1)]		Brief of Variables	OFW Table QW-352	SMAW Table QW-353	SAW Table QW-354	GMAW [Note (2)] Table QW-355	GTAW Table QW-356	PAW Table QW-357
			Essential					
QW-402 Joints	.4	− Backing		X		X	X	X
	.7	+ Backing	X					
QW-403 Base Metal	.2	Maximum qualified	X					
	.16	ϕ Pipe diameter		X	X	X	X	X
	.18	ϕ P-Number	X	X	X	X	X	X
QW-404 Filler Metals	.14	± Filler	X				X	X
	.15	ϕ F-Number	X	X	X	X	X	X
	.22	± Inserts					X	X
	.23	ϕ Filler metal product form					X	X
	.30	ϕ t Weld deposit		X	X	X	X	X
	.31	ϕ t Weld deposit	X					
	.32	t Limit (s. cir. arc)				X		
QW-405 Positions	.1	+ Position	X	X	X	X	X	X
	.3	ϕ ↑↓ Vert. welding		X		X	X	X
QW-408 Gas	.7	ϕ Type fuel gas	X					
	.8	− Inert backing				X	X	X

Paragraph [Note (1)]			Essential					
		Brief of Variables	OFW Table QW-352	SMAW Table QW-353	SAW Table QW-354	GMAW [Note (2)] Table QW-355	GTAW Table QW-356	PAW Table QW-357
QW-409 Electrical	.2	ϕ Transfer mode				X		
	.4	0 Current or polarity					X	

Welding Processes:

OFW	Oxyfuel gas welding
SMAW	Shielded metal-arc welding
SAW	Submerged-arc welding
GMAW	Gas metal-arc welding
GTAW	Gas tungsten-arc welding
PAW	Plasma-arc welding

Legend:

ϕ Change	t Thickness
+ Addition	↑ Uphill
− Deletion	↓ Downhill

NOTES:

(1) For description, see Article IV.

(2) Flux-cored arc welding as shown in Table QW-355, with or without additional shielding from an externally supplied gas or gas mixture, is included.

1) Highly reactive metals such as aluminum oxidize easily and rapidly. When aluminum is to be welded, it is common first to etch it using a phosphoric acid compound immediately prior to welding. If such a process is not followed, porosity in the weld can form due to combination of hydrogen in the air with the aluminum oxide film on the surface of the joints.
2) For some materials such as 2.25Cr-1Mo steel, a flame cut edge is not an acceptable weld preparation due to the formation of oxides that cause delayed hydrogen cracking in the weld.
3) Section UW-32 of ASME BPVC Section VIII, Division 1 requires that cast materials be machined, chipped, or ground to expose sound material prior to welding.

8.4 Weld Joint Design and Process to Reduce Stress and Distortion

Because of the high temperatures and thermal gradients involved in the welding process, weld residual stress and distortion are common problems. The weld nugget and its immediate surroundings start out very hot, while the rest of the weldment is at a lower temperature. During cooling, shrinkage varies with the change in temperature, so parts of the weldment that start at different temperatures will shrink by different amounts. If this occurs, the results are residual stresses and distortion. Industry and individual companies within industry have developed various ways to deal with these issues.

8.4.1 Reduced heat input

Since heat is the problem, much effort goes into minimizing total heat input. Some welding processes involve more heat than others. GTAW, for example, while one of the cleanest welding processes, is also one of the slowest and hottest. Weld design is often the greatest determinant of heat input, but careful fit-up and control of the welding procedure, including voltage, current, and travel speeds, are also important. A wide weld will have more shrinkage than a narrow one. Thus, in addition to reducing the cost of welding by requiring less of it, use of a relatively narrow J-groove weld rather than a simpler V-groove also typically reduces stress and distortion. The lower total heat input helps, as does the more consistent width of the weld.

8.4.2 Lower temperature differential

Another way of reducing residual stress and distortion is reduction of thermal gradients by preheating the parent material of the weldment. Preheating often has other favorable effects as well, but the simple reduction of temperature differential between the parent material and the molten weld metal can be a significant factor.

8.4.3 Choice of weld process

As noted, some welding processes, such as GTAW, are hotter than others. Some such as EBW may be very hot, but heat such a small volume of material, and of a very consistent width, that the stress and distortion are minimized. FSW has generally reduced stress and distortion because of lower temperatures and consistent width of the weld.

8.4.4 Weld configuration and sequencing

The only changes that can be made in configuration of fillet welds are size and whether they are continuous or intermittent. Distortion can sometimes be minimized, though, by using multiple passes, rather than a single pass.

In groove welds, however, a number of changes can be made for efficiency and to control undesirable effects. Making the weld narrower reduces total weld volume, and to the extent that the weld sidewalls get steeper, the difference in shrinkage from the root to the surface is reduced. Even a weld with parallel sidewalls, however, will be subject to differential shrinkage. In a multi-pass weld, the first passes will already be solid as the following passes are laid in. Shrinkage of the later layers will bring the early ones into compression.

If all the welding is done from one side, then as the groove fills up and the weld shrinks (which occurs in all directions), the component will tend to bend toward the side with the weld. For this reason, welds in thick materials are often performed from both sides in an attempt to balance the tensile loads on the two sides of the material.

Producing a full penetration weld typically requires back-gouging after welding or partial welding one side, so as to remove slag, blow through, and lack of penetration at the root. For relatively thin materials, if the weld prep is symmetrical around the midpoint of the thickness, then after back-gouging much more weld will be placed on the second side than that remains on the first. This will cause bending toward the second side. This is avoided by using an asymmetric weld prep. A one-third, two-thirds prep is often used, with welding performed first on the side with a deeper groove. In this way, the second side weld goes approximately to the center of the material.

Switching welding from one face of the weld to the other to balance stresses on the two sides is also effective in reducing distortion. The effect on stress may be less than that of reducing total weld because while balancing stresses can reduce distortion, much of the stress remains. This technique, however, is typically used in conjunction with others so that often minimal straightening is needed.

Sometimes a procedure calls for skipping around during the weld process. The result is less heat buildup in the area of the weld, with the goal of forcing more of the deformation due to shrinkage to happen in the weld itself.

There are also innovative ways of using heat of later weld passes to relieve stress in some of the earlier ones.

It can be seen that there are many approaches to reduce stress and deformation due to welding. Some require changes at the design stage, while others allow and depend on skill of the welder and the care that he or she takes in placing the weld. To the extent that use of these techniques eliminates the need for remedial steps such as jacking or pressing the structure back into shape, they can improve both cost and quality.

8.5 Weld Preheat and Interpass Temperature

A minimum parent material temperature required before welding begins is referred to as preheat. The temperature that must not be exceeded during welding is referred to as the interpass temperature. If the interpass temperature is exceeded, then welding must stop until the requirement is again met.

The requirement for preheat is part of the welding procedure specification, WPS, as shown in Table 8.1. It is determined by the fabricator as part of an approved welding procedure. It may be an essential or nonessential variable depending on the welding process.

Preheat and interpass temperature monitoring and control: When weld preheat and weld interpass temperature must be controlled, monitoring is usually performed using either a series of crayons (temp sticks) calibrated to melt at specific temperatures or sometimes with thermocouples attached to the component. If calibrated crayons are used, then marks are made using successive crayons, going up in temperature until the one that does not melt is found. The temperature has then been determined to be within the temperature range designated by the last crayon that melts and the first one that does not. Crayons are typically available every 25°F (14°C) from 125°F (52°C) to 800°F (427°C), and every 50°F (28°C) up to 2000°F (1093°C). Thermocouples, with their associated readout circuit, usually provide a continuous readout of actual temperature. It should be pointed out that they are only as accurate as their calibration. Sometimes the precision of a readout is not justified by its accuracy.

Preheat is often achieved either by heating the vessel in a furnace, removing it just before welding begins, or by rotating the vessel on turning rolls with a large torch (often referred to as a rosebud) trained on the weld area. At times it is necessary to perform welding on the inside of a preheated vessel. In this case, it is usual to keep heat on the outside, while at the same time blowing large quantities of cool air through the vessel for the comfort of the welder.

8.6 Welder Versus Welding Operator

8.6.1 Welders

A welder is one who performs manual or semiautomatic welding.

8.6.1.1 Manual welding

Manual welding is welding wherein the entire welding operation is performed and controlled by hand. SMAW and GTAW, when performed by hand, are examples. The welder sets the current, voltage, etc., controls how fast material is fed into the weld, and controls the advance of the welding in accordance with an established WPS.

8.6.1.2 Semiautomatic arc welding

In semiautomatic arc welding, the equipment controls only the filler metal feed. All other aspects of the weld are controlled by the welder. Typical examples are FCAW and GMAW in which the welder sets the welding parameters, holds the gun, and determines the advance of the weld and any weaving in accordance with an established WPS.

8.6.2 Welding operators

A welding operator is one who operates machine or automatic welding equipment.

8.6.2.1 Machine welding

Machine welding is welding in which the torch, gun, or electrode holder is held by a mechanical device, but controlled by the welder in response to changes in the welding conditions. Thus, the welder sets the welding parameters in accordance with the WPS and can change the machine settings provided he remains within the limits established in the WPS. The machine just holds the torch, gun, or electrode holder. SAW is an example.

8.6.2.2 Automatic welding

Automatic welding is welding performed by the equipment without adjustment of the controls by the welding operator. EBW and orbital GTAW are examples. The equipment is preprogrammed. Once a weld is started, the equipment goes through its cycle, including ramp-up, welding, perhaps multiple passes, any oscillation (called weave if performed manually), and ramp-down.

8.6.3 Differences in qualifications

For WPSs and PQRs, differences between qualifications for welders versus welding operators are in general encapsulated in the requirements for the WPS and PQR. For SMAW, SAW, GMAW, FCAW, GTAW, and PAW, a change from manual to automatic is a nonessential variable, requiring requalification only when impact testing is required. Detailed requirements are spelled out in ASME Section IX beginning with section QW-250.

For welder qualifications, the essential variables requiring requalification are addressed beginning with section QW-350.

8.7 Weld Repair

Welds may have to be repaired during fabrication for a number of reasons such as the following:

1) Slag inclusions occurring during welding.
2) Surface indications after cooling of the welds.
3) Delayed hydrogen cracking after welding.
4) Cracks occurring subsequent to PWHT.

8.7.1 Slag inclusion during welding

Slag inclusion and defects occurring during welding are removed by either thermal (e.g., carbon arc) or mechanical means. The area is then rewelded using an approved WPS procedure for groove welds.

8.7.2 Surface indications after cooling of welds

Crack indications may appear at the surface of welds subsequent to welding due to thermal residual stresses. These indications are usually removed by mechanical means. The repaired area may be blended and left as is if the remaining thickness is equal to or larger than the minimum required thickness. Otherwise the area must be repaired following an approved WPS for groove welds.

8.7.3 Delayed hydrogen cracking after welding

Cracking in the weld occurs sometimes due to hydrogen generated by the welding process. The cracking occurs during or shortly after welding is completed. Normally, ultrasonic testing or radiographic examination is required to determine the extent of cracking. Repairs of the weld must be done in accordance with an approved WPS procedure.

8.7.4 Cracks occurring subsequent to PWHT

Cracks sometimes develop in welds subsequent to PWHT especially in thick sections and are usually caused by thermal or residual stresses. The procedure for repairing such cracks is outlined in the ASME code and consists of items such as the following:

1) Informing the user of cracks, proposing a method of repair, and getting approval from the user prior to proceeding with the repair.
2) Selecting the appropriate welding procedure in accordance with an approved WPS.
3) Determining whether PWHT will be required subsequent to repairing the cracks in accordance with the ASME Code rules.
4) Deciding which NDE method to use for inspecting the repaired area.

8.8 Post Weld Heat Treating

Most carbon steel welds, some stainless steel welds, and some nonferrous welds require post weld heat treating (PWHT), Figures 8.1 and 8.2. PWHT is primarily used to reduce the hardness of the heat-affected zone of the weld and reduce residual welding stresses.

The following is a summary of some of the PWHT requirements in Section VIII, Division 1, of the ASME code.

8.8.1 PWHT of carbon steels

The weld designations for these steels are P1 Groups 1, 2, and 3. PWHT is required for thicknesses over 1.5 in. (38 mm). PWHT is performed at a minimum temperature of 1100°F (595°C). Section UCS of the ASME Code Section VIII, Division 1, gives full details regarding the PWHT of these materials as well as some exemptions.

Figure 8.1 Inserting a pressure vessel into a furnace (*Source:* Nooter Construction)

Figure 8.2 Pressure vessel subsequent to PWHT (*Source:* Nooter Construction)

8.8.2 PWHT of low alloy steels

For 1/2Cr–1/2Mo steels (P3 material), PWHT is done at 1100°F (595°C) for all thicknesses except as specified in Section UCS.

For low alloy steels such as 1.25Cr–1/2Mo (P4 materials), PWHT is done at 1200°F (650°C) for all thicknesses except as specified in Section UCS.

For low alloy steels such as 2.25Cr–1Mo and 9Cr–1Mo (P5 materials), PWHT is done at 1250°F (675°C) for all thicknesses except as specified in Section UCS.

8.8.3 Some general PWHT requirements for carbon steels and low alloy steels

The holding time for PWHT at the above specified temperatures is one hour/inch of thickness with a 15 minutes minimum time. In cases where the minimum temperature cannot be maintained, the ASME code allows a reduction in the temperature with a substantial increase in holding time. For example, a 100°F (55°C) drop in minimum temperature requires an increase in holding time from 1 hours/in. to 4 hours/in. This requirement is given in Table UCS 56.1 of Section VIII, Division 1 of the ASME code.

The temperature of the furnace cannot exceed 800°F (425°C) when the vessel component is inserted for PWHT. The heating rate above 800°F (425°C) is restricted to 400°F/hour (220°C/hour) divided by the thickness of the component but in no case greater than 400°F/hour (220°C/hour).

Above 800°F (425°C), the cooling rate is 500°F/hour (275°C/hour) divided by the thickness of the component but in no case greater than 500°F/hour (275°C/hour).

The ASME code Section VIII, Division 1, gives other details regarding PWHT such as permissible variation in metal temperature during PWHT and segmental PWHT.

8.8.4 PWHT of stainless steel

Generally, PWHT is not required for stainless steels. Exceptions are ferritic stainless steels such as types 405, 410, and 430. Under certain conditions specified in part UHA of Section VIII, Division 1, these materials require PWHT with special cooling rates.

8.8.5 PWHT of nonferrous alloys

PWHT of nonferrous alloys is not desirable except for the following materials:

1) PWHT is required for zirconium alloy R60705 within 14 days after welding. The minimum PWHT temperature is 1000°F (540°C). The minimum hold time is 1 hour for thicknesses up to 1 inch (25 mm) plus 1/2 hour for each additional 1 in. (25 mm) of thickness.
2) Welds of nickel alloys 800, 800H, and 800 HT shall be PWHT when the design temperature is higher than 1000°F (540°C). The minimum PWHT temperature is 1625°F (885°C). The holding time is 1.5 hours for thicknesses up to 1 in. (25 mm) plus 1 hour for each additional 1 in. (25 mm) of thickness.

8.9 Cladding, Overlay, and Loose Liners

Cladding, overlay, and loose liners are often used on the inside surface of carbon steel vessels where the contents are too corrosive for the carbon steel. The following sections provide descriptions of these three methods.

8.9.1 Cladding

Titanium, zirconium, and tantalum reactive alloys cannot be readily welded to carbon steel. The construction choices when one of these alloys is required are to build a solid wall vessel using the required material, or, when the required thickness results in prohibitive cost, either to construct a carbon steel vessel with the alloy clad to the inside surface of the carbon steel or to construct a vessel with a loose liner. There are two methods for cladding these reactive alloys to carbon steel.

The first method is to explosively bond the cladding material to the carbon steel. This process consists of placing the thin clad material over a flat carbon steel plate with a very small gap between them. Explosives are then placed on the top of the cladding material. The explosives are detonated, resulting in the reactive clad material bonding to the carbon steel. The flat plate is then formed into a shell or head for construction of the vessel.

The second method of cladding the reactive alloy to the carbon steel shell is by hot rolling. The process consists of starting with a thick carbon steel plate and a thick plate of the reactive alloy. The two are positioned one on top of the other, heated, and placed in a mill roll. The assembly is then rolled back and forth to reduce the thickness to the desired value. The hot rolling bonds the reactive alloy to the carbon steel plate.

The clad material is normally stripped back prior to welding any joint in the parent carbon steel or other alloy material. The cladding is normally stripped back by one of two methods. The first method is by chiseling. This method is used when welding a small area such as nozzles and lugs. The second method is by machining the cladding back. This method is used for welding large areas such as circumferential and longitudinal welds. Care must be exercised in the depth of chiseling or machining to ensure getting down to uncontaminated base material for welding purposes without

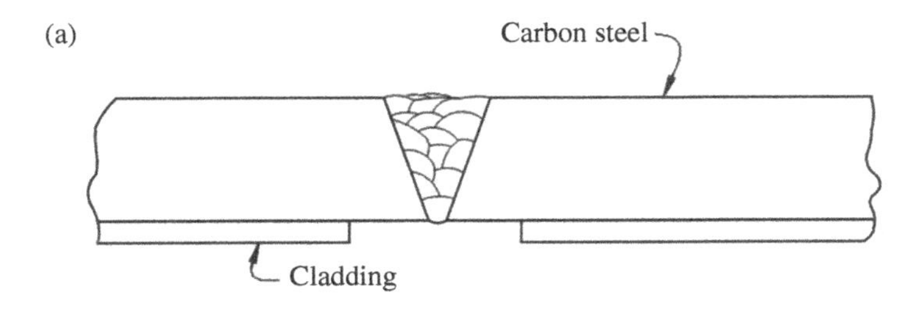

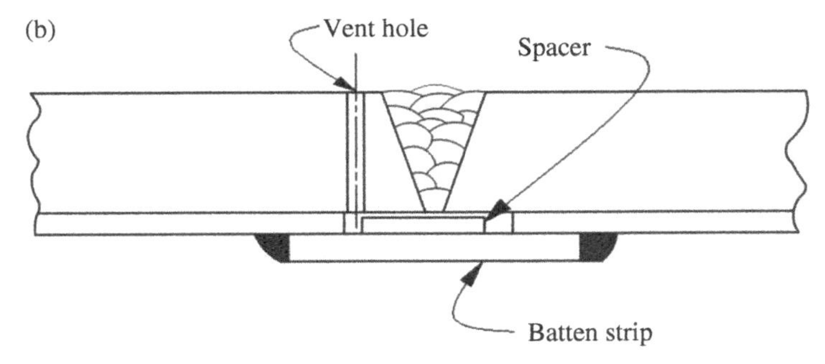

Figure 8.3 Welding of clad plates

reducing the required base metal thickness below minimums. Some fabricators add a small amount of extra material, about 1/32 in. (0.8 mm) or more, to the base material thickness, in order to assure an uncontaminated surface for welding and a minimum base material thickness based on design requirements.

Welding of clad pieces is illustrated in Figure 8.3. Figure 8.3(a) shows two carbon plates welded together with the cladding stripped back from the joint. A vent hole is then drilled at the joint as shown in Figure 8.3(b). A copper spacer is inserted and a batten strip of the same material as the cladding is welded to seal the joint as shown in Figure 8.3(b). The vent hole serves several important purposes. During welding, the vent allows heated gases to escape from the cavity, facilitating welding. Later during fabrication, pressure can be applied through the vent hole to ascertain that the joint is properly sealed. Finally, during operation, the vent hole is monitored and the vessel is shut down if leakage is detected. Repairs can then be made on the inside surface before any damage to the carbon steel shell occurs.

8.9.2 Weld overlay

Weld overlay is a practical method of constructing a thick-walled vessel using carbon steel for the pressure boundary and a weld overlay, normally of stainless steel or nickel alloy as a corrosion barrier. Such construction can be more economical than constructing the total thickness using the more expensive stainless steel or nickel alloy. The fabrication procedure consists of melting strips of stainless steel or nickel alloy onto the inside surface of a formed shell or head plate. Either electroslag or submerged arc welding is normally used to melt and lay down the strips.

Figure 8.4 shows the detail of a typical weld overlay. The chemistry of the first layer is normally contaminated by carbon steel. Accordingly, many users require two layers to assure proper chemistry in contact with the vessel contents. Overlay welding is performed with a single wire, dual wire, or a ribbon of up to about 2 in. (50 mm) wide.

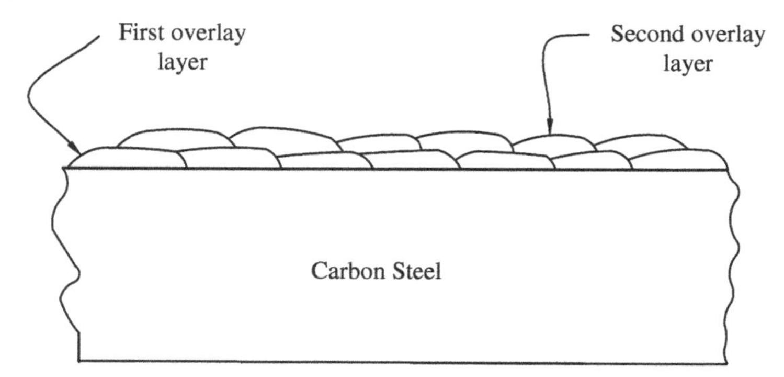

Figure 8.4 Weld overlay

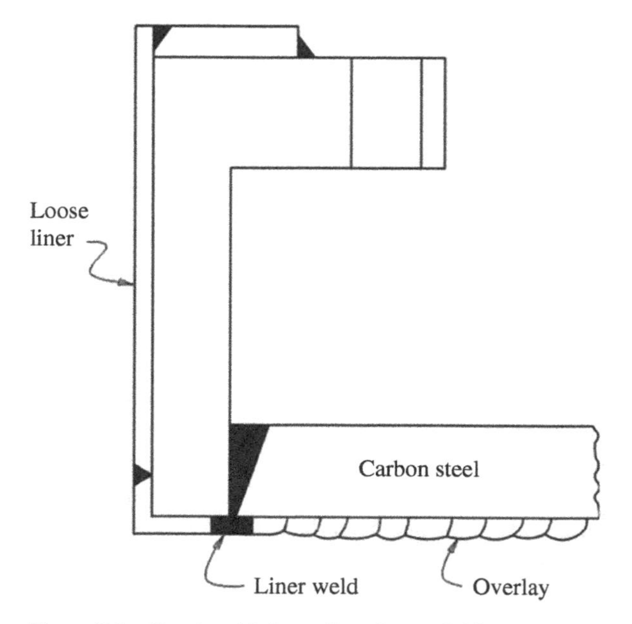

Figure 8.5 Nozzle with loose liner in overlaid vessel

8.9.3 Loose liners

Small nozzles and openings of overlaid vessels and clad vessels are normally loose lined when it is impractical to overlay them. Figure 8.5 shows a schematic of a typical detail in an overlaid vessel, while Figure 8.6 shows a schematic of a typical detail in a clad vessel.

8.10 Brazing

Brazing is the process of joining two or more metal components using a filler metal with a melting point ("liquidus") above 840°F (450°C) and below the solidification temperature ("solidus") of the base materials. The process differs from welding in that the parent materials are not melted. It is

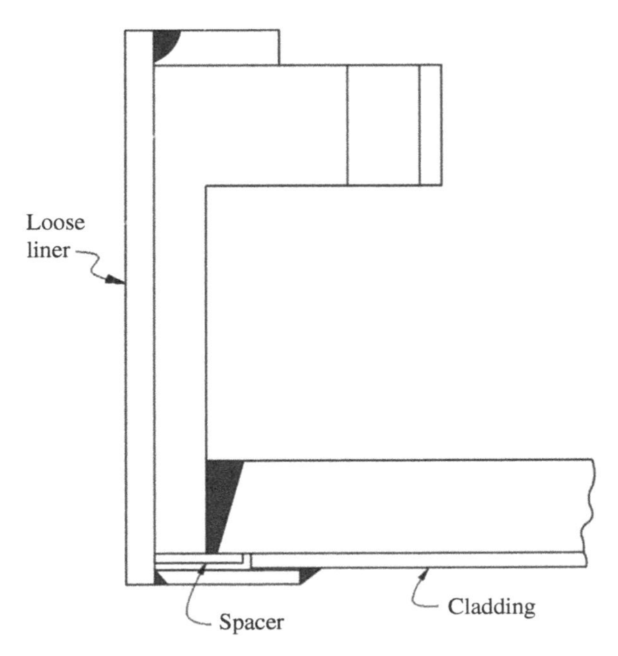

Figure 8.6 Nozzle with loose liner in a clad vessel

sometimes referred to as silver soldering because silver is used in a number of brazing alloys, but many such alloys contain little or no silver, and the temperature range at which brazing occurs is above that for soldering.

8.10.1 Applications

Brazing is often used to join dissimilar metals, in cases in which welding would result in unacceptable distortion, and when part tolerance is better achieved with snug fitting parts. Brazed pressure vessels are not common because of the tight tolerances required to facilitate flow and full coverage of the filler metal.

8.10.2 Filler metal

In brazing, the filler metal is most often flowed into the joint using capillary action. It is also sometimes plated on one or both of the surfaces to be joined, or combined with the flux and applied at the same time. Premanufactured filler metal inserts can also be used. Cleanliness of the parts to be joined is important, as the filler will not flow well on a contaminated surface. While the surface should be cleaned before application of the flux, the flux also serves to remove small amounts of impurities and oxidation.

8.10.3 Heating

Heating is accomplished using a gas torch, in a furnace, using a dip process, by induction, and very occasionally using a heli-arc (GTAW) torch. An advantage of furnace brazing over torch brazing is that the whole assembly is heated together and there is less tendency toward residual stress and distortion. It typically requires a special furnace with a vacuum, an inert atmosphere, or a reducing atmosphere to prevent oxidation, and often must use fixturing to hold parts in their correct relative locations and orientations.

8.10.4 Flux

A flux is typically used to prevent formation of oxides during heating and serves as a cleaning agent for small amounts of contamination. It is also possible to braze without flux in some cases. Furnace brazing with an inert or reducing atmosphere, or in a vacuum, can permit brazing without flux if parts are thoroughly cleaned prior to insertion in the furnace, and a few braze alloys are self-fluxing with certain metals. When flux is used, removal of residual flux after brazing is important to avoid corrosion.

There are a number of ways of applying flux to the joint. The most familiar is probably simply painting the joint surfaces with a brush, using a paste or liquid flux. Sometimes it is combined in a paste with the filler metal, or applied with the filler metal either using coated rods or flux-cored rods.

8.10.5 Brazing procedures

As with welding procedures, the use of brazing procedures is critical to the quality of the braze and the final brazed product. The production and use of brazing procedures are entirely parallel to those of welding procedures shown in Section 8.2 and ASME Section IX, Part QB [1].

Reference

1 ASME. 2021. "*Boiler and Pressure Vessel Code, Section IX*". American Society of Mechanical Engineers, New York.

9

Fabrication of Pressure Equipment Having Unique Characteristics

9.1 Introduction

In this chapter, four types of pressurized equipment with unique features are discussed: heat exchangers, dimpled jackets, layered vessels, and rectangular vessels. The fabrication of these types of equipment requires special attention to some of the details discussed in what follows.

9.2 Heat Exchangers

Heat exchangers are the workhorses in refineries and chemical plants. They have been used for cooling or heating fluids needed for many varied processes. In some large refineries and chemical plants, a hundred or more heat exchangers might be used. There are essentially three types of shell and tube heat exchangers: U-tube, fixed, and floating head (also referred to as floating tubesheet), Figure 9.1.

U-tube heat exchangers, Figure 9.1(a), consist of a shell encasing a U-tube bundle. A pass partition, Figure 9.1(a), is inserted in the channel section of U-tube heat exchangers and is normally welded to the tubesheet.

Fixed heat exchangers, Figure 9.1(b), consist of a tube bundle attached to tubesheets at both ends. An expansion joint is normally provided to handle thermal expansion difference between the tubes and the shell.

A floating head exchanger is shown in Figure 9.1(c). The tubes are welded at one end to a fixed tubesheet and welded at the other end to a tubesheet assembly that allows a specified amount of movement to accommodate the temperature differential between the tubes and the shell.

In all three illustrations, baffles are used both to support the tubes and to ensure efficient heat transfer by directing flow of the fluid back and forth across the tubes. While this is a common arrangement, Figure 9.2 illustrates some other baffle configurations.

Figure 9.2 shows some configurations of various components of heat exchangers. These components can be divided into three categories: front end, shell, and back end. Sketches A, B, C, N, and D for the front end show various configurations of channels for U-tube and floating head heat exchangers. Sketches E, F, G, H, J, and X show possible locations of inlet and outlet nozzles in the shell as well as various baffle configurations. Sketch K shows a kettle-type vessel where a tube bundle is inserted through the body flange. Sketches L, M, and N show the back side of some fixed tubesheet configurations, while sketches P, S, T, and W show details of some floating head configurations.

Fabrication of Metallic Pressure Vessels, First Edition. Owen R. Greulich and Maan H. Jawad.

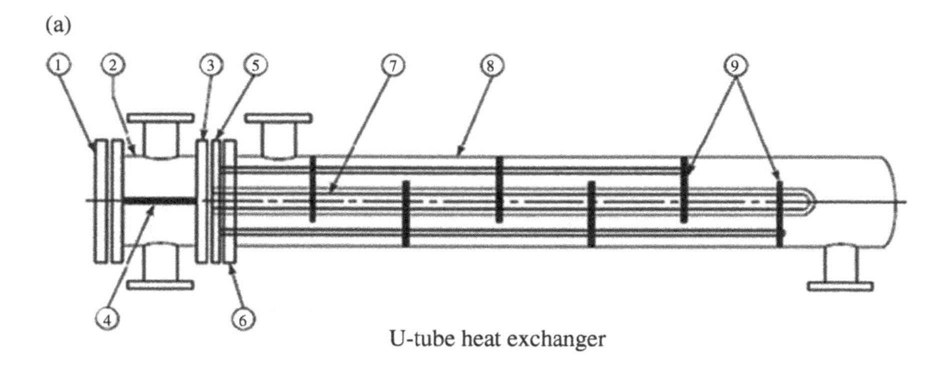

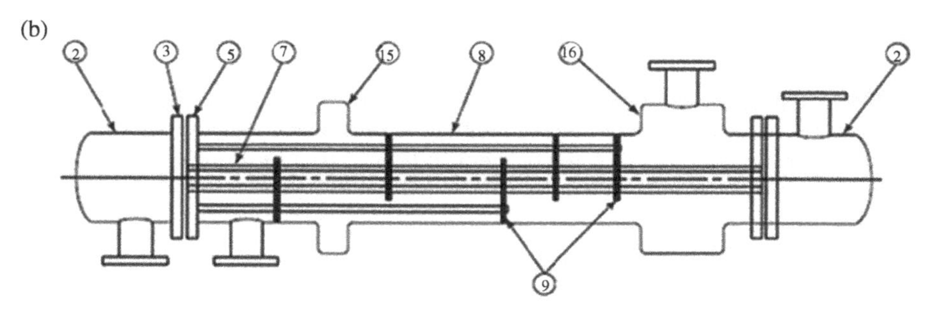

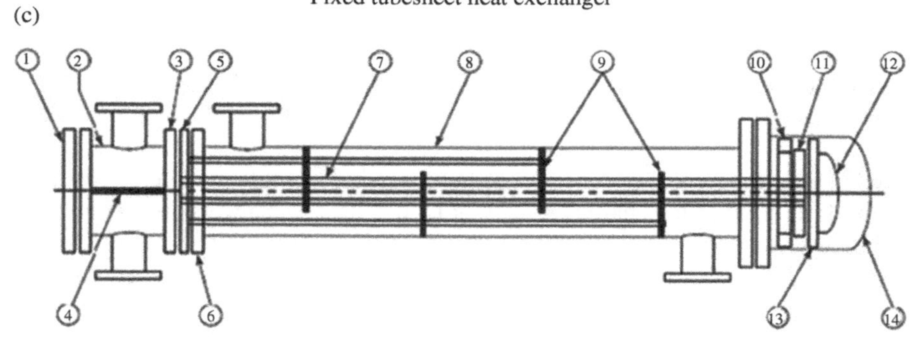

① Channel cover (bolted flat cover)
② Channel
③ Channel flange
④ Pass partition
⑤ Stationary tubesheet
⑥ Shell flange
⑦ Tubes
⑧ Shell
⑨ Baffles or support plates
⑩ Floating head backing device
⑪ Floating tubesheet
⑫ Floating head
⑬ Floating head flange
⑭ Shell cover
⑮ Expansion joint
⑯ Distribution or vapor belt

Figure 9.1 Various types of heat exchangers (*Source:* ASME VIII-1 [1])

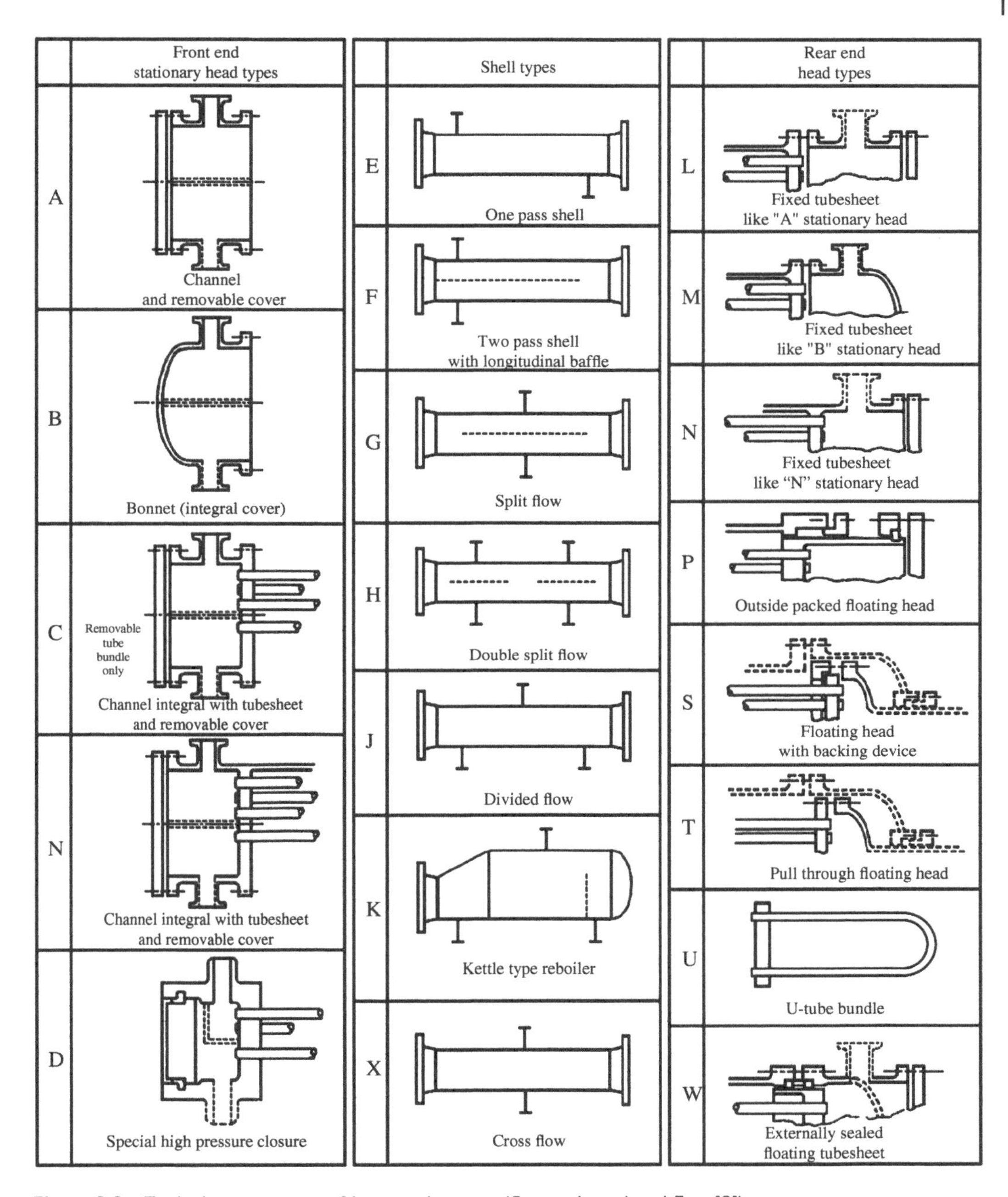

Figure 9.2 Typical components of heat exchangers (*Source:* Jawad and Farr [2])

9.2.1 U-tube heat exchangers

U-tube heat exchangers use only one tubesheet. The tube bundle is inserted through baffles, which act as tube spacers as well as direct fluid flow for efficient heat transfer. For horizontal units, the baffles also transfer the load of the tubes to the shell. The spacing of the baffles is essentially determined during the design stage by vibration and buckling considerations for the tubes. The tubes are attached to the tubesheet as illustrated in Figures 9.3 and 9.4, and discussed in Section 9.2.4. Special consideration needs to be given to the U-bend of tubes during the U-forming process due to thinning of the outer walls. Appendix F describes issues associated with tube bending and thinning in detail.

Many standards such as TEMA and ASME B31.3 set limits on the acceptable value of the minimum thickness at the bend region and the fabricator needs to be aware of those limits when purchasing material for forming into U-tubes.

9.2.2 Fixed heat exchangers

Assembly of the shell and tube bundle of a fixed heat exchanger requires careful planning. The process consists of holding the two tubesheets and the intermediate baffles with rods extending from one tubesheet to the other to form a skeleton assembly. The tubes are then pushed, one at a time, through the first tubesheet, through all of the baffles, and then through the second tubesheet. The tubes are first rolled and/or welded to one tubesheet. They are then trimmed, rolled, and welded to the second tubesheet. It is common in some heat exchangers to have the spacing of the baffles relatively large, while the tubes are relatively small in diameter and the tube spacing is small. With such a configuration, the cantilevered tubes have a tendency to hit the baffles during insertion through the tubesheets and baffles, rather than aligning with the holes. In such situations, plastic bullets, Figure 9.5, are inserted at the ends of the tubes to guide them through the designated holes in the baffles and tubesheets. Another situation that can occur when the baffle spacing is large and the tube diameter is small is for the tube to "jump" holes. If this situation is not caught early, the consequences will be costly since a portion of the tube bundle must be disassembled and reassembled correctly. Accordingly, special procedures need to be in place for preventing such occurrences.

9.2.3 Floating head heat exchangers

Heat exchangers with floating heads use packing glands in the sliding area to prevent mixture of tubeside and shellside fluids. A detail of one such joint is shown in Figure 9.6. The choice of the packing gland is a function of temperature, pressure, and amount of calculated movement. Many fabricators have their own proprietary packing gland details.

9.2.4 Attachment of tubes-to-tubesheets and tubes-to-headers

Details of tube-to-tubesheet or tube-to-header attachments differ substantially between the ASME Boiler and Pressure Vessel Code, Section 1, Rules for Construction of Power Boilers and Section VIII, Pressure Vessels. In the boiler code, the tubesheet and tube materials are normally made of carbon or low alloy steels and the service is usually water or steam. Accordingly, experience has shown that the tubes can satisfactorily be attached to the tubesheet by rolling then either flaring or beading the ends as shown in Figure 9.3, sketches (a) and (b). The welding option is sometimes used as shown in sketches (c) and (d). Grooves in the tubesheet, Figure 9.3(e), may also be used to improve securing of the tubes to the tubesheet.

Rolling of the tubes into the tubesheet is accomplished by using tube rollers. Examples of standard tube rollers are shown in Figure 9.7. An arrangement of tapered mandrel and counter-tapered rollers ensures that as the mandrel is moved axially into the tube, the tube is expanded as a cylinder

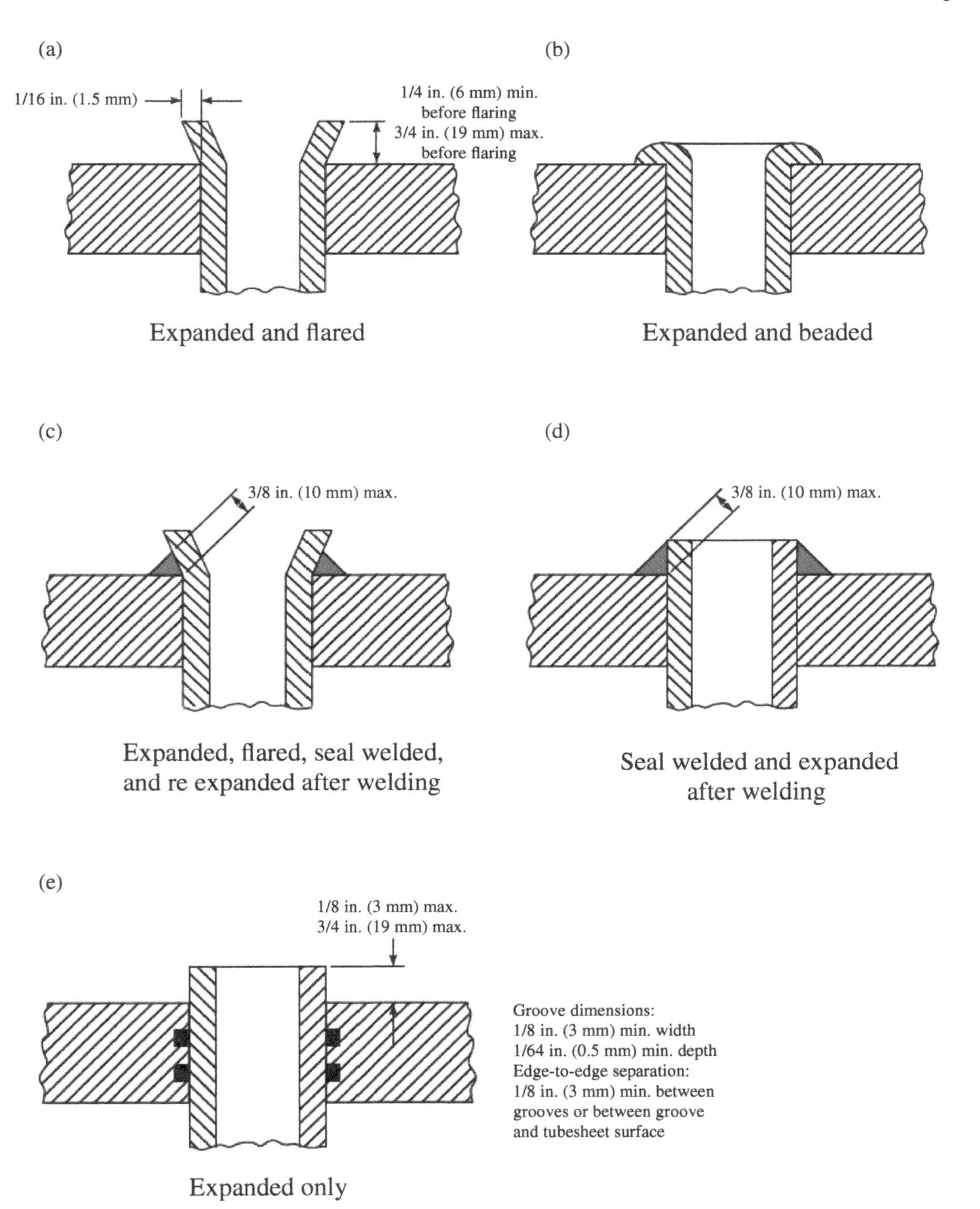

Figure 9.3 Typical tube-to-tubesheet junctions in water tube boilers (*Source:* ASME-I [3])

over the length of the rollers, making even contact with the tubesheet, Figure 9.8. When a flared tube as shown in Figure 9.3(a) is required, a tube roller with additional rollers for flaring is used.

Tube-to-tubesheet junctions in the ASME Pressure Vessel Code almost always require welded tube ends as shown in Figure 9.4. This is because the tubes and tubesheets are often made of

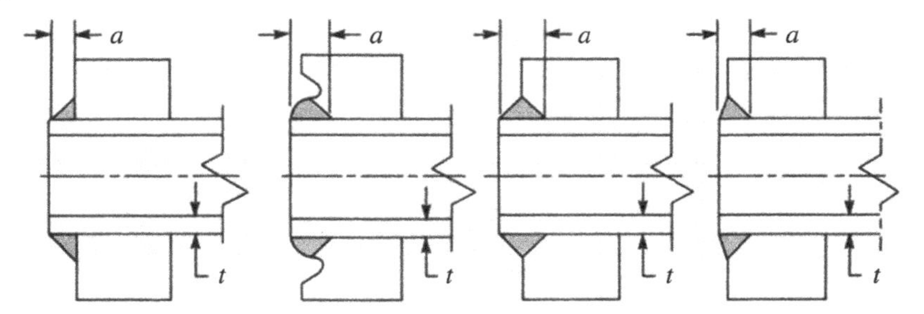

Some acceptable weld geometries where a is not less than $1.4t$

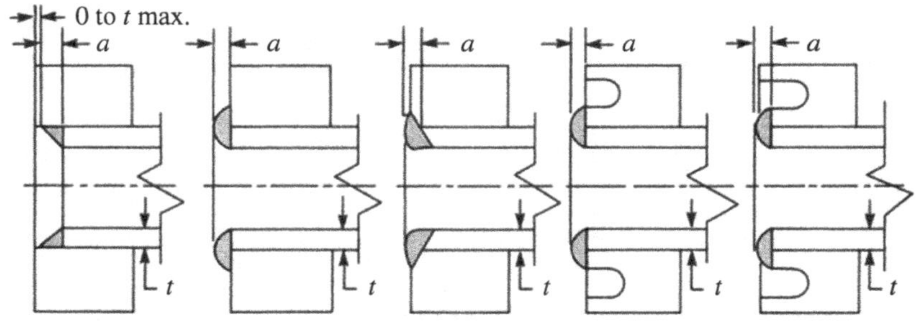

Some acceptable weld geometries where a is less than $1.4t$

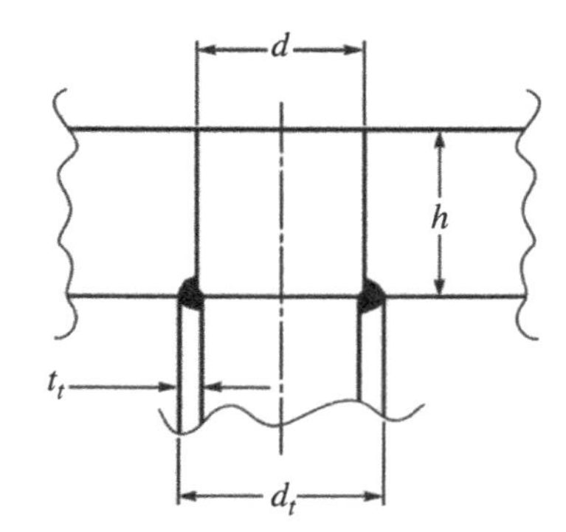

Tubes welded to back side of tubesheet

Figure 9.4 Some typical tube-to-tubesheet junctions in pressure vessel applications (*Source:* Jawad and Farr [2])

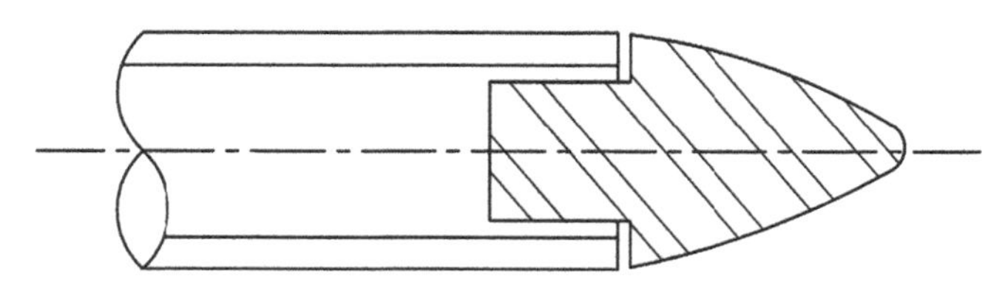

Figure 9.5 Plastic alignment bullet at the end of a tube

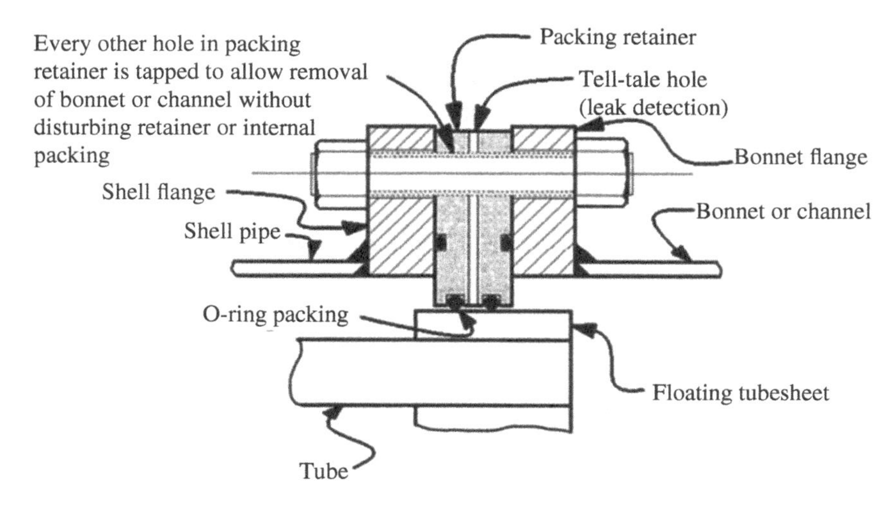

Figure 9.6 An example of a floating tubesheet

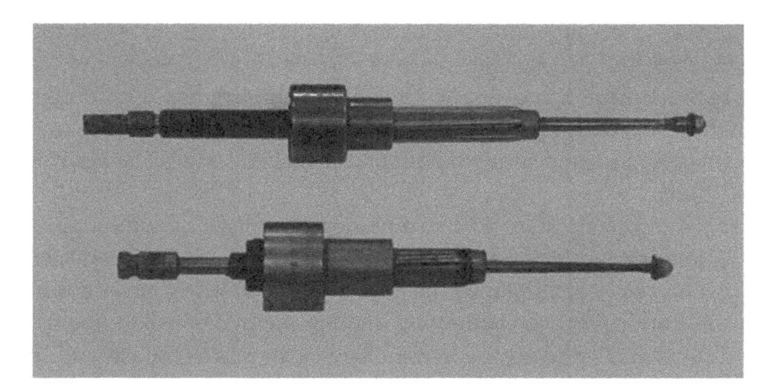

Figure 9.7 Examples of tube rollers (*Source:* Marks Brothers, Inc., ASME Fabricator)

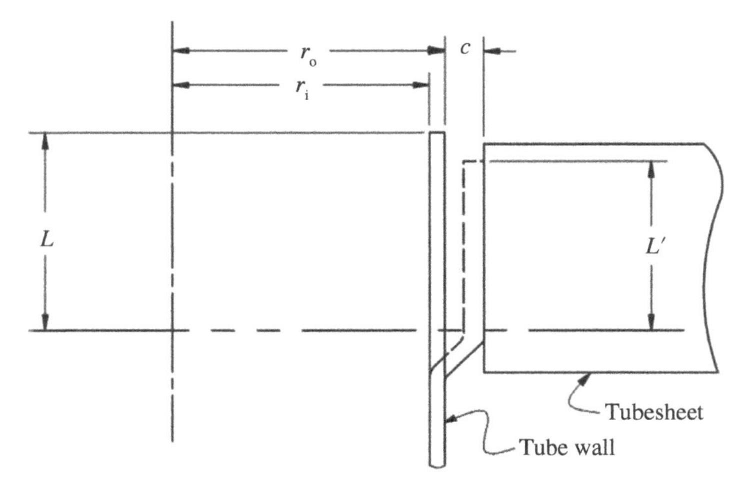

Figure 9.8 Expansion of the tube diameter

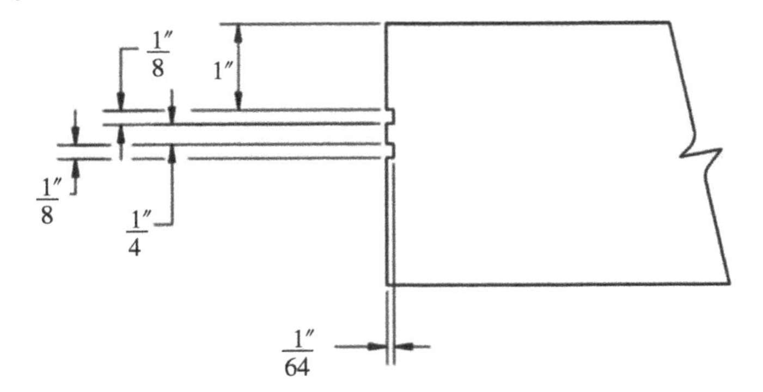

Figure 9.9 Typical grooved tubesheet configuration

different materials, with different properties, with unmatched moduli of elasticity and coefficients of thermal expansion. Such differences often render rolling alone inadequate and a weld is needed to ensure the strength of the joint. Also, in the pressure vessel industry, it is often desirable to avoid even slight leakage between the tubeside and the shellside fluids due to the nature of the chemicals used in heat exchangers. The tubesheet may or may not be grooved, Figure 9.9, based on design conditions. The grooves are beneficial in improving the strength of the joint when the tube and tubesheet materials have approximately the same strength and modulus of elasticity. The grooves also reduce leakage of unwelded joints.

As noted above, rolling the tubes to attach them to the tubesheets involves circumferential stretching of the tubes as well as longitudinal extension due to the pressure of rolling the tube against the tubesheet hole [4]. This biaxial stretching may increase the probability of stress corrosion cracking in certain process applications. In such situations, another method of expanding the tubes is used to minimize total distortion. The tubes are hydraulically expanded as illustrated in Figure 9.10. The expander head consists of a steel block with a hole and a neoprene sleeve. A hydraulic hose is attached to the block to supply pressure. The pressure, often at 40,000 psi (275 MPa) or higher, causes the neoprene sleeve to expand forcing the tube into intimate contact with the tubesheet hole. The process is completed without stretching the tubes longitudinally. Total distortion is thereby reduced; hence, stress corrosion cracking is minimized in the tubes.

Another method of attaching tubes to tubesheets is by using explosives as shown in Figure 9.11. This method is more expensive than roller or hydraulic expansion and is used in areas with limited

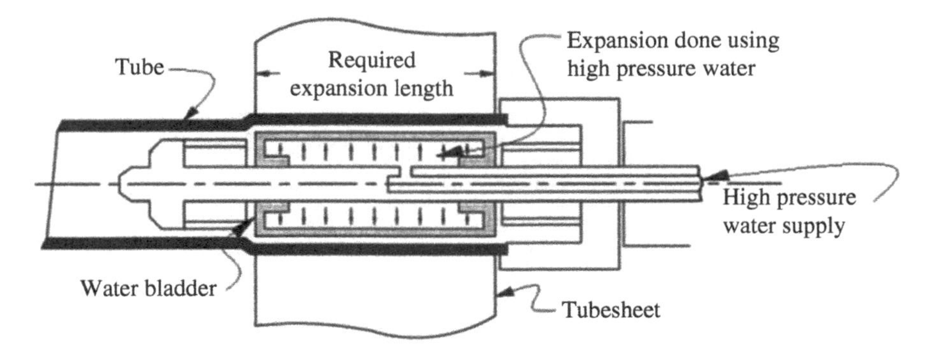

Figure 9.10 Hydraulic expander

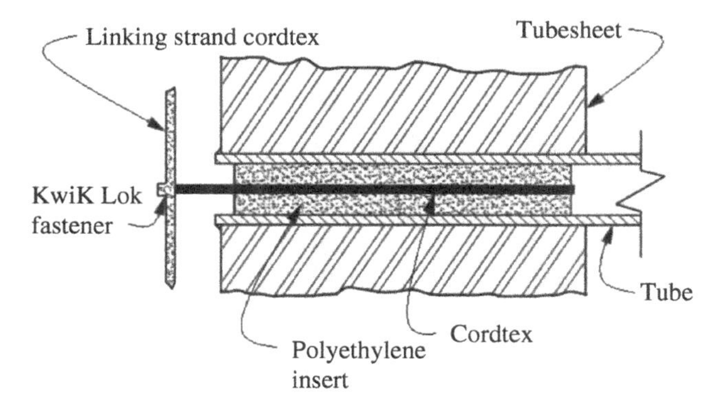

Figure 9.11 Typical arrangement of explosive expansion prior to detonation

access such as commercial and naval nuclear heat exchanger applications. The expansion mechanism of the tubes is the same as that for hydraulic expansion.

The properties of the three tube expansion methods discussed above are as follows:

Roller expansion

1) Easy to use.
2) Requires minimum amount of equipment.
3) Stretches the tube longitudinally, which may increase the likelihood of stress corrosion cracking during operation.
4) In fixed tubesheets, the tubes will be in compression.

Hydraulic expansion

1) Requires people with special skills.
2) Requires expensive equipment.
3) Shortens the tube due to radial expansion.
4) In fixed tubesheets, the tubes will be in tension.

Explosive expansion

1) Requires people who are skilled in explosives.
2) Application limited to commercial and naval nuclear applications.
3) Shortens the tube due to radial expansion.
4) In fixed tubesheets, the tubes will be in tension.

The following are some general characteristics of tube end expansion:

1) Expansion of carbon steel tubes into carbon steel tubesheets normally results in a strong structural joint (not necessarily leak-free).
2) Welding the ends of carbon steel tubes into carbon steel tubesheets subsequent to rolling does not appreciably increase the strength of the joint but prevents leakage.
3) Expansion of titanium or zirconium tubes into carbon steel tubesheets normally results in a very weak structural joint due to spring back of the tubes and also may not be leak-free.

4) Welding the ends of titanium or zirconium tubes into the titanium or zirconium cladding of carbon steel tubesheets appreciably increases the strength of the joint and prevents leakage.
5) Putting grooves in carbon steel tubesheets helps strengthen the joint when using carbon steel tubes. The grooves are less effective at retaining titanium tubes.

9.2.5 Expansion joints

Expansion joints are fabricated either using corner joints, as flanged and flued, or a combination, Figure 9.12, or as bellows, Figure 9.13. Flanged-and-flued expansion joints are normally manufactured by the heat exchanger supplier although some companies specialize in fabricating flanged-and-flued joints for use by others. Forming of the various pieces requires only a few rolling machines.

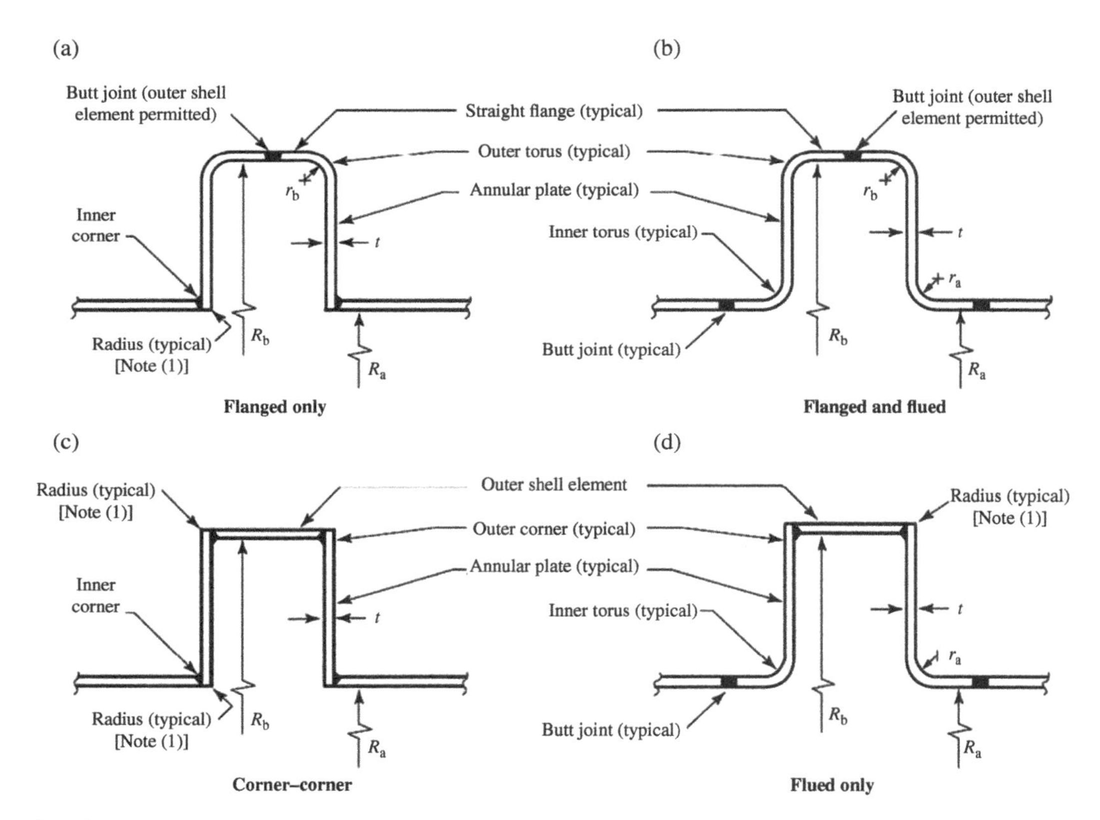

Legend:

R_a, R_b = inside radius of expansion joint straight flange, shell, or outer shell element
t = thickness of expansion joint flexible element

General note: $r_a, r_b \geq 3t$.

Note:
(1) Where the term "Radius" appears, provide a 1/8-in. (3 mm) minimum blend radius.

Figure 9.12 Some types of expansion joints (*Source:* ASME VIII-1 [1])

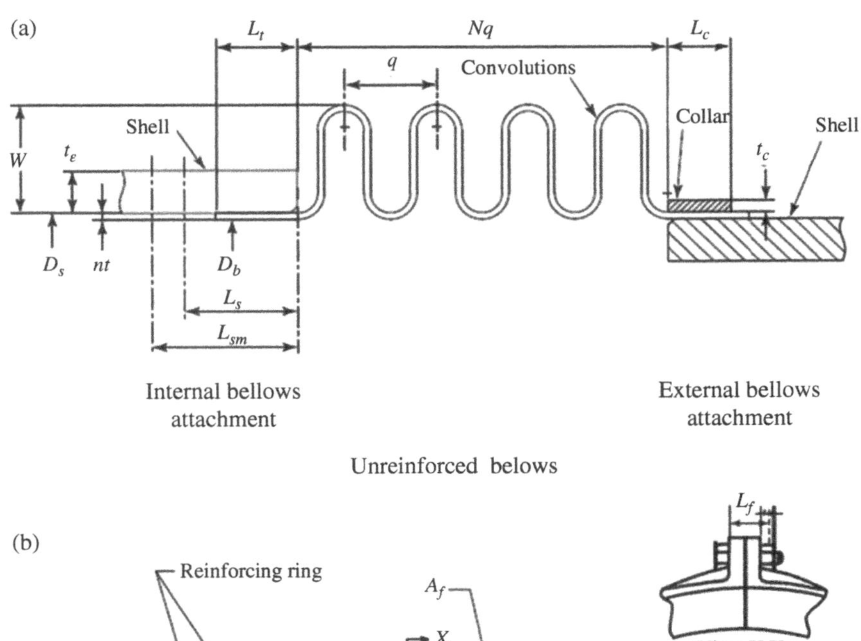

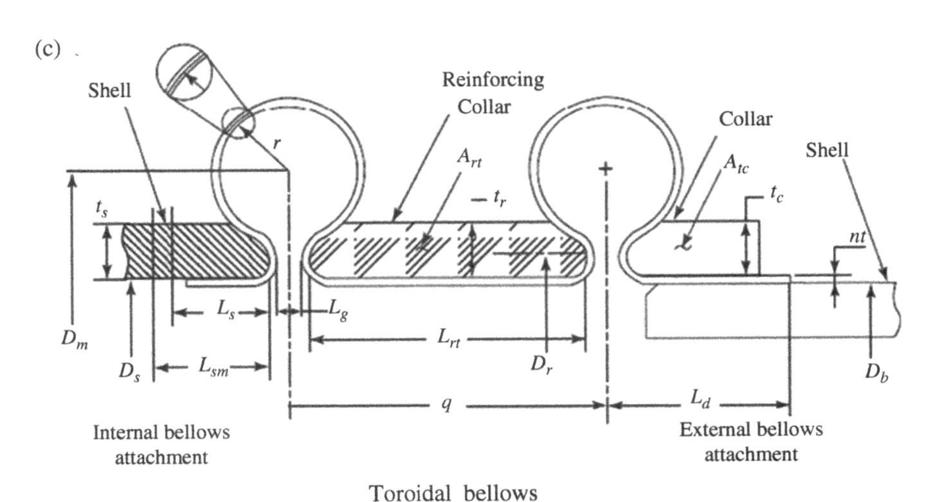

Figure 9.13 Bellows expansion joints (*Source:* ASME VIII-1 [1])

Table 9.1 Differences between flanged-and-flued and bellows expansion joints

	Flanged and flued	Bellows
Advantages	1) Economical for large diameters 2) Do not tend to squirm with angular displacements	1) Economical for large deflections 2) Economical for small diameters
Disadvantages	1) Economical for small deflections only	1) Tendency to squirm with angular displacements or excessive length

The manufacturing of bellows expansion joints requires specialized forming equipment. Accordingly, most manufacturers of heat exchangers buy bellows from companies who specialize in manufacturing them.

The choice of using a flanged-and-flued versus bellows expansion joint depends on multiple factors. Some of these factors are given in Table 9.1.

9.2.6 Assembly of heat exchangers

This section discusses some general issues and concerns related to heat exchanger fabrication.

9.2.6.1 Bowing of tubesheets

Rolling the tubes into the tubesheets of large-diameter fixed heat exchangers causes the tubes to elongate when using rollers and to shorten when using hydraulic or explosive expansion. The tubes are normally expanded and welded at one tubesheet and then trimmed and expanded and welded at the second tubesheet. Elongation or shortening of the tubes as they are expanded into the second tubesheet may cause the tubesheets to deflect since the tubes are already fixed at the first tubesheet. The sequence of tube rolling on thin tubesheets may need to be considered to minimize potential bowing of the tubesheet.

9.2.6.2 Attachment of tubesheets to shells or channels

Welding a thin tubesheet to a thin channel or shell does not pose any problem. However, welding a thick tubesheet, such as 10 in. (255 mm) thick, to a thick channel, such as 6 in. (150 mm) thick, may cause serious problems during manufacturing. The following issues should be considered:

1) Welding of a thick tubesheet to a thick shell or channel, Figure 9.14(a), may result in edge cracks in the tubesheet or channel due to weld shrinkage because of possible microscopic laminations in the tubesheet or channel. Accordingly, the following may need to be done:
   a) Use ultrasonic or dye penetrant test to inspect the edges of the tubesheet or channel prior to welding.
   b) Require transverse tensile testing of the tubesheet or channel to ensure good properties in the transverse direction, and through thickness tensile testing near its edge to identify potential weakness in that direction due to the forging process.

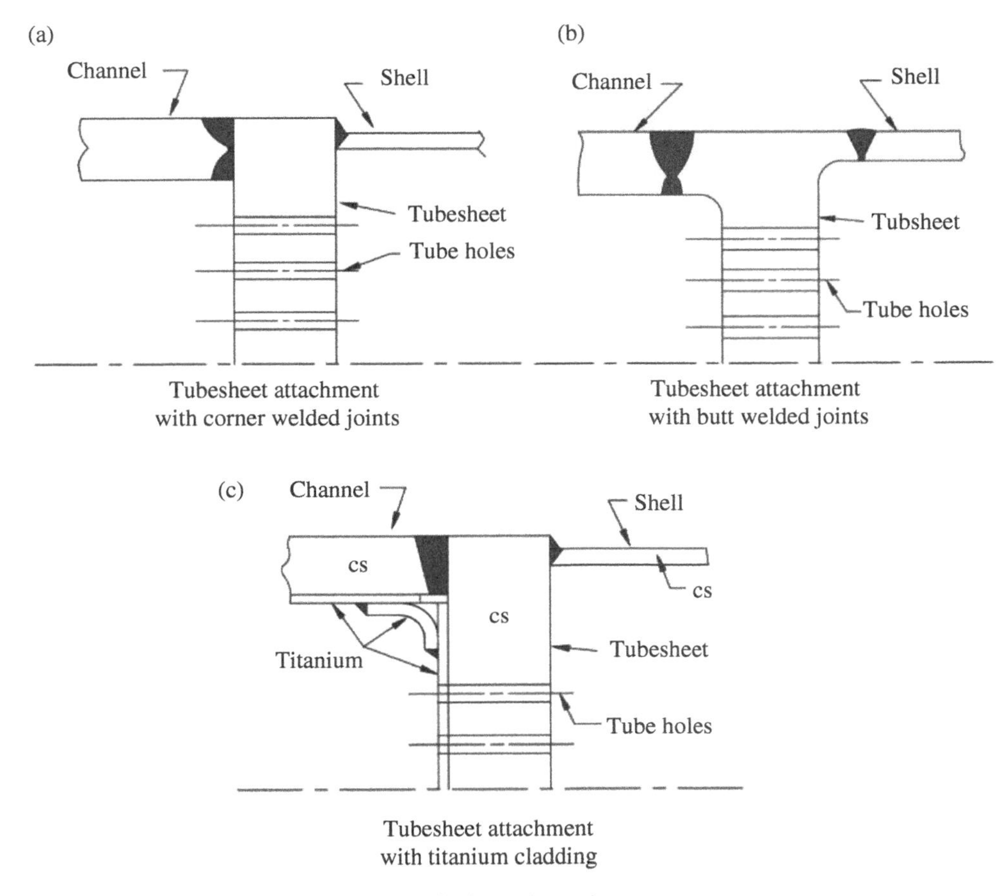

Figure 9.14 Some tubesheet-to-shell and -channel attachments.

c) Use a weld process with minimum heat input.
d) Use a weld design with minimum volume.

2) To eliminate the possibility of cracking, the fabricator may purchase a tubesheet with an extension on the shell side, channel side, or both as shown in Figure 9.14(b). This detail is more expensive than that shown in Figure 9.14(a) since the fabricator must pay for the extra material most of which will be machined off. But this may be more economical in the long run than the potential of having to repair a cracked tubesheet.
3) One possible detail for welding a titanium clad tubesheet to a titanium clad channel is shown in Figure 9.14(c). The batten strip shown must be strong enough to resist the internal pressure in the component.

9.2.6.3 Rails for sliding tube bundles in shells

In many cases, the tube bundle, including tubesheet, tubes, and rods, is assembled outside the shell and then inserted into the shell. To avoid twisting of the tube bundle as it is inserted, grooves are cut

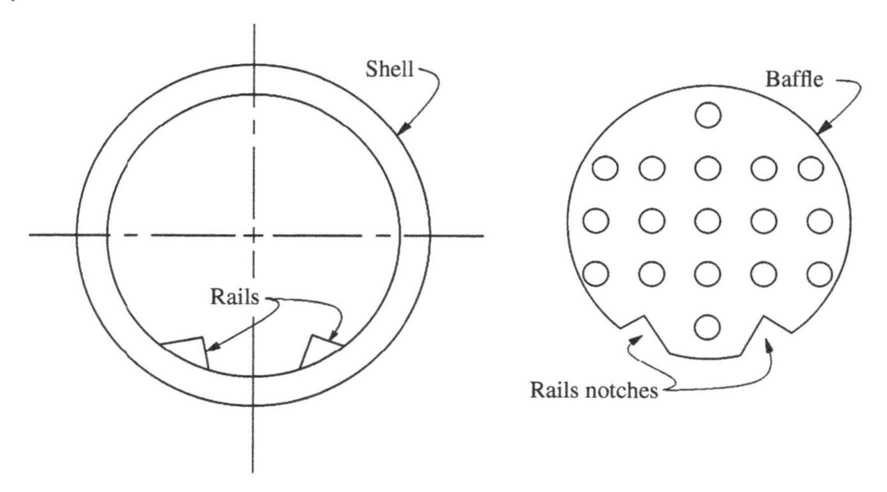

Figure 9.15 Tube bundle sliding rails

in the baffles and rails are provided in the shell to ease insertion and alignment as shown in Figure 9.15. Other methods involving lugs instead of rails are used as well.

9.3 Dimpled Jackets

Dimpled jackets, Figure 9.16, are supplied as components built for strapping on the inside or outside surfaces of pressure vessels, Figure 9.17. They are used for either heating or cooling depending on the service.

The fabrication of a dimpled jacket consists of placing a thin plate on top of a thicker plate and spot or plug welding the two plates together. Pressure is then applied between the two plates. The pressure is increased until the stress in the thinner plate exceeds the yield point and the thinner plate bulges to form the jacket space. The design pressure of the jacket is determined by proof tests in accordance with the ASME code.

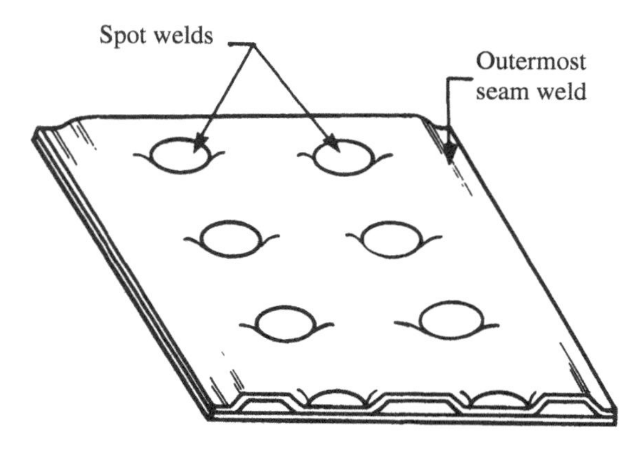

Figure 9.16 Dimpled plate welded to plain plate (*Source:* ASME VIII-1 [1])

Figure 9.17 Dimpled plate jacket welded to a pressure vessel (*Source:* Nooter Construction)

9.4 Layered Vessels

9.4.1 Introduction

Layered vessels, Figure 9.18, are generally used in high pressure service with small temperature gradients. They are found in various industrial applications ranging from petrochemical process reactors to aerospace gas storage vessels to hydrostatic presses for forming complicated parts. A recent use has been destruction of chemical weapons using explosives contained within a layered vessel. The development and use of layered vessels started in earnest around World War II. In Germany, wide corrugated steel strips were helically wrapped around an inner shell to form a layered "wickeloffen" vessel used for coal gasification. At about the same time, A. O. Smith Corporation in the United States developed a method for wrapping thin steel plates over an inner shell to build layered vessels.

Layered vessels offer some unique advantages over solid wall vessels in certain applications. Some of them are as follows:

1) Better mechanical properties of thin materials due to additional rolling, refining the grain structure.
2) Purchasing thin layers with good physical and chemical properties to construct a thick shell is more expeditious and cost effective than purchasing a thick shell with the same properties.
3) Rolling multiple thin layers is easier than rolling a single thick layer, and it can usually be accomplished without heating.
4) A crack through the inner shell during operation can be detected through the vent system in the layered vessel, allowing timely repairs without damaging the rest of the shell.

9.4.2 Fabrication of layered shells

There are many approaches to fabricating layered shells, some of which are shown in Figure 9.18, taken from the ASME code. Sketch (a) shows concentrically wrapped shell using thin layers typically in the range of 3/16 in. (5 mm) to 7/16 in. (11 mm). Sketch (b) shows a coil layered shell

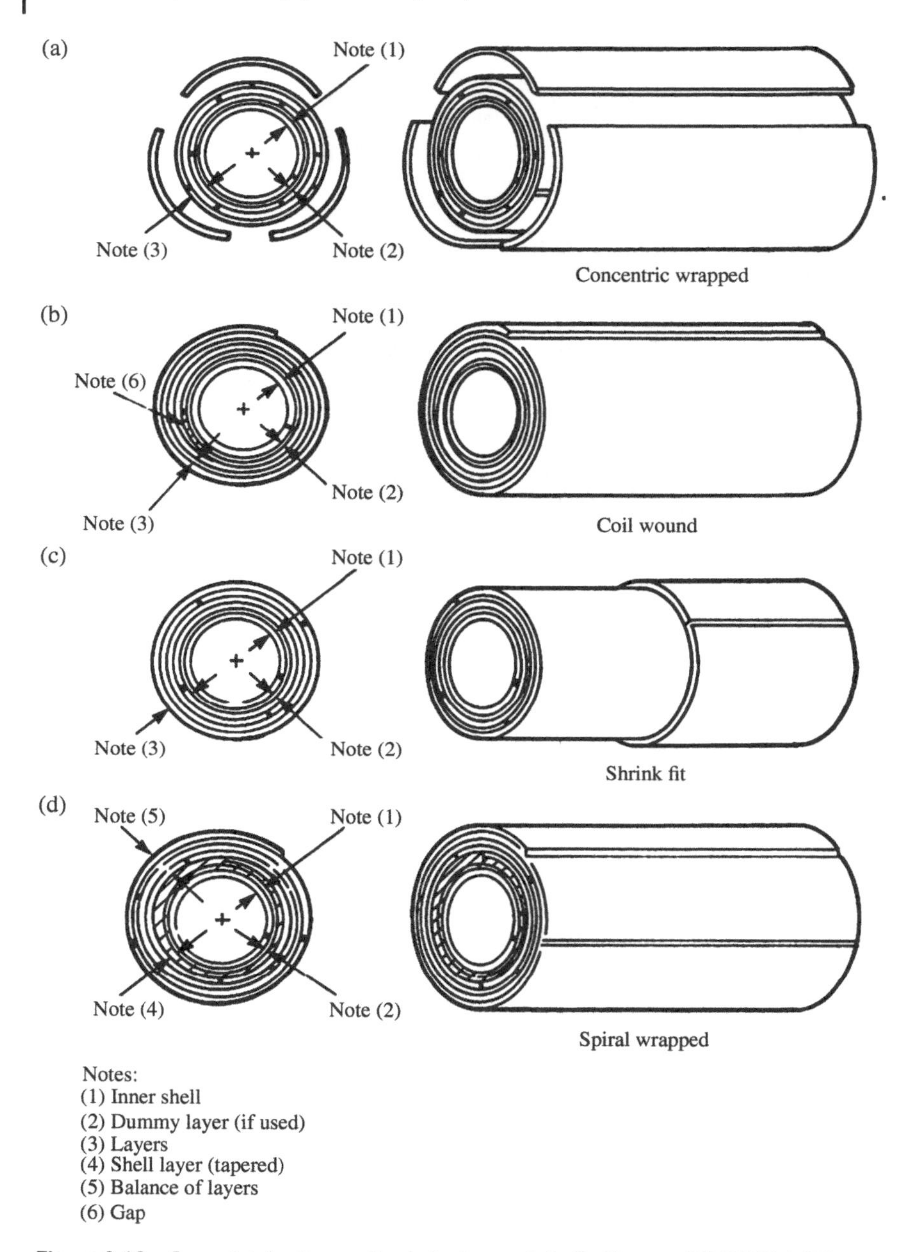

Figure 9.18 Some fabrication methods for layered shells (*Source:* ASME VIII-1 [1])

consisting of an inner cylinder with a thin gauge material, about 0.02 in. (0.5 mm) thick, continuously wrapped around it. Sketch (c) shows a layered vessel fabricated from shrunk-fit cylinders. Sketch (d) is a layered vessel consisting of an inner shell with a tapered plate and spirally wound shell segments in a thickness range of 3/16 in. (5 mm) to 7/16 in. (11 mm), shown in greater detail in Figure 9.19.

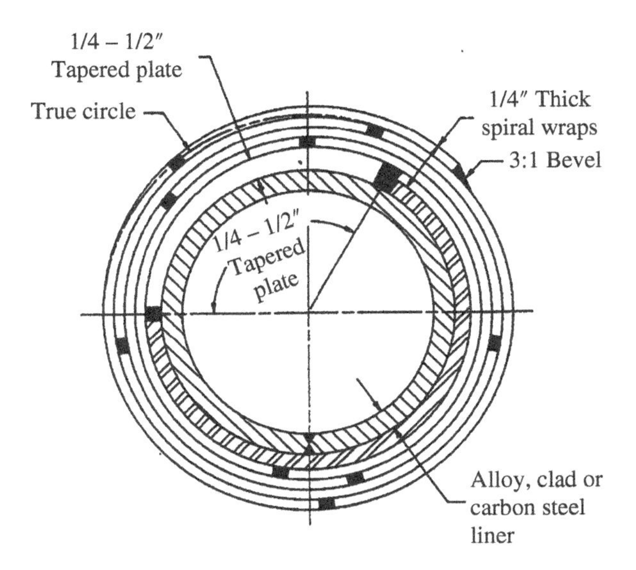

Figure 9.19 Spiral-wrapped shell with tapered plate [5]

The fit up for ensuring tight wraps for both concentric thin layer shells, Figure 9.18(a), and spiral thin layer shells, Figure 9.18(d), is illustrated in Figure 9.20. The inner shell (A) is slid over the mandrel (B). The first layer (C) consists of one or more plates depending on the shell diameter. The first plate of the first layer is placed on top of the inner shell and held tight by two or more cables (D) that are attached to frame (E) by hydraulic cylinders. Plate (C) is tack welded along both longitudinal edges. A second plate (C) is added and tightened using the cables and is tack welded. The cables are then removed and the two plates are full penetration welded to the inner cylinder (A). The process is repeated until all plates of the first layer (C) are welded to the inner shell. The process is repeated for subsequent layers, with welds offset. Rollers (F) are used to rotate the cylinder so welding of the joints is done in the down position. Rollers (G) are attached to frame (E), so the frame can travel along the length of the cylinder to place the cables at the required location.

In some applications such as hydrostatic presses, the wrapping process can be adjusted to produce large compressive stress in the inner cylinder and tension in the outside layer [6]. This process produces a result similar to autofrettaging of solid wall cylinders. This operation consists of applying a large force in the cables (D) to press plate (C) tightly against the previous plate and using wide longitudinal welds to increase the weld shrinkage.

Figure 9.18(c) shows a layered shell fabricated by shrinking cylinders around the liner and other layers. The thickness of the individual cylinders ranges between 0.75 in. (19 mm) and 2 in. (50 mm). The fabrication process consists of first forming and welding the inner shell and measuring the inside and outside radii R_1 and R_2, Figure 9.21(a). The first layer is then fabricated with the inside radius R_3 about the same dimension as the outside radius R_2 of the inner shell. The inside and outside radii R_3 and R_4 are then measured, Figure 9.21(b). The first layer cylinder is then heated to a temperature high enough so it can be slid over the inner shell. When the first layer cools down, it shrinks and locks onto the inner shell. The process is repeated until enough layers are shrunk-fit together to obtain the desired thickness.

Figure 9.18(b) shows a coil layered shell fabricated using a spirally wound thin sheet. This method of construction is not currently prevalent in industrial use. However, it remains viable for construction.

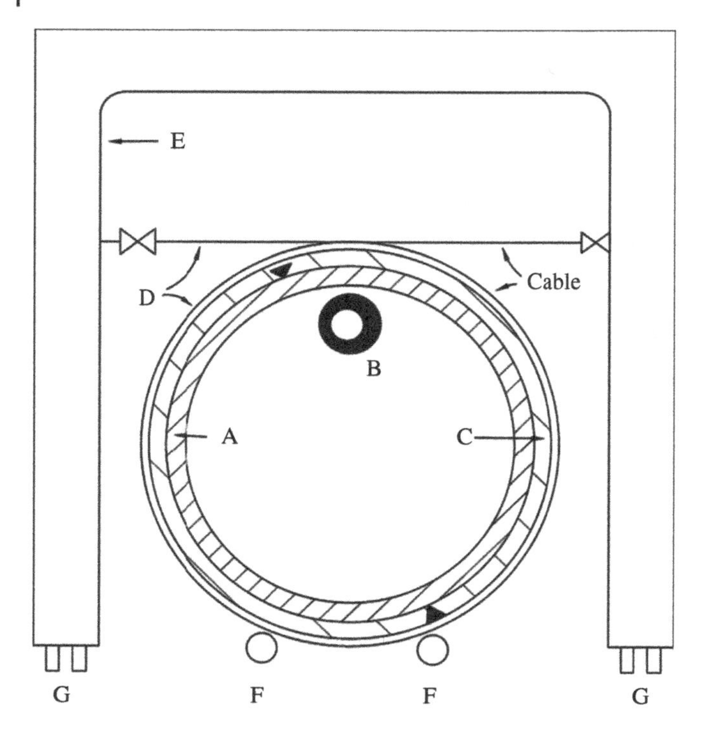

Figure 9.20 Wrapping of thin layer concentric and spiral layered vessels

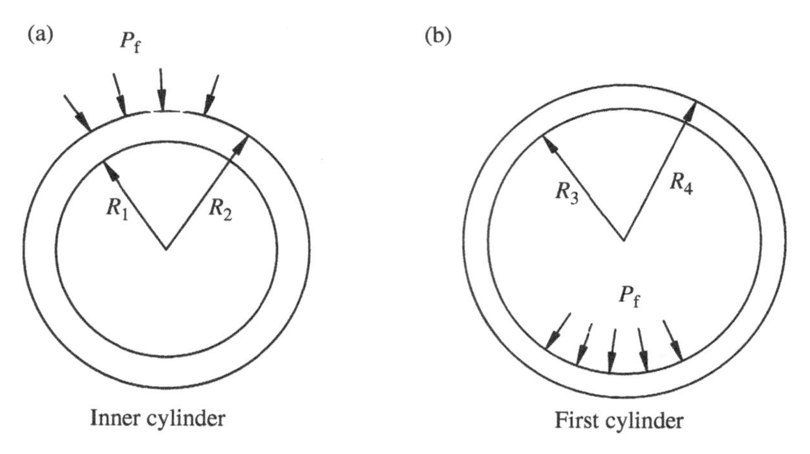

Figure 9.21 Components of shrink-fit layered vessel

Some layered vessels are built from plate strips helically wound to form the shell. The German Wickeloffen is one such a method. Some fabricators in China use strips to helically wind layered vessels. These two methods are not currently addressed by the ASME code.

Some details of attaching the layered shell to thinner heads are shown in Figure 9.22 and for thicker heads are shown in Figure 9.23. The heads are normally buttered with a thin layer of weld

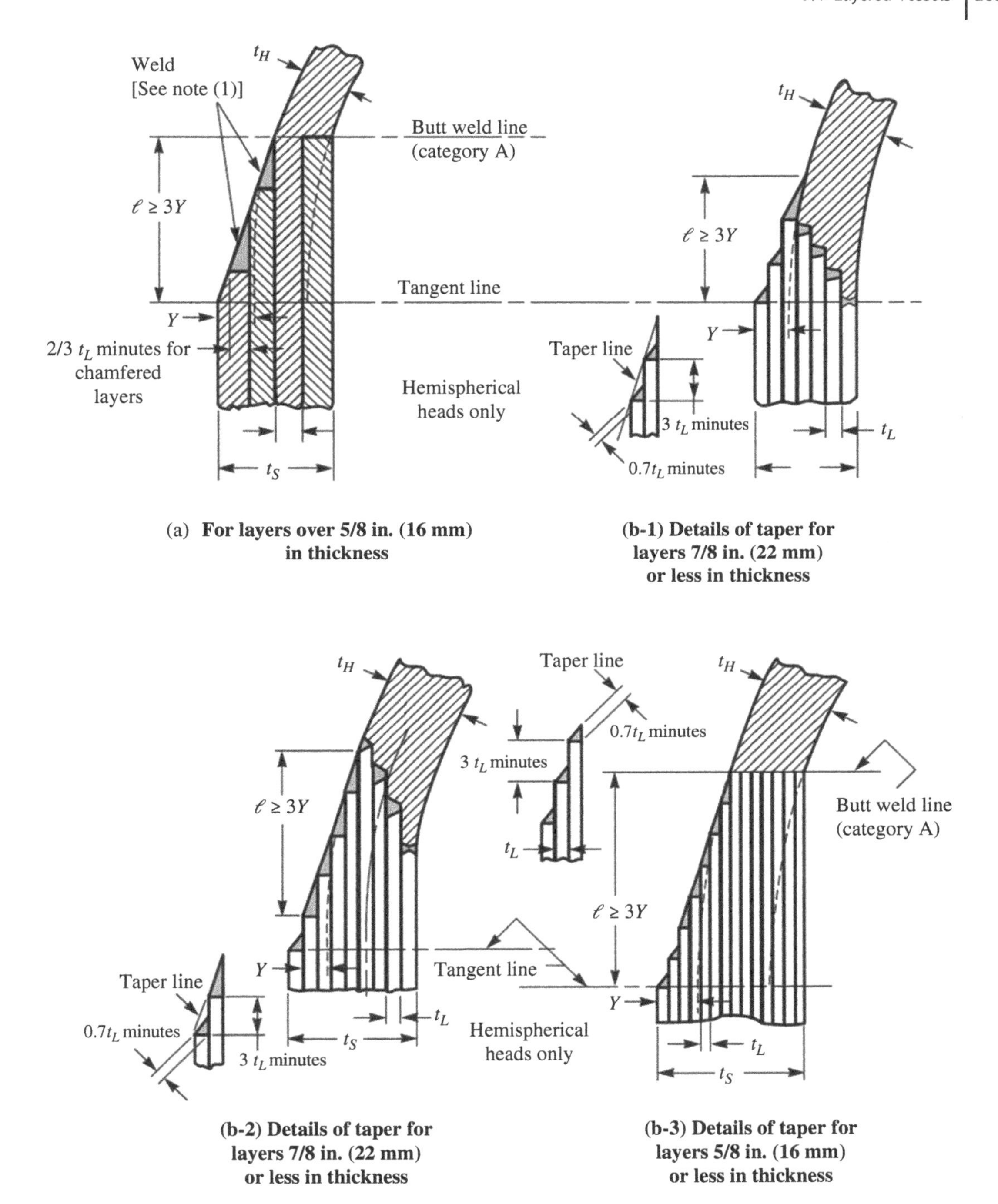

Figure 9.22 Layered shell-to-thinner head attachments (*Source:* ASME VIII-1 [1])

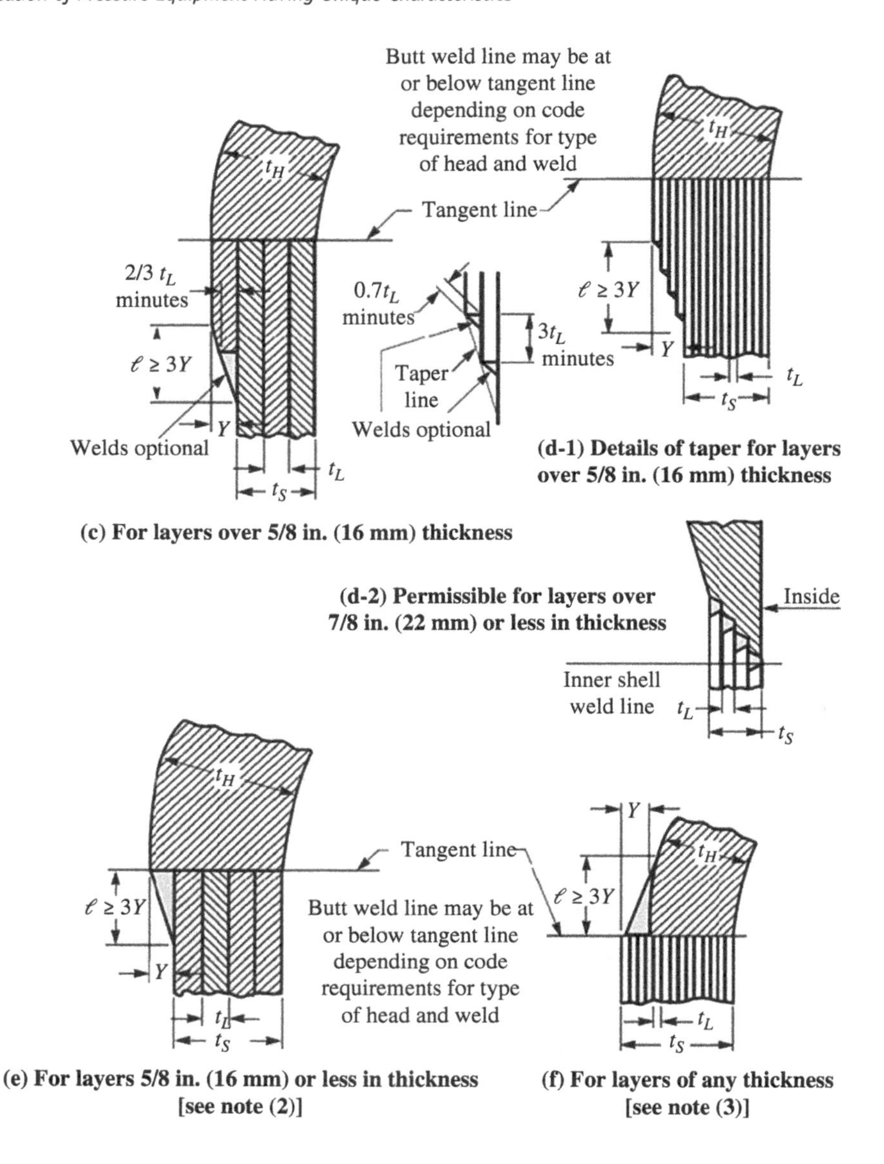

Legend:

t_H = thickness of head at joint
t_L = thickness of one layer
t_S = thickness of layered shell
Y = offset

General note: In all cases, ℓ shall not be less than 3Y. The shell centerline may be on either side of the head centerline by a maximum of 1/2 (t_S–t_H). The length of required taper may include the width of the weld.

Notes:

(1) Actual thickness shall not be less than theoretical head thickness.
(2) In sketch (e), Y shall not be larger than t_L.
(3) n sketch (f), Y shall not be larger than 1/2 t_S.

Figure 9.23 Layered shell-to-thicker head attachments (*Source:* ASME VIII-1 [1])

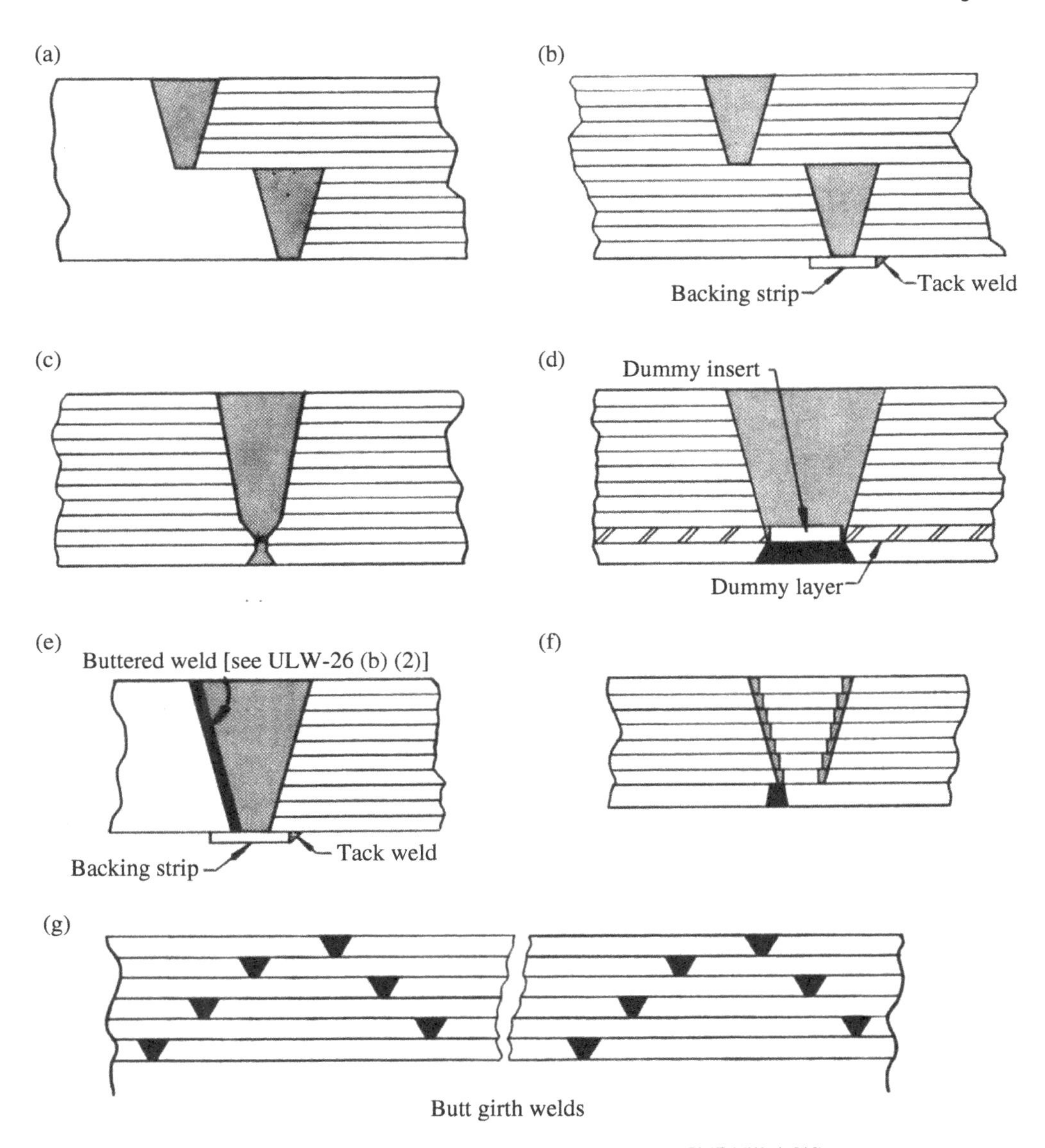

Figure 9.24 Layered pressure vessel shell-to-shell weld (*Source:* ASME VIII-1 [1])

and post weld heat treated prior to welding to the shell. The weld edge of layered shells constructed of thin layers is also buttered with a thin layer of weld prior to welding to the heads.

Some details for welding layered shell course to layered shell course are shown in Figure 9.24. Details of supports for layered shell cylinders are shown in Figure 9.25.

9.5 Rectangular Vessels

Rectangular pressure vessels are used in hospitals, medical laboratories, and other institutions where a pressure vessel with a flat surface is needed to support trays for sterilizing purposes. The pressure is normally close to 45 psi (0.31 MPa) and the temperature is about 300°F (150°C).

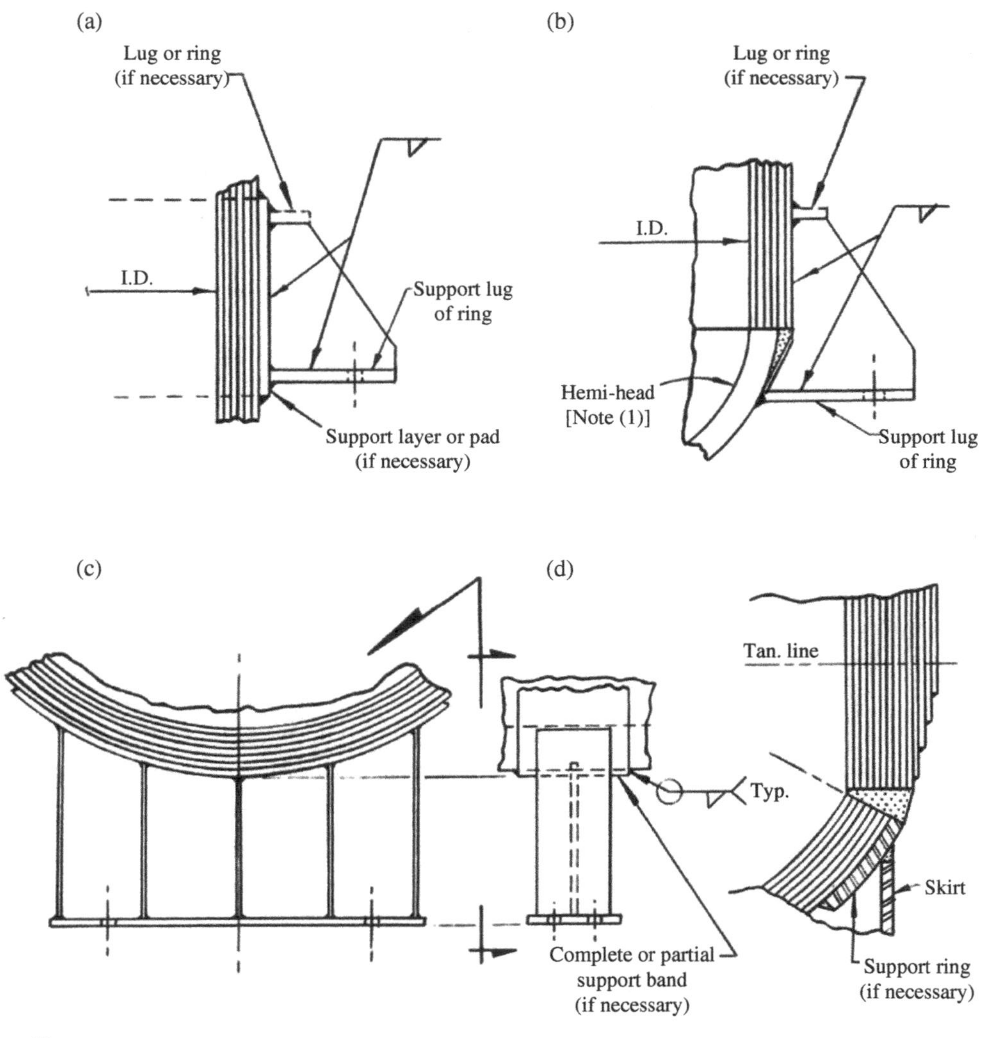

Note:
(1) For other than hemi-heads, special consideration shall be give to the discontinuity stress.

Figure 9.25 Layered shell supports (*Source:* ASME VIII-1 [1])

Sizes vary from about 16 in. (400 mm) high by 16 in. (400 mm) long to about 36 in. (900 mm) high by 48 in. (1200 mm) long, Figures 9.26 and 9.27.

9.6 Vessels with Refractory and Insulation

Vessels operating at high temperatures require refractory and/or insulation to protect the metal from high temperatures and possible erosion and or corrosion. A summary of some of the decisions that need to be made if the insulation or refractory is installed at the pressure vessel manufacturer's plant is given as follows:

Figure 9.26 Rectangular vessel fabricated of two formed C sections (*Source:* Marks Brothers, Inc., ASME Fabricator)

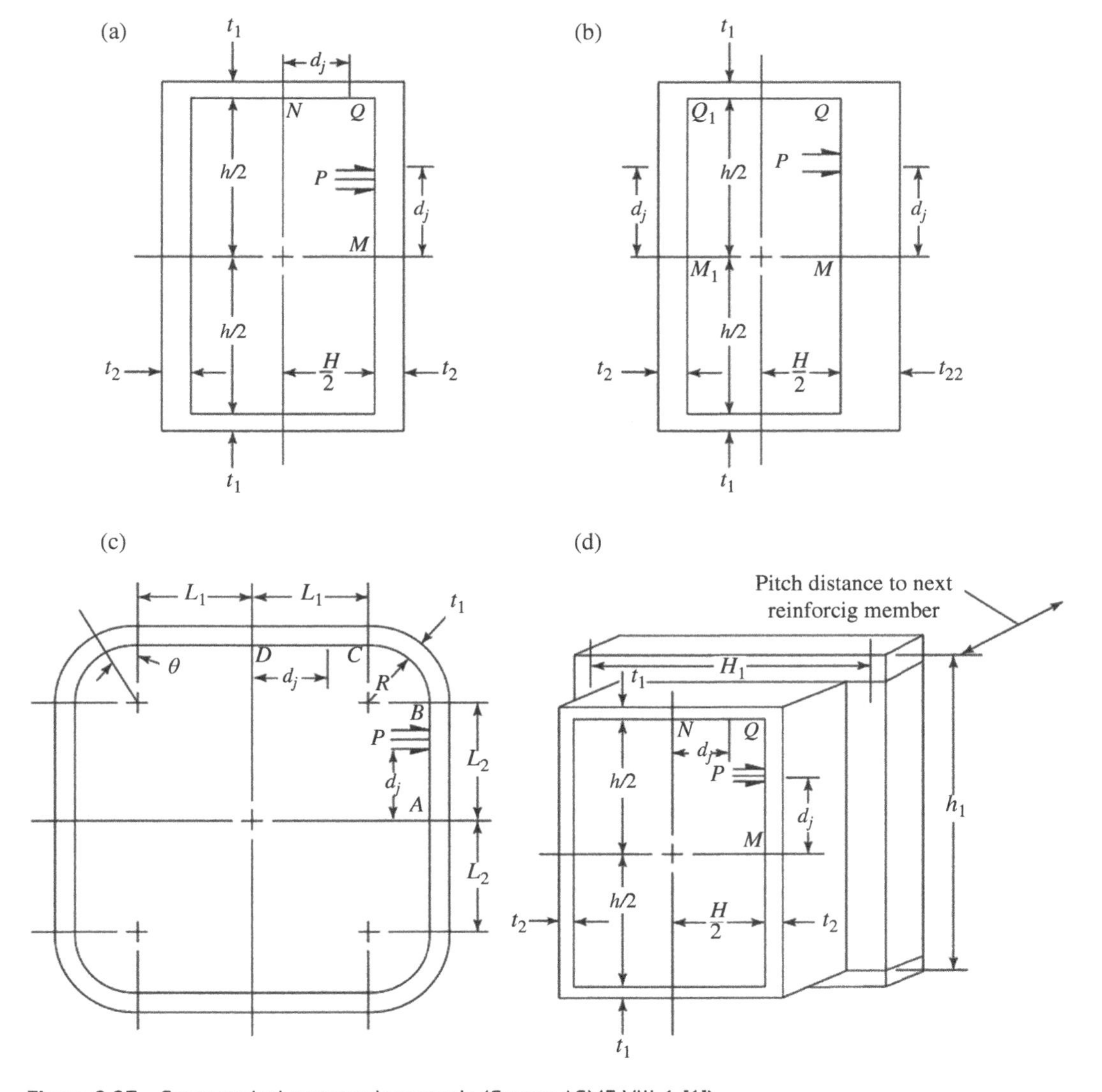

Figure 9.27 Some typical rectangular vessels (*Source:* ASME VIII-1 [1])

1) If anchors are needed for the refractory/insulation and the vessel requires post weld heat treating, then a decision needs to be made as to whether to install the anchors prior or subsequent to post weld heat treating.
2) Since refractory and insulation need to be kept dry, a decision must be made whether hydrotesting should be performed prior to installing the refractory/insulation or pneumatic test is to be performed subsequent to installation. The requirements for hydrotesting and pneumatic testing can be vastly different regarding cost and safety.
3) Time must be considered for dry-out (also called bake-out) of the insulation prior to moving or shipping the vessel.

9.7 Vessel Supports

Pressure vessels are supported in a variety of ways, as shown in Figure 9.28. Small lightweight vessels are normally supported by structural legs as shown in Figure 9.28 sketch (b). This type of support is the most economical one to fabricate. For heavier vessels it is more practical to use pipe leg supports with their centerline aligned with the shell as shown in Figure 9.29. This construction eliminates the bending moment in the shell and is especially economical for thin shells, which might otherwise require a stiffening ring.

For large heavy vessels, it is more practical to use skirt supports as shown in Figure 9.28 sketch (a). The dead weight and wind and seismic loads are better distributed into the shell and are transferred more uniformly to the skirt and down to the foundation. The cost of fabricating the skirt and

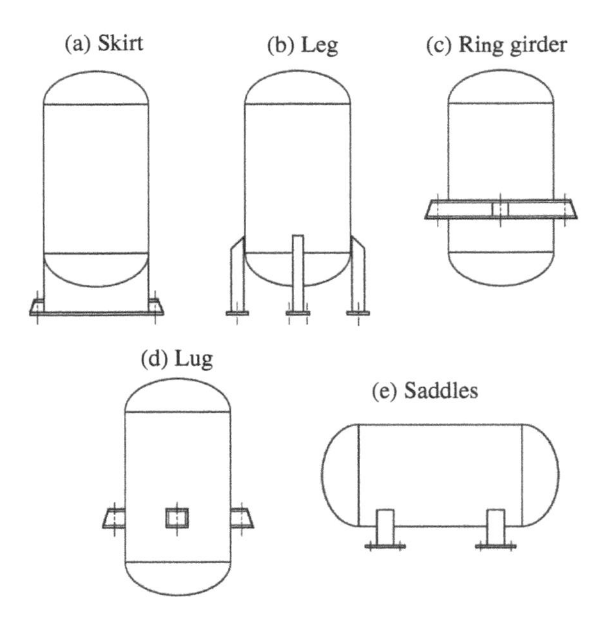

Figure 9.28 Vessel supports

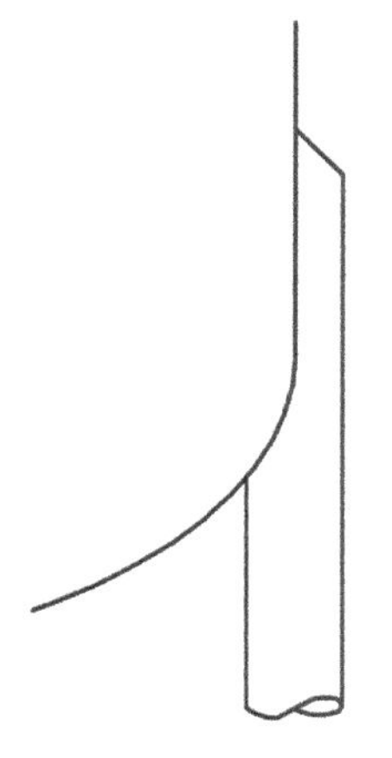

Figure 9.29 Pipe leg support

base ring is less than that of heavy pipe legs. Also, the foundation for skirt-supported vessels tends to be more economical than that for individual legs supporting heavy loads.

Many vessels are located in structural multistory frames. In such cases, lugs are used to support the vessel as shown in Figure 9.28 sketch (d). The shell is subjected to a bending moment due to the eccentricity of the support. The local bending stresses due to the bending moment may result in a thicker shell at the lug location unless the original thickness is adequate to resist the additional loads. For thin shells, it is more economical to fabricate a ring girder for a support as shown in Figure 9.28 sketch (c). Fabricating a ring girder is more expensive than lugs but may still be more economical than increasing the shell thickness in the vicinity of the support. This is particularly true if the shell material is an expensive alloy.

Horizontal vessels are normally supported on saddles as shown in Figure 9.28 sketch (e). If the vessel wall is thin, then stiffening rings are often required around the cylinder in the vicinity of the saddle, and possibly between the saddles. The cost of the rings, especially if the shell is thin, may represent a substantial portion of the total cost of the horizontal vessel.

9.8 Summary

The previous discussion highlights a few types of equipment that are out of the ordinary for many shops. In addition, special applications, especially in the research community, often call for unique configurations or materials. Figure 9.30 illustrates a unique vessel that required specially machined internal passages and unconventional structural accommodations. When the need for such products arises, designers and fabricators must work together to develop appropriate configurations, weld details, and fabrication sequences to fill the need.

Figure 9.30 Unusual pressure vessel configuration and view of portion of internals prior to final assembly

References

1 ASME VIII-1. 2021. "*Boiler and Pressure Vessel Code, Section VIII, Division 1.*" American Society of Mechanical Engineers, New York.
2 M. Jawad, J. Farr. 2019. "*Structural Analysis and Design of Process Equipment*". Wiley, Hoboken, NJ.
3 ASME. 2021. "*Boiler and Pressure Vessel Code, Section I.*" American Society of Mechanical Engineers, New York.
4 M. Jawad, E. Clarkin, and R. Schuessler. "Evaluation of Tube-to-Tubesheet Junctions". *ASME Journal of Pressure Vessel Technology*, February 1987, Volume **109**, Pages 19–26.
5 M. Jawad. "Layered Vessels in the Petrochemical Industry". May 1979. API Proceedings, Volume **58**.
6 M. Jawad. "Wrapping Stress and Its Effect on Strength of Concentrically Formed Plywalls". September 1972. ASME 72-PVP-7.

10

Surface Finishes

10.1 Introduction

Most pressure vessels have finishes applied to external surfaces either for protection of the vessel from the surrounding environment or for appearance. External finishes may also include advertising for the vessel manufacturer or for the company that assembles the vessel into a system, such as a tank mounted air compressor system. Many vessels also have finishes applied to internal surfaces for protection from the media that they contain.

Almost all vessels fabricated of materials likely to corrode in an outside environment (e.g., steels) receive a surface treatment. Even stainless steel vessels, which are generally impervious to external environmental conditions, are often painted for reasons of appearance.

In many cases, the external environment of a pressure vessel is not highly aggressive. For vessels operated near salt water or in plants where chemicals drift from other processes, however, the environment may be a concern.

In some cases, the internal environment is dealt with by just adding a corrosion allowance, then accepting the fact that after a certain number of years the vessel will be corroded below its minimum wall and will need to be replaced. If the medium is significantly more aggressive than just moisture saturated air, this approach may not be cost effective. Often in locations such as fertilizer and process plants, the media of a vessel presents conditions that are much more aggressive than the outside air. The means of dealing with the environment vary depending on the problems it presents.

10.2 Types of Surface Finishes

A number of ways have been developed to protect the surface of a vessel. Categories of surface finishes include the following:

1) Surface characteristics, unfinished.
2) Finishes involving a reduction in the reactivity of the material itself, without any coating.
3) Applied coatings.

Some involve changing the surface properties to reduce the tendency of the metal surface to react with the environment and are typically referred to as passivating. Others depend on separating the surface from the environment and include applied coatings.

Fabrication of Metallic Pressure Vessels, First Edition. Owen R. Greulich and Maan H. Jawad.

Almost all of the techniques used to treat material surfaces pose environmental issues and challenges.

10.2.1 Surface characteristics, unfinished

For some applications, the finish is defined largely by surface roughness and contour characteristics, with no additional properties or coatings. The finish of machined or polished surfaces is typically measured as an average roughness, either microinches (μin.) or micrometers (μm). It is the average of the absolute values of deviation from a theoretical, perfectly flat surface. A specified surface roughness still allows for a significant variety in surface appearance, however, and even in how much the surface deviates from "perfectly flat." Since this measurement is an average, it can allow significant highs and lows to occur, provided the average still remains at a low number.

A surface finish comparator gauge, Figure 10.1, is often used to determine the finish by performing a visual comparison. Sometimes a profilometer is used to take actual measurements using a high-resolution probe similar to the needle on a phonograph, or by measuring optical or ultrasonic scattering or changes in capacitance.

Part of surface finish is the "lay," or the direction of any marks. For example, a flat surface machined on a lathe will have a spiral cut, with (roughly) circular marks spiraling in from the outside to the center. A milled surface will have circular marks that progress across the piece, but with a radius equal to that of the cutter. If a 1/2 in. (12.7 mm) endmill is used, then the radius will be 1/4 in. (6.4 mm), while a 10 in. (254 mm) circular cutter will leave marks with a 5 in. (127 mm) radius.

Machined surfaces vary in surface roughness depending on the material, lubricant, type of cutter, shape of the cutting tool, tool sharpness, number of cutters, feed rate, and cutting speed. A common rough machined surface is often about 125–250 μin. (3.2–6.3 μm), though it could be outside this range on either side. A machined flange face will usually have a concentric circular finish in this range so as to retain the gasket well and minimize leakage.

Polished surfaces such as stainless steel also have a range. A "brushed" finish on stainless steel may be as rough as about 16 μin. (0.4 μm), while a polish that will reflect an image is about 2 μin. (0.05 μm). When appearance is a factor, the "lay" of the finish is important. A sheet that has a grain going all one direction appears uniform and pleasing. Such a finish is usually characterized by

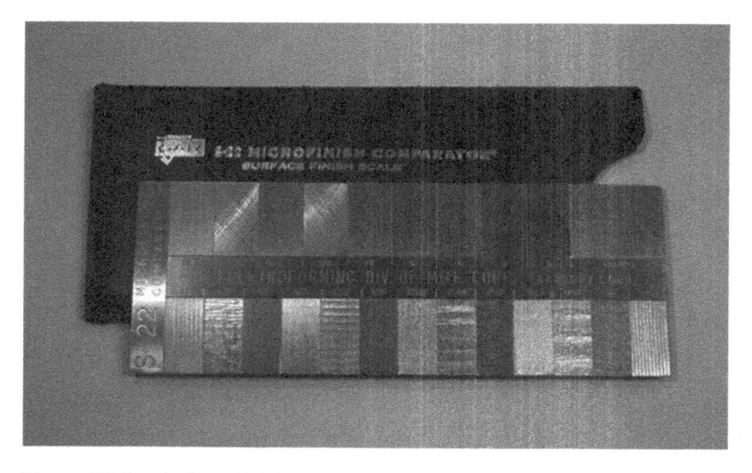

Figure 10.1 Surface finish comparator gauge

marks all running in the same direction, resulting from passing the sheet under a wide sanding belt. If random scratches at various angles occur on the surface, it will likely appear marred, even though the material may be no less smooth.

Pharmaceutical, hospital, biotech, food processing, and other industries often require polished surfaces to minimize bacterial growth and to facilitate cleaning. For these applications, the surface roughness is likely in the range of 16 μin. (0.4 μm) to 2 μin. (0.05 μm). Other applications may require such finishes for the sake of appearance. Sometimes electropolishing is specified.

Abrasive blasting is often used to clean and provide a surface that will hold paint or other coatings. The "standard" abrasive blast used to be sandblasting, typically with silica, but environmental and health concerns have reduced the popularity of this material. Also, a sand or grit blasted surface is not particularly easy to clean, because of the sharpness of the contours. For paint and other coatings, the sharper contours of a grit blasted surface are often an advantage since they provide more surface area and a contour for the coating to grip. Bead blasted stainless steel, on the other hand, shows a light satiny finish that is easy to clean. Blasting with various other less aggressive materials, including walnut shells and dry ice, is used to clean metallic surfaces without producing significant texture.

Grinding, polishing, and grit blasting of surfaces, regardless of the type of grit or particle, are normally conducted in a specially confined area with particle recovery and filtering to prevent environmental damage and to reduce health concerns. Some grits, such as garnet for sandblasting, can often be used more than once, while materials such as glass beads and walnut shells are probably not recycled.

10.2.2 Passivation

Passivation includes those processes used to reduce reactivity of the material surface itself. These processes can be applied to a number of metals and alloys, including austenitic and martensitic stainless steels and other ferrous metals, steels, aluminum, and nickel, among the materials often used for fabrication of pressure vessels.

Passivation is accomplished by a chemical reaction with the base metal, either electrolytically, by application of chemicals to facilitate the reaction, or by allowing a reaction to occur naturally with oxygen in the air. The effect is to produce a very thin and relatively inert corrosion layer that prevents further corrosion from developing.

Austenitic stainless steels are frequently passivated. The process typically includes cleaning, a hot nitric acid bath that causes the surface reaction, a neutralizing rinse in an aqueous sodium hydroxide bath, and drying. ASTM A967/A967M Standard Specification [1] for Chemical Passivation Treatments for Stainless Steel Parts provides a standard for these operations.

Martensitic stainless steels are passivated less frequently and with more difficulty, and proprietary processes are used.

Ferritic steels can be made passive for mild nitric acid environments by placing them first in concentrated nitric acid. Passivation achieved in this way can be lost almost instantaneously, however, by contact with unpassivated steel, and the process is rarely used.

Ferrous materials including steel are also sometimes protected by allowing formation of oxidation (rust), which is then treated with phosphoric acid to create a more inert metalophosphate. This is sometime referred to as a conversion coating process.

Nickel is passivated by formation of a layer of nickel fluoride, making it useful for handling elemental fluorine, which is present in water and sewage treatment applications.

Some aluminum alloys react with oxygen in the atmosphere to form a natural thin surface oxide layer that prevents further oxidation or corrosion in certain environments. Some aluminum alloys do not form such a layer to the extent of protecting themselves and require the use of other processes to produce the passive layer. The two common ones are anodizing and chromate conversion coating.

Anodizing uses an electrolytic process to produce an oxide layer, while chromate conversion produces a thin aluminum chromate coating chemically.

Chromate conversion can also be used for other materials such as cadmium, copper, magnesium, silver, tin, and zinc, though most of these are not frequently used in construction of pressure vessels.

Many of the chemicals used in passivation processes are corrosive or otherwise aggressive. When no longer useful, they must typically be treated or disposed of as hazardous waste.

10.2.3 Applied coatings

10.2.3.1 Plating

Plating is the process of chemically or electrolytically bonding a very thin layer of another metal or alloy to the surface of a metal. Its use is not common for pressure vessels, though electroless nickel plating has occasionally been used on vessel interiors. Metal gaskets, however, are often plated with a soft material, commonly silver, because it more readily adapts to any imperfections in the adjacent seating surface.

The process of plating requires a clean surface, and the finish will be no better than the finish on the unplated product. The acids used for etching and the electrolytes used in the plating process must be contained and properly disposed of.

10.2.3.2 Polymeric coatings

Polymers are chemical compounds formed of repeating units. Paints are almost all polymers, but the category extends well beyond paints. Many of these products, though not considered paints, are used as coatings on vessels either to protect the surface or to give it other desirable properties. Among these polymeric materials are ETFE (ethylene tetrafluoroethylene), FEP (fluorinated ethylene propylene), PFA (perfluoroalkoxy alkane), PTFE (polytetrafluoroethylene, or Teflon), and others.

Whether dealing with paints or other polymers, both compatibility with the media and compatibility with other surrounding materials must be considered. For example, reactions resulting in corrosive byproducts have occurred between paints and insulating materials. In addition, solvents, reaction byproducts, and excess product from these processes are usually harmful to the environment and must be contained and disposed of as hazardous waste.

10.2.3.2.1 Paint

Paints are available in a wide variety of formulations, designed for specific applications. Some are intended to produce the absolutely best appearance, while others are designed for durability, resistance to moisture, fading, or chalking, or for easy application, or long life in harsh environments.

Brush painting of process equipment is currently almost unheard of except for touch up or for painting field joints that could not be done in the shop. Most paints used in industry use solvents that must evaporate to dry or cure the paint. These solvents present environmental hazards. As a result, most industrial coating is performed in specialized facilities, either in a dedicated portion of a fabricator's plant or at the facility of a company dedicated to such coating. Environmental factors

Figure 10.2 Abrasive blasting room (*Source:* Clemco Industries Corp.)

are only some of the issues. Another reason is that coatings have become sophisticated products that require special knowledge, experience, and equipment for proper application.

Proper surface preparation is important to ensure a good bond between the paint and the vessel surface. A white or near-white blast is recommended for many paints. Abrasive blasting of large components is preferably done in self-contained abrasive blast rooms, Figure 10.2, while small components are blasted manually. This is often accomplished in a small abrasive blasting cabinet, using protective rubber gloves to handle the components and blasting nozzle. One or more prime coats and finish coats are common. Minimum build thicknesses are specified. Users should consult with their paint suppliers regarding what paint to specify for each application.

When plate sections and other products are prepared for field fabrication, at least a prime coat of paint is usually applied in the shop. Sometimes a finish coat is applied as well. When this is done, painters must mask weld areas to prevent paint from getting mixed in the weld, causing porosity and other problems, Figure 10.3. The masked areas and the welds will then be touched up in the field.

In addition to environmental issues, field painting is minimized because of the risk of overspray drifting onto adjacent surfaces. Even on a calm day, it is difficult to avoid entirely this problem. In at least one case, even with the use of low volatile organic compound paint, over three hundred cars required repainting or significant detailing after the wind picked up and carried paint from what was planned as a "simple" paint job involving a large pressure vessel.

10.2.3.2.2 Other polymers

Polymeric coatings designed to resist aggressive chemical or other operating conditions have also been designed. These are most often applied to the interior of pressure vessels and systems to reduce corrosion, erosion, or other types of chemical or physical attack. Table 10.1 provides a brief summary of some industrial coating characteristics.

Figure 10.3. Large equipment paint room (*Source:* BlastOne International)

Table 10.1 Properties of some industrial coatings

Material \ Application	Corrosion resistance	Abrasion resistance	Chemical resistance	Nonstick	Baked on
ECTFE/Halar	x	x	x		x
Epoxy	x	x	x		
ETFE	x	x	x	x	x
FEP	x		x	x	x
PFA	x		x	x	x
PPS/Ryton	x	x	x		x
PTFE	x		x	x	x
PVDF/Dykor	x	x	x		x
Xylan	x		x	x	x

Even more than for paint, users must consult with experts in the coating field to ensure correct selection and application of coatings for critical applications.

10.2.3.3 Thermally applied coatings

When other coatings are insufficient, thermal application of metal alloys, ceramics, and polymers can be used. The application process typically involves heating and spraying a powdered material onto the surface to be coated. The processes used are referred to by various names, including flame spraying, metal spraying, plasma spraying, etc.

Heat sources vary, including air and oxygen fueled gas mixtures as well as electric arc, some involving a plasma at 20,000°F (11,100°C) or more. The part temperature, however, typically remains below about 250°F (120°C), typically eliminating concerns regarding heat treatment.

The material to be deposited is sometimes supplied as a powder, sometimes as a wire. Materials used include metals such as aluminum, copper, nickel, carbon and stainless steels, Inconel®, Hastelloy®, Stellite®, various ceramics such as alumina, titania, tungsten carbide, and zirconia, mixtures of metals and ceramics, and corrosion resistant polymers.

While thermally applied coatings can be used in the harshest of environments, a number of factors affect the success of the coating process. Feedstock, temperature, means of heating, gas mixture, flow and deposition rates, and other variables affect coating quality. As with other processes, users of these processes should consult with product and equipment suppliers regarding the best product for their application.

A factor that must be considered with flame sprayed coating is that surface finish is often fairly rough. If a fine finish is needed, then a thicker coating must be applied to allow for machining, grinding, and/or polishing to the required degree.

Reference

1 ASTM (2017). *A967/A967M Standard Specification for Chemical Passivation Treatments for Stainless Steel Parts*. ASTM International, West Conshohocken, PA

11

Handling and Transportation

11.1 Introduction

Handling of vessels and vessel components at the fabrication plant and transporting them to their final destination requires a fair amount of planning. Some of the logistical problems encountered are discussed in this chapter.

11.2 Handling of Vessels and Vessel Components Within the Fabrication Plant

Moving a vessel or components of a vessel within the plant requires cranes when the movement is within the bay or dollies and trucks if the movement is from one bay to another. In any of these operations, an overhead crane is most likely involved. A typical setup is shown in Figure 11.1. Two vertical cables are slung symmetrically around the vessel and attached to a spreader beam. The beam is attached to the hook of an overhead crane by two slanted cables as shown. In most cases, the two vertical cables are kept equally loaded and symmetrically attached to the spreader beam.

- In some cases, the vertical cables are not equally loaded. This occurs when part of the vessel is heavier than the other part such as in the case of U-tube heat exchangers, vessels with partial jackets, and vessels with large nozzles toward one end of the vessel. In such cases, careful consideration is given to the attachment of the vertical cables to the spreader beam as illustrated in Example 11.1.

Example 11.1
The U-tube heat exchanger is supported by two vertical cables as shown in Figure 11.2. A 20 ft spreader beam is used to lift the exchanger. The hook cables are at 45° angle. Determine the best location for attaching the vertical cables to the spreader beam.

Solution

Case 1. Vertical cables are equidistant from the beam centerline
From Figure 11.3(a), the forces in the hook cables are

$F_{1V} = 14.4$ kips	$F_{1H} = 14.4$ kips	$F_1 = 20.3$ kips
$F_{2V} = 10.6$ kips	$F_{2H} = 10.6$ kips	$F_2 = 14.8$ kips

Fabrication of Metallic Pressure Vessels, First Edition. Owen R. Greulich and Maan H. Jawad.

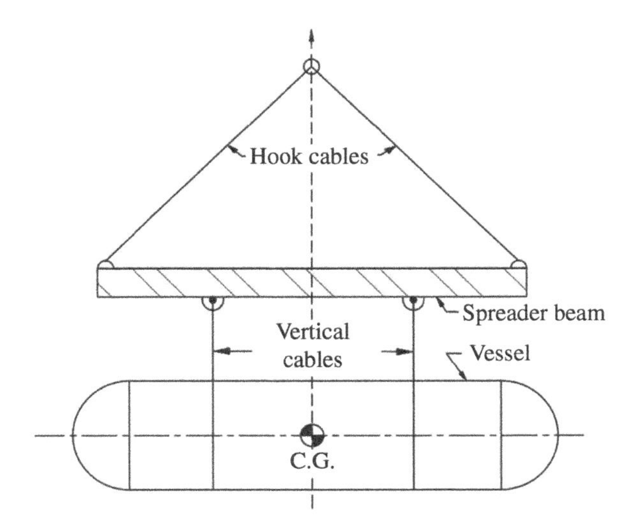

Figure 11.1 Lifting a pressure vessel using a spreader beam

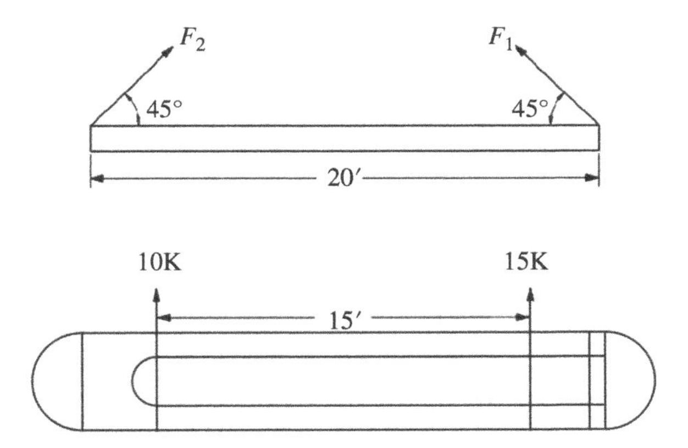

Figure 11.2 Lifting a U-tube heat exchanger

The analysis indicates that the horizontal forces in the hook cables result in an unbalanced force of 3.8 kips (14.4 – 10.6). Since there is no way of reacting this unbalanced load, the beam would tilt until the centroid of the load was located below the crane hook, thereby balancing the horizontal reactions. This, of course, also tilts the vessel, a condition that is normally unacceptable. Since this would happen anyway, a better design is to locate the centroid of the vessel at the centerline of the beam.

Case 2. Centroid of vertical cables at the beam centerline.

From Figure 11.3(b), the forces in the hook cables are

$F_{1V} = 12.5$ kips	$F_{1H} = 12.5$ kips	$F_1 = 17.7$ kips
$F_{2V} = 12.5$ kips	$F_{2H} = 12.5$ kips	$F_2 = 17.7$ kips

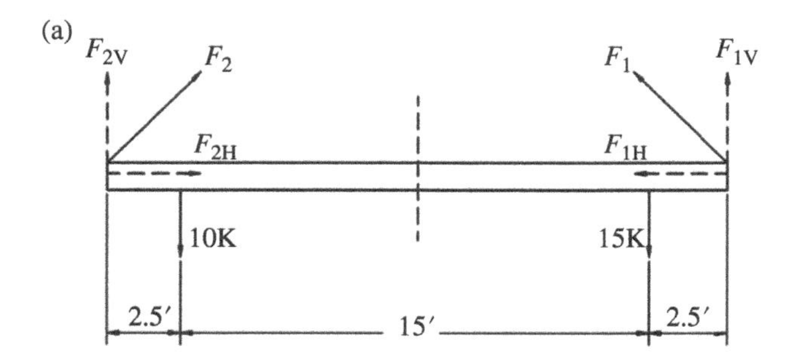

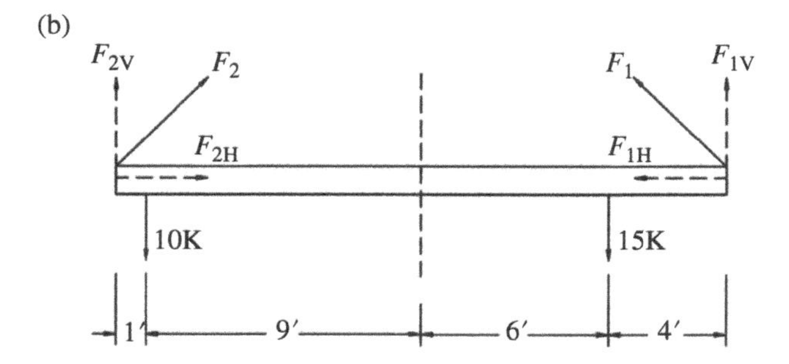

Figure 11.3 Attachment of cables to spreader beam. (a) Applied loads are equidistant from centerline of beam. (b) Applied loads are not equidistant from centerline of beam

This case results in a zero unbalanced horizontal force in the overhead crane and a maximum load on both hook cables of 17.7 kips.

This example shows the need to attach the vertical cables at various locations along the spreader beam. Accordingly, spreader beams are usually designed with multiple pick-up points as shown in Figure 11.4.

Occasions arise where the weight of a vessel is greater than the overhead crane capacity. In such cases, the manufacturer may choose to lift one end of the vessel at a time and crib underneath the vessel so that a dolly or truck can be inserted underneath it for moving. The location of the cribbing underneath the vessel has to be determined accurately in order not to exceed the crane capacity.

11.3 Transportation of Standard Loads

The vast majority of pressure vessels can be shipped as standard loads on trucks. This has several advantages. No special permits are required, cost is reasonable, and shipping times are typically much less than for rail shipments. On-time delivery still usually requires planning, unless the product is so small that it can be handled by a package delivery service such as UPS or FedEx.

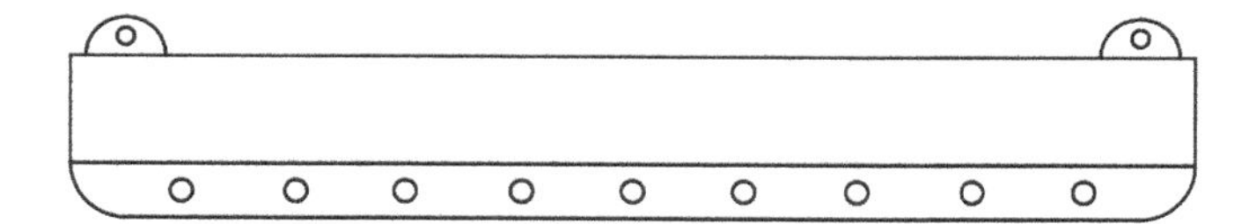

Figure 11.4 Typical spreader beam

Figure 11.5 Light load on an exclusive truck

Some vessels come near to the capacity of a truck due to their size or their weight. If this happens, it is usual to ship on an "exclusive" truck, Figure 11.5. This means that no other products will be loaded on the truck and it can be expected to leave the facility of the shipper and drive essentially directly to the facility of the purchaser. This option is attractive even for smaller loads, if either the cost of the vessel or the criticality of schedule warrants it. Planning ahead and making arrangements with the trucker ensures on-time pick-up and delivery.

Otherwise, smaller loads are frequently shipped "less than truckload," or "LTL." The product is packaged to protect it during the journey, typically by placing it on a pallet. Usually, when this is done, it will be picked up by the trucking company and routed to a local trucking depot or distribution center. There, if the pressure vessel is being shipped some distance, it will be loaded onto a long-haul truck with products from other shippers going the same direction, and brought to another trucking depot nearer the final destination. At this point, it may either go directly to the purchaser, if sufficiently close, or it may ship to another depot that is closer to the final destination. Of course, such shipments take longer than using an exclusive truck, but even allowing for one or more transfers on the way, delivery within a couple of weeks is usual. With careful coordination by truckers, even cross-country delivery within a week is feasible.

Sometimes tight schedules require air shipment. A small pressure vessel might be shipped using a package delivery service, but such services typically have weight limits of about 70 lb. Major airlines often carry freight on commercial passenger flights. If the shipment is delivered to the air carrier at the airport, weight limits are often significantly higher – in hundreds of pounds or more – than those for package delivery services, and shipments can be planned based on regular passenger flights. Such shipments can normally be coordinated either directly with the airline or with freight forwarders who specialize in such coordination and in finding advantageous costs.

Larger shipments can be coordinated with airfreight companies and airfreight forwarders. The larger the shipment, the more important advance coordination becomes.

In all cases, packaging is an important consideration, and the choice of carrier is sometimes made by selecting the carrier whose service permits the least expensive packaging. How a vessel will be secured is critical. Chains or straps or lugs attached to the vessel or the vessel supports may be used to resist sliding and overturning forces. The addition of several inexpensive lugs at the design stage may more than pay for itself when it is time to ship.

In any case, consideration of shipping, and shipping costs, should be considered at the bid stage on a project. In addition to the cost of shipping, the cost of pallets and cribbing can easily run into hundreds or thousands of dollars. While in most cases failure to include such costs in a bid is not an insurmountable error, it may make the difference between a profitable and an unprofitable project.

11.4 Transportation of Heavy Vessels

Moving oversize or heavy loads requires special considerations by the fabricator. Means of handling and transportation are as follows:

1) Cranes.
2) Trucks.
3) Railroad cars.
4) Barges and ships.
5) Airplanes.

Planning for shipping a vessel using one of the above five methods starts at the bidding stage. Factors such as diameter, thickness, length, weight, center of gravity, geometric configuration, how the product will be lifted, and transport permits have to be considered at the beginning of the project.

The cost of shipping oversize or overweight loads is often disproportionate to the weight because of the need for special equipment, special routing, or limitations on time of day or week when shipping is permitted by jurisdictions. The cost of a permit is not typically significant, but the cost of compliance with permit requirements often is.

Some of the requirements needed for each of these transport methods are discussed in this chapter.

11.4.1 Handling heavy vessels using specialty cranes

Specialty cranes, besides the overhead cranes commonly used in fabrication, are frequently needed to move heavy loads within the fabricator's facility or at the user plant. Figure 11.6 shows a stationary crane at a river dock of a fabricator used to load vessels on barges.

Specialty mobile cranes, Figure 11.7, lifting up to 1000 ton (900 tonne) loads or more can be leased to lift heavy weights at refineries and chemical plants during turnarounds or shutdowns. The load lift points become extremely critical and need to be thoroughly evaluated prior to lifts.

11.4.2 Shipping by truck

In the United States, the maximum length of a standard flat-bed trailer varies by state between 48 and 53 ft (15 and 16 m). The maximum standard width set by federal regulations is 8.5 ft (2.6 m). The dead weight of a standard semitruck and trailer is usually kept below 35,000 lb (16,000 kg) and the maximum gross weight of a loaded standard semitruck is 80,000 lb (36,000 kg). This leaves about 45,000 lb (20,000 kg) for carrying live loads in order to stay below the 80,000 lb (36,000 kg) maximum limit. In addition, certain bridges have limits that may not be exceeded, and there are often axle weight limits. Dimensions and weights beyond these typically require a permit

Figure 11.6 A 300 ton (270 tonne) stationary crane (*Source:* Nooter Construction)

Figure 11.7 Heavy-lift crane in the field (*Source:* Nooter Construction)

issued by the state department of transportation. A "permit load" traversing a number of states will often require a number of permits, each issued by the respective state.

Hence, a fabricator may more easily ship a vessel by truck when the length is not greater than 53 ft (16 m), the width is less than 8.5 ft (2.6 m), and the weight does not exceed about 45,000 lb (20,000 kg). This results in a maximum cylindrical shell thickness of about 0.70 in. (18 mm). A greater thickness will require adjustments in the diameter and/or the length. The following approximate equation gives the weight versus dimensions for a cylindrical vessel with spherical heads:

$$W = (\pi/12)\left[(D)(L)(t_s) + \left(D^2\right)(t_h)\right](490)(\gamma) \tag{11.1}$$

$$W \leq 45,000\ \text{lb} \tag{11.2}$$

where

D = diameter of vessel, ft
γ = 1.0 for carbon and stainless steel
= 1.14 for copper and copper alloys
= 0.35 for aluminum and aluminum alloys
= 0.58 for titanium and titanium alloys
L = length of shell, ft
t_h = thickness of head, inches
t_s = thickness of shell, inches
W = weight of vessel

Example 11.2

What is the maximum length of a steel pressure vessel that can be shipped on a standard trailer? The vessel dimensions are $D = 6$ ft, $t_h = 0.75$ in., and $t_s = 1.25$ in.

Solution

From Eqs. (11.1) and (11.2),

$$45,000 = (\pi/12)\left[(6.0)(L)(1.25) + \left(6.0^2\right)(0.75)\right](490)(1.0)$$

Solving this equation for L results in

$$L = 43.0\ \text{ft}.$$

When the weight and/or dimensions of a vessel exceed the standard limits, then special trailers are used. The trailers can be small such as those shown in Figure 11.5 or fairly sophisticated such as those shown in Figures 11.8–11.10. The use of these specialty trailers requires calculations to determine the correct weight distribution on the axles. Also, and as shown in Figure 11.10, these trailers are often fairly wide and require special permits to operate on the interstate highways and local state and city roads. Special routing may be required and particular daytime or nighttime hours may be allocated. Attention must be given to bridge clearances, overhead power lines, maximum weight on bridges, and turning radii. Depending on the size, escort vehicles are also required to accompany the load.

Permit costs for oversize or overweight loads are not typically excessive, but cost is driven up by the high cost of special equipment, limitations on time and day when large loads may be shipped,

Figure 11.8 Specialized transport (*Source:* Nooter Construction)

Figure 11.9 Sophisticated dolly for transporting large cylindrical shells (*Source:* NASA)

Figure 11.10 Heavy load trailers (*Source:* Bigge)

the cost of pilot cars, and the cost of finding and following routes with sufficient overhead and turning clearance. Processing of "standard" permits up to a certain width and length may be fairly routine, but highly unique, extremely large, or extremely heavy loads can require weeks or more.

11.4.3 Shipping by rail

Shipping by rail is economical when both the fabricator and user have access to railroad tracks. Standard specialized flat beds operated by the major railroad companies in the United States come in lengths of 60 and 89 ft (18 and 27 m) with a width of about 9.5 ft (2.9 m). Load capacity ranges between 145,000 and 160,000 lb (66,000 and 73,000 kg), much greater than that of a standard truck trailer. This load capacity, however, may substantially be reduced if the route includes bridges that cannot support such weights. Rerouting may be necessary with associated additional costs.

There are numerous heavy weight flat-bed cars owned by major railroad companies as well as by private companies that can carry heavier loads than the standard flat beds. Many of these heavy weight flat beds are available for lease. Large widths and heights may be carried on routes that are not limited by tunnels, overpasses, or turns.

Many of the loads shipped by rail tend to be long and overhang beyond the car as shown in Figure 11.11. In such cases, idler cars are placed behind or in front of the vessel car to provide sufficient length during shipment.

For very long vessels, three railroad cars are used as shown in Figures 11.12 and 11.13. The two end cars have saddles for supporting the vessel while the middle (idler) car provides needed length. One of the saddles is designed to swivel, while the other is designed to slide and swivel.

This loading condition requires special attention when the train is going through a curve with a bridge or other width restriction on it. This is illustrated in Figure 11.13 where the train cars tend to follow the curve centerline of the track while the vessel follows the chord between the saddles. The distance H between the arc and the chord needs to be determined and kept below the actual measured clearance between the track and any obstructions, including trains on adjacent tracks.

The amount of swivel of the saddles considered in the design depends on the length and number of the railroad cars and the radius of the curve. In the United States, the minimum track radius for

Figure 11.11 Idler flat railroad car behind a vessel (*Source:* Nooter Construction)

Figure 11.12 A long load requiring three railroad cars (*Source:* Nooter Construction)

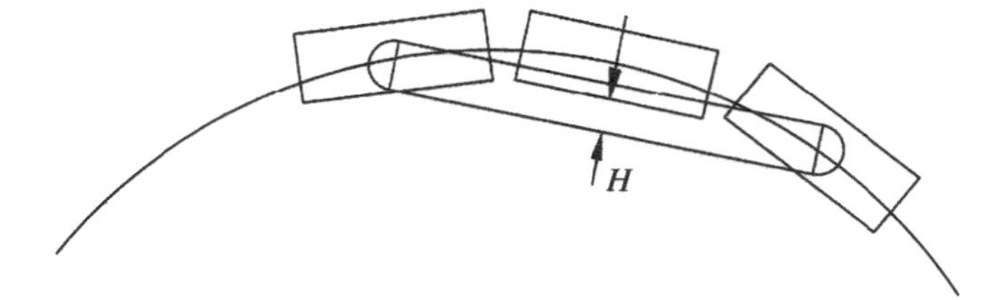

Figure 11.13 Position of railroad cars and vessel along a curve

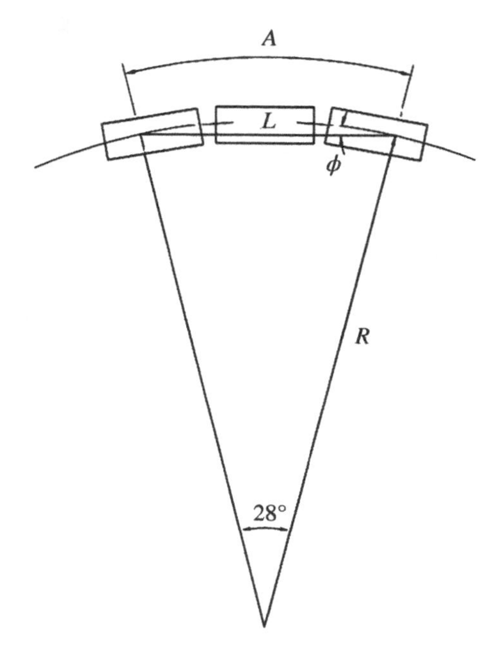

Figure 11.14 Swivel angle and saddle motion

commercial trains on major railroads is $R = 574$ ft (175 m). The approximate swivel angle, Figure 11.14, is

$$\theta = \sin^{-1}\left(\frac{L}{2R}\right) \tag{11.3}$$

where

L = distance between saddles, ft
θ = swivel angle, degrees
R = radius of track curve, ft

The approximate sliding distance, Δ, of the movable saddle to be considered in the design is

$$\Delta = A - L \tag{11.4}$$

where

Δ = sliding distance of movable saddle
A = arc length = $(\theta)(R)(\pi/90)$

Example 11.3

A pressure vessel is transported using a three railroad car system as shown in Figure 11.12. Each car is 89 ft long. The saddles are stationed in the middle of the leading and trailing cars. The distance between the saddles is $L = [(89/2) + 89 + (89/2)] = 178$ ft. The train track has a curve of $R = 574$ ft. Determine the magnitude of the swivel angle needed to be incorporated in the design of both saddles and the amount of movement incorporated in the design of the sliding saddle.

Solution

From Eq. (11.3), the swivel angles are

$$\theta = \sin^{-1}\left(\frac{178}{2(574)}\right)$$
$$= 8.92^\circ$$

From Eq. (11.4), the approximate total movement of the sliding saddle is

$$\Delta = \frac{(8.92)(574)(\pi)}{90} - 178$$
$$= 0.722 \text{ ft} = 8.66 \text{ in.}$$

Longitudinal, vertical, and transverse acceleration and deceleration loads are considered in the design of the straps and tie-downs. Figure 11.15 shows straps and rods used for tie-down, while in Figure 11.16 a cradle is used to support the vessel. Special attention must be given to vessel tie-downs when shipping by rail due to the practice of "humping." Humping refers to a process used in the railyard to sort cars to make up trains. It involves sending cars down a slope (one side of the "hump") to give them speed to roll through the switch to the required location in a train. It sometimes results in a significant jolt as the humped car bumps into the rest of the train.

Figure 11.15 Strap and rod tie-downs (*Source:* Nooter Construction)

Figure 11.16 Cradle support (*Source:* Nooter Construction)

Shipping of partial components, Figure 11.17, by train or truck requires special attention to the center of gravity of the load as well as tie-downs and supports to prevent tipping of the component or overturning of the flat bed.

11.4.4 Shipping by barge or ship

Barges can carry massive loads. Standard river barges in the United States are 195 ft (59 m) long by 35 ft (11 m) wide with a load capacity of 3,000,000 lb (1,400,000 kg). Newer barges of 209 ft (64 m) long by 50 ft (15 m) wide are being constructed. These will be able to carry up to 6,000,000 lb (2,700,000 kg).

The load can be driven onto the barge, Figure 11.18, or loaded by a crane, Figure 11.6. If the load is driven on, then provisions are made to prevent the front of the barge from tipping down as the load is moved on. This is normally done by having a tug boat push the barge against the shore allowing the front rake of the barge to rest on solid ground during loading to avoid excessive downward motion of the front of the barge.

Because of their size and high load capacity, barges can often accommodate multiple loads as shown in Figure 11.19. The additional draft of the barge due to the mass of the cargo is approximated by the equation

$$h = \frac{F}{62.4(\lambda)(L)(W)} \tag{11.5}$$

where

F = applied load, lb
h = additional barge draft due to applied load, ft
  ≈ 1.0 for fresh water
  ≈ 1.025 for salt water

Figure 11.17 Shipping partial components by rail (*Source:* Nooter Corporation)

Figure 11.18 Drive on barge load (*Source:* Nooter Construction)

Figure 11.19 Multiple loads on a barge (*Source:* Nooter Construction)

L = length of barge at waterline, ft
W = width of barge, ft

Water ballast is normally used in selected compartments of the barge to level it in the event that the center of gravity of the load is placed off-center. Large ocean-going barges normally have Plimsoll, or load limit, marks, Figure 11.20, along both sides of the barge to verify level and to prevent overloading. The L-R symbol in Figure 11.20 indicates the barge certifying agency, in this case Lloyd's Register. The horizontal line in the circle is for checking the top deck level of the barge.

Example 11.4

Two pressure vessels, each weighing 478,000 lb, are loaded on a river barge. The barge width is 35 ft and the length is 175 ft. What is the approximate additional draft of the barge due to the load?

Solution

From Eq. (11.5),

$$h = \frac{(2)(478,000)}{62.4(1.0)(175)(35)}$$

$$h \approx 2.5 \text{ ft.}$$

Vessels are also transported by ships. The shallow draft of a barge compared to its width limits its tendency to rock. On a ship, tie-downs become an important consideration due to the possible rolling action of the ship caused by ocean waves. Weight and size limits of pressure vessels are normally not a consideration aboard ships.

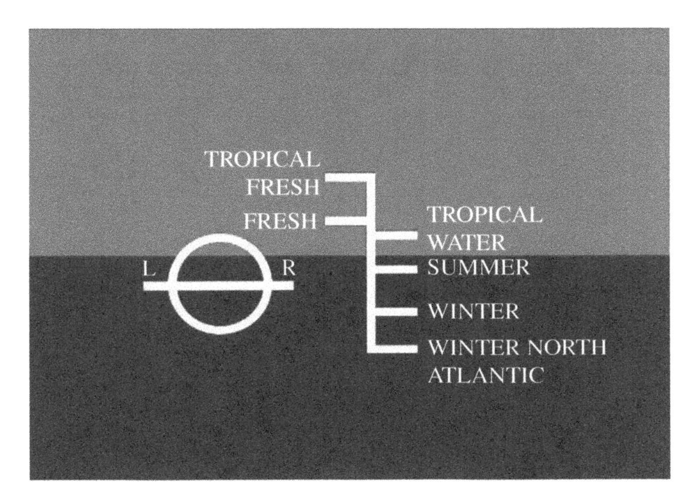

Figure 11.20 Plimsoll marks

Figure 11.21 Heavy vessel being loaded onto an Antonov An-124 aircraft (*Source:* Nooter Construction)

11.4.5 Shipping by air

Pressure vessels occasionally have to be shipped by air when a refinery, chemical plant, or other facility has an urgent need for the vessel and shipping cost becomes secondary. For heavy pressure vessel loads, the most commonly used commercial plane is the Antonov An-124, Figure 11.21. The payload capacity for this airplane is around 340,000 lb (154,000 kg). Internal cargo dimensions are approximately 21 ft wide × 14 ft high × 119 ft long (6.4 × 4.3 × 36 m).

Special consideration must be given to tie-downs when shipping by plane. The pressure vessel designer needs to know the airplane tie-down locations and capacity to make provisions for them at the design stage.

11.5 Summary

A number of choices are often available for shipping, even for large loads. Costs of the various options can vary significantly, as can delivery times and the required preparation for the product being shipped. While the purchaser specifies shipping in some cases, it is incumbent on the fabricator to be aware of the various possibilities and their relative costs, and to plan for these up front. A good shipping choice made and planned for in advance of the project may make the difference between receiving and not receiving a contract or between a profit and a loss on a contract received.

12

ASME Code Compliance and Quality Control System

12.1 Need for ASME Code Compliance

Over 100 years ago, a long history of failed boiler and pressure vessels led to the development of the ASME Boiler and Pressure Vessel Code. Following the creation of this code, there was a precipitous drop in boiler failures. As a result of this history, the ASME BPVC is enshrined in laws or regulations in almost all 50 states, plus the provinces of Canada, as well as in the regulations of the United States Occupational Safety and Health Administration (OSHA).

In spite of this history and the legal requirements, pressure vessels not constructed in accordance with this standard continue to fail, though with much less regularity.

It is well understood that no one constructs a vessel with the intent of having it fail. Causes of failure are typically lack of adequate design, failure to construct the vessel properly, or misapplication of the vessel. Compliance with the code ensures that these issues have been addressed effectively.

12.2 What the ASME Code Provides

The ASME BPVC is a comprehensive standard providing design, fabrication, test, and inspection guidance and requirements. In addition, it provides qualification requirements for material, to ensure that a good design is not implemented using less than adequate material. And finally, it has requirements for relief devices to ensure that a vessel is not over-pressurized. These, together with the conformity assessment requirements described in the next section, ensure that a vessel is safe for its intended application.

12.3 Fabrication in Accordance with the ASME Code

Fabricators intent on building pressure vessels and other equipment in accordance with the requirements of the ASME code must have three items in place:

1) A Quality Control System, QC, approved by ASME demonstrating the ability of the fabricator to follow the ASME rules. This is described in Section 12.4.
2) An independent Authorized Inspector, AI, who usually works for an insurance company not associated with the fabricator. The inspector reviews the fabrication documents and the vessel fabrication to ascertain compliance with the ASME code. Section 12.5 describes the function of the AI.
3) An ASME code stamp for stamping vessels indicating they meet the code requirement. In addition to the stamping, the fabricator must furnish specific fabrication documents as outlined in Section 12.4.

Fabrication of Metallic Pressure Vessels, First Edition. Owen R. Greulich and Maan H. Jawad.

Fabricators may need to provide additional stamps to an ASME vessel to meet various customer and jurisdictional requirements. Some of these stamps are discussed in Section 12.7.

12.4 ASME Code Stamping

Pressure vessels built in accordance with the ASME Pressure Vessel Code, Section VIII Division 1, must comply with many requirements such as those regarding design, materials, fabrication, and inspection. The following are some of the documents generated during fabrication and that the fabricator is obligated to keep for a minimum of five years after fabrication of the vessel. Some users request copies of these documents from the fabricator to store at their facilities.

1) Design calculations
2) Fabrication drawings
3) Material mill test reports
4) WPS for the vessel welds
5) Records of radiographic or ultrasonic examination
6) Record of PWHT
7) Record of hydrotesting
8) Manufacturer's Data Report, U-1 form
9) Manufacturer's Partial Data Report, U-2 form
10) Facsimile of name plate

12.4.1 Design calculations

The design calculations show the minimum required thicknesses and configurations needed for various components of the pressure vessel. These calculations are normally performed either in-house by the fabricator or by an engineering firm on behalf of the fabricator. On some occasions, the calculations are produced by the customer/user. If the customer/user produces them, they need to be checked and accepted by the fabricator. The design calculations form the basis for constructing the pressure vessel and are an essential part of the vessel documentation.

12.4.2 Fabrication drawings

The fabrication drawings form the road map for fabricating the pressure vessel. They show dimensions and orientation details of various components. They also list materials of construction and other information such as pressure, temperature, and some required nondestructive examination.

12.4.3 Material mill test reports

All materials associated with the pressure boundary of the vessel are supplied with mill test reports. These reports, Table 2.23, give the chemical composition of the material and physical properties such as tensile strength, yield stress, and elongation. Other information may also be in the mill test reports such as Charpy impact values and nondestructive examination.

There are instances in which the fabricator needs to use a small piece of material for which the mill test report is lost or not available. In such cases, the fabricator may generate its own test report if it has a qualified laboratory on-site or access to such a laboratory.

12.4.4 WPS for the vessel welds

All welds in a vessel pertaining to the pressure boundary are required to be performed in accordance with a WPS as described in Chapter 8. Records of these procedures are maintained by the manufacturer as part of the vessel documentation.

12.4.5 Records of nondestructive (NDE) examination

Records of all NDE examinations are maintained by the fabricator. These records are either digital or on film or paper. Some of the NDE methods used are given as follows:

1) Radiographic (RT). In this method a radioactive source is used to take a picture, either with film or digitally, of the weld volume. RT is the main method used in the United States for checking the integrity of welded joints. Portable RT machines can be used for thin welds. However, stationary machines or radioisotopes such as iridium or cobalt in thick-walled rooms or large access-restricted areas are needed for thick welds.
2) Ultrasonic (UT). This method is normally used when RT is difficult to use due to the location or configuration of the weld. The size and angle of the probes have to be matched with the thickness and configuration of the weld. Conventional UT inspections are recorded on reader sheets, but more current equipment, especially that for phased array ultrasonic testing (PAUT), provides a visual plot of the data that can be revisited for reinterpretation if needed.
3) Liquid penetrant (PT). This test is used to detect surface cracks in ferrous and nonferrous alloys. After the surface is cleaned to ensure that no extraneous materials or oils might mask a defect, a penetrant dye is applied to the surface to be checked. After allowing time for the dye to flow into any crevices, excess dye is wiped off and a white developer is applied. Time is then allowed for the dye to bleed back out into the white developer, providing vivid indications of any rejectable indications. Sometimes fluorescent penetrant is used, with viewing under ultraviolet light. This is typically a more sensitive inspection.
4) Magnetic particle (MT). This method is used for ferrous alloys. Iron powder is applied on the surface and then subjected to a magnetic field. Any indications on the surface will result in disruption of the magnetic lines, causing the iron powder to cluster in those areas. As with PT, fluorescent materials are sometimes added, usually in a liquid, for greater sensitivity.
5) Visual (VT). VT is essential in assuring overall integrity of fabrication. Items such as gaps, weld contour, and surface finishes can only be evaluated by visual inspection.
6) Dimensional inspection. Dimensional inspections verify accuracy of critical items such as roundness and other physical characteristics such as diameter, length, nozzle locations, alignments, and orientations of flange surfaces and bolt holes.
7) Leak testing (LT). Leak testing is used to assure integrity of gaskets and other connections and to identify defects such as pin holes in welded joints.
8) Bubble test. This test is used in components subjected to pressure from gas. A solution, or immersion, is used on the surface to check for bubbles in case of a leak.

12.4.6 Record of PWHT

Records of any PWHT done on any part are maintained. They consist of charts or digital output showing ramping-up and ramping-down of temperature and the actual time and temperature maintained during PWHT.

12.4.7 Record of hydrotesting

These records show the pressure cycles during hydrotesting or pneumatic testing of the pressure vessel. Included in the record is the hold time at various pressures.

12.4.8 Manufacturer's Data Report, U-1 Form

The ASME U-1 form is probably the most important document for a pressure vessel. It provides a summary of critical information regarding the vessel. It lists important vessel information such as orientation, thicknesses, and number of nozzles including their size and location. It lists major components of the vessel as well as other important information such as design temperature and pressure. Table 12.1 shows items that ASME [1] requires to be entered on a U-1 form. The form is signed by an independent Authorized Inspector, as described in Section 12.5.

Table 12.1 Manufacturer's Data Report, U-1 Form

Page ______ of ______

FORM U-1 MANUFACTURER'S DATA REPORT FOR PRESSURE VESSELS
As Required by the Provisions of the ASME Boiler and Pressure Vessel Code Rules, Section VIII, Division 1

1. Manufactured and certified by ______ (1) ______ (Name and address of Manufacturer)
2. Manufactured for ______ (2) ______ (Name and address of Purchaser)
3. Location of installation ______ (3) ______ (Name and address)
4. Type ______ (4) ______ (Horizontal, vertical, or sphere) ______ (5) ______ (Tank, separator, jkt. vessel, heat exch., etc.) ______ (8) ______ (Manufacturer's serial number)
______ (9) ______ (CRN) ______ (10) ______ (Drawing number) ______ (12) ______ (National Board number) ______ (Year built)
5. ASME Code, Section VIII, Div. 1 ______ (13) ______ [Edition and Addenda, if applicable (date)] ______ (14) ______ (Code Case number) ______ (15) ______ [Special service per UG-120(d)]

Items 6–11 incl. to be completed for single wall vessels, jackets of jacketed vessels, shell of heat exchangers, or chamber of multichamber vessels.

6. Shell: (a) Number of course(s) ______ (16) ______ (b) Overall length ______ (17) ______

Course(s)			Material	Thickness		Long. Joint (Cat. A)			Circum. Joint (Cat. A, B & C)			Heat Treatment	
No.	Diameter	Length	Spec./Grade or Type	Nom.	Corr.	Type	Full, Spot, None	Eff.	Type	Full, Spot, None	Eff.	Temp.	Time
	(18)	(19)	(20)	(21)	(22)	(23)	(24)		(25)	(26)		(27)	

Body Flanges on Shells												
									Bolting			
No.	Type	ID	OD	Flange Thk	Min Hub Thk	Material	How Attached	Location	Num & Size	Bolting Material	Washer (OD, ID, thk)	Washer Material
						(20)			(32)	(32)		

7. Heads: (a) ______ (20) ______ (27) ______ (Material spec. number, grade or type) (H.T. — time and temp.) (b) ______ (Material spec. number, grade or type) (H.T. — time and temp.)

	Location (Top, Bottom, Ends)	Thickness		Radius		Elliptical Ratio	Conical Apex Angle	Hemispherical Radius	Flat Diameter	Side to Pressure		Category A		
		Min.	Corr.	Crown	Knuckle					Convex	Concave	Type	Full, Spot, None	Eff.
(a)		(28)	(22)	(29)	(30)								(31)	
(b)														

Body Flanges on Heads												
									Bolting			
	Location	Type	ID	OD	Flange Thk	Min Hub Thk	Material	How Attached	Num & Size	Bolting Material	Washer (OD, ID, thk)	Washer Material
(a)												
(b)							(20)		(32)	(32)		

8. Type of jacket ______ (33) ______ Jacket closure ______ (34) ______ (Describe as ogee and weld, bar, etc.)
If bar, give dimensions; if bolted, describe or sketch ______
9. MAWP ______ (35) ______ (Internal) ______ (External) at max. temp. ______ (36) ______ (Internal) ______ (External) Min. design metal temp. ______ (37) ______ at ______.
10. Impact test ______ (38) ______ [Indicate *yes* or *no* and the component(s) impact tested] at test temperature of ______ (38) ______.
11. Hydro., pneu., or comb. test pressure ______ (39) ______ Proof test ______ (40) ______

Items 12 and 13 to be completed for tube sections.

12. Tubesheet ______ (20) ______ [Stationary (material spec. no.)] ______ (18) ______ [Diameter (subject to press.)] ______ (21) ______ (Nominal thickness) ______ (22) ______ (Corr. allow.) ______ [Attachment (welded or bolted)]
______ (20) ______ [Floating (material spec. no.)] ______ (18) ______ (Diameter) ______ (21) ______ (Nominal thickness) ______ (22) ______ (Corr. allow.) ______ (Attachment)
13. Tubes ______ (20) ______ (Material spec. no., grade or type) ______ (O.D.) ______ (Nominal thickness) ______ (Number) ______ [Type (straight or U)]

(07/17)

Table 12.1 (Continued)

FORM U-1

Page ______ of ______

Manufactured by ______ (68)

Manufacturer's Serial No. ______ (8) ______ CRN ______ (9) ______ National Board No. ______ (12) ______

Items 14–18 incl. to be completed for inner chambers of jacketed vessels or channels of heat exchangers.

14. Shell: (a) No. of course(s) ______ (b) Overall length ______

Course(s)			Material	Thickness		Long. Joint (Cat. A)			Circum. Joint (Cat. A, B & C)			Heat Treatment	
No.	Diameter	Length	Spec./Grade or Type	Nom.	Corr.	Type	Full, Spot, None	Eff.	Type	Full, Spot, None	Eff.	Temp.	Time

Body Flanges on Shells

No.	Type	ID	OD	Flange Thk	Min Hub Thk	Material	How Attached	Location	Bolting: Num & Size	Bolting: Bolting Material	Bolting: Washer (OD, ID, thk)	Bolting: Washer Material
						(20)			(32)	(32)		

15. Heads: (a) ______ (b) ______
(Material spec. number, grade or type) (H.T. — time and temp.) (Material spec. number, grade or type) (H.T. — time and temp.)

	Location (Top, Bottom, Ends)	Thickness		Radius		Elliptical Ratio	Conical Apex Angle	Hemispherical Radius	Flat Diameter	Side to Pressure		Category A		
		Min.	Corr.	Crown	Knuckle					Convex	Concave	Type	Full, Spot, None	Eff.
(a)														
(b)														

Body Flanges on Heads

	Location	Type	ID	OD	Flange Thk	Min Hub Thk	Material	How Attached	Bolting: Num & Size	Bolting: Bolting Material	Bolting: Washer (OD, ID, thk)	Bolting: Washer Material
(a)							(20)		(32)	(32)		
(b)												

16. MAWP ______ (Internal) ______ (External) at max. temp. ______ (Internal) ______ (External) Min. design metal temp. ______ at ______ .

17. Impact test ______ [Indicate *yes* or *no* and the component(s) impact tested] at test temperature of ______ (38) ______ .

18. Hydro., pneu., or comb. test pressure ______ (39) ______ Proof test ______ (40) ______

19. Nozzles, inspection, and safety valve openings:

Purpose (Inlet, Outlet, Drain, etc.)	No.	Diameter or Size	Type	Material: Nozzle	Material: Flange	Nozzle Thickness: Nom.	Nozzle Thickness: Corr.	Reinforcement Material	Attachment Details: Nozzle	Attachment Details: Flange	Location (Insp. Open.)
				(20)	(20)						
(41)		(42)	(43)	(44)	(45)	(46)		(47)	(48) (49)	(48) (49)	(50)

20. Supports: Skirt (51) (Yes or no) Lugs (51) (Number) Legs (51) (Number) Others ______ (51) (Describe) Attached ______ (51) (Where and how)

21. Manufacturer's Partial Data Reports properly identified and signed by Commissioned Inspectors have been furnished for the following items of the report (list the name of part, item number, Manufacturer's name, and identifying number):
(52)
______

22. Remarks ______
(53)
______

(07/17)

Table 12.1 (Continued)

FORM U-1 Page ____ of ____

Manufactured by ____ (58) ____

Manufacturer's Serial No. ____ (8) ____ CRN ____ (9) ____ National Board No. ____ (12) ____

(59) CERTIFICATE OF SHOP COMPLIANCE

We certify that the statements in this report are correct and that all details of design, material, construction, and workmanship of this vessel conform to the ASME BOILER AND PRESSURE VESSEL CODE, Section VIII, Division 1.

U Certificate of Authorization number ____ Expires ____

Date ____ Name ____ (Manufacturer) Signed ____ (Representative)

(60) CERTIFICATE OF SHOP INSPECTION

I, the undersigned, holding a valid commission issued by the National Board of Boiler and Pressure Vessel Inspectors and employed by ____ of ____

have inspected the pressure vessel described in this Manufacturer's Data Report on ____, and state that, to the best of my knowledge and belief, the Manufacturer has constructed this pressure vessel in accordance with ASME BOILER AND PRESSURE VESSEL CODE, Section VIII, Division 1. By signing this certificate neither the Inspector nor his/her employer makes any warranty, expressed or implied, concerning the pressure vessel described in this Manufacturer's Data Report. Furthermore, neither the Inspector nor his/her employer shall be liable in any manner for any personal injury or property damage or a loss of any kind arising from or connected with this inspection.

Date ____ Signed ____ (60) (Authorized Inspector) Commissions ____ (61) (National Board Authorized Inspector Commission number)

(62) CERTIFICATE OF FIELD ASSEMBLY COMPLIANCE

We certify that the statements in this report are correct and that the field assembly construction of all parts of this vessel conforms with the requirements of ASME BOILER AND PRESSURE VESSEL CODE, Section VIII, Division 1. U Certificate of Authorization number ____ Expires ____.

Date ____ Name ____ (Assembler) Signed ____ (Representative)

(63) CERTIFICATE OF FIELD ASSEMBLY INSPECTION

I, the undersigned, holding a valid commission issued by the National Board of Boiler and Pressure Vessel Inspectors and employed by ____

of ____, have compared the statements in this Manufacturer's Data Report with the described pressure vessel and state that parts referred to as data items ____ (64) ____, not included in the certificate of shop inspection, have been inspected by me and to the best of my knowledge and belief, the Manufacturer has constructed and assembled this pressure vessel in accordance with the ASME BOILER AND PRESSURE VESSEL CODE, Section VIII, Division 1. The described vessel was inspected and subjected to a pressure test of ____. By signing this certificate neither the Inspector nor his/her employer makes any warranty, expressed or implied, concerning the pressure vessel described in this Manufacturer's Data Report. Furthermore, neither the Inspector nor his/her employer shall be liable in any manner for any personal injury or property damage or a loss of any kind arising from or connected with this inspection.

Date ____ Signed ____ (60) (Authorized Inspector) Commissions ____ (61) (National Board Authorized Inspector Commission number)

(07/17)

Source: American Society of Mechanical Engineers, 2021.

12.4.9 Manufacturer's Partial Data Report, U-2 form

Many manufacturers buy subassemblies from other manufacturers when building a pressure vessel, such as welded heads, tube-to-tubesheet subassemblies, and expansion joints. These subassemblies are furnished with a Manufacturer's Partial Data Report, U-2 form. The form is similar to the U-1 form but is more limited in scope since the part's manufacturer is not responsible for the completed vessel. The U-2 form(s) is attached to the U-1 form as a supplementary document.

12.4.10 Name plate

Every ASME code stamp vessel is required to have a stamping showing certain critical vessel information. This is most often accomplished by application of a name plate, though on a vessel of sufficient thickness it may be stamped directly on the shell or head. The stamping or name plate,

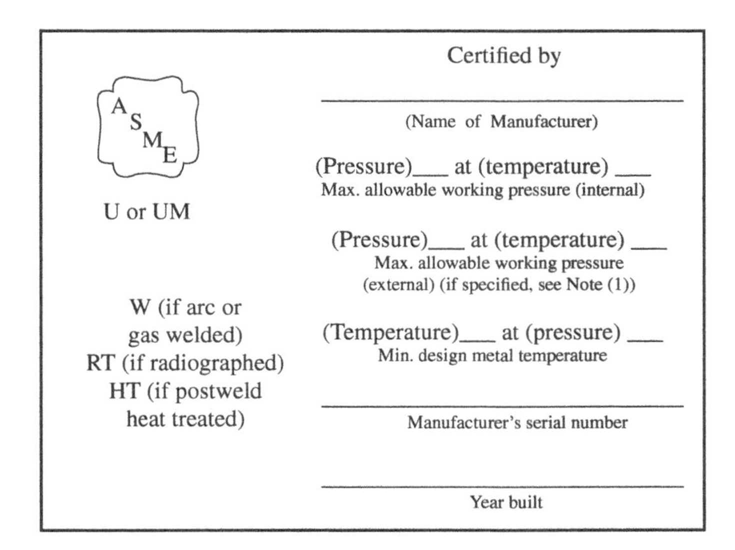

General note: Information within parentheses is not part of the required marking. Phrases identifying data may be abbreviated; minimum abbreviations shall be MAWP, MAEWP, MDMT, S/N, FV, and year respectively. See L-11 for sample Nameplate markings.

Notes:
(1) The maximum allowable external working pressure is required only when specified as a design condition.

Figure 12.1 Sample name plate (*Source:* American Society of Mechanical Engineers [1])

Figure 12.1, includes the ASME code stamp and in addition provides essential data such as the following:

1) Name of manufacturer
2) Maximum allowable working pressure and temperature
3) Maximum allowable external pressure, if applicable
4) Minimum design metal temperature (MDMT)
5) Serial number
6) Welding, PWHT and radiographic examination, if performed
7) Year built
8) National Board number, if applicable

Many manufacturers show additional information on the name plate. As an example, Figure 12.2 shows a name plate for a propane gas tank and lists the following information:

1) Name of manufacturer: American Welding & Tank
2) Location of various fabrication facilities of the manufacturer
3) National Board NB stamp indicating the vessel is registered with the National Board
4) ASME U stamp
5) W symbol indicating a welded vessel

6) RT4 indicating a partially radiographed vessel
7) Maximum allowable working pressure: MAWP = 250 psi
8) Maximum temperature: 400°F
9) Minimum design metal temperature, MDMT: −20°F at 250 psi
10) Vessel was fabricated at plant number 5
11) Canadian Registration Number CRN L-4709.5C. Vessels installed in a given Canadian Province must obtain a Canadian Registration Number, CRN and be approved by that province. The "C" at the end of this number indicates that this vessel has been approved by all provinces
12) Manufacturer's vessel serial number: 7SE 095282
13) Year built: 2009
14) Length: 119 in.
15) Outside diameter: OD 37.4 in.
16) Head thickness: 0.185 in.
17) Shell thickness: 0.218 in.
18) Above ground type: BS-AP
19) Surface area: 97.5 ft^2
20) Type of head: Hemi
21) Water capacity: 500 gallons
22) Underwriters Lab stamp, UL, with "Listed container assembly for LP gas 695A."
23) Statement regarding limitation on media content: "This container shall not contain a product having a vapor pressure in excess of 215 psi at 100°F"
24) Statement: "Dip tube length – 80% full at 40°F. D.T. = 10.7 in."

12.5 Authorized Inspector and Authorized Inspection Agency

Fabricators of ASME pressure vessels are required to have a contract with an Authorized Inspection Agency, AIA, to supply an Authorized Inspector, AI. The AIA can be an insurance company such as Hartford Steam Boiler Insurance Company, Onecis Insurance Company, or Factory Mutual Insurance Company or it can also be a jurisdiction willing to provide AIs. The AI provides numerous services at the fabricator's plant such as

1) Reviewing various documents associated with fabrication such as mill test reports, welding WPS and PQR records, radiographic inspection records, PWHT charts, and hydrostatic testing
2) Signing the U-1 form
3) Inspecting various pressure vessel components during the fabrication process
4) Having been assigned by ASME and functioning as a member of the team reviewing the fabricator's Quality Control System

12.6 Quality Control System for Fabrication

Manufacturers of pressure vessels built in accordance with the ASME code are required by ASME to be audited periodically to assure compliance with the ASME rules. Auditing is done by an ASME team or a team designated by ASME such as from the National Board. The Authorized Inspector is also included as a part of the team. Auditing is performed to assure the manufacturer has an effective Quality Control System in place for handling various aspects of manufacturing. The following are some of the items reviewed by the Auditing team.

12.6.1 Organizational chart

This chart shows the flow of line of authority and line of communications between the various managers responsible for fabrication. The managers usually identified include quality assurance, chief engineer, plant, production control, purchasing, and information systems.

12.6.2 Authority and responsibility

This section describes the responsibilities and authority of various personnel regarding the quality assurance and quality control within the organization as pertains to fabrication.

12.6.3 Quality control system

This part details the method of implementing the Quality Control, QC, System. It identifies the manager responsible for maintaining and updating the quality control system. It also details the distribution system of the QC document and its tracking method.

12.6.4 Design and drawing control

The QC system includes controls on the design calculations and generation of the drawings. The system includes a procedure for tracking revisions of the calculations and revisions and distribution of the drawings to various departments within the fabrication plant.

12.6.5 Material control

Material control requires material tracking procedures that are applicable prior to fabrication, during fabrication, and subsequent to fabrication. Controls prior to fabrication include material requisition, material quality verification upon receipt, and material stocking procedures. Controls during fabrication include usage of customer furnished material and usage of material from stock. Controls subsequent to fabrication address restocking of material remnants and matching mill test reports with the remnants.

12.6.6 Production control

This area includes the coordination between production and areas such as inspections, tests, and examination.

12.6.7 Inspection

Quality control of inspection involves in-house inspectors and the coordination between the fabricator and other inspectors such as the Authorized Inspector, AI, and the user inspector.

12.6.8 Hydrostatic and pneumatic testing

Controls for hydrostatic and pneumatic testing cover test equipment, pumps, hoses, valves, gauges, and recording equipment. Also included are such items as additional code nondestructive testing subsequent to hydrostatic and pneumatic testing.

12.6.9 Code stamping

Code stamping includes the area where all documents must be accounted for. These include gathering such items as mill test reports, PWHT charts, inspection reports, hydrostatic test records, and design calculations. The code stamp cannot be applied until all of these are in compliance with requirements.

12.6.10 Discrepancies and nonconformances

The quality system covers control and disposition of discrepancies and nonconformances. These include material, weld, and fabrication deficiencies, and what must be done in each case.

12.6.11 Welding

This section addresses the controls applicable to personnel responsible for the WPS, PQR, and WPQ of the welds in the vessel.

12.6.12 Nondestructive examination

This section defines the system for controlling nondestructive examination, NDE, operations.

12.6.13 Heat treatment control

Control of heat treatment is described in this section. It includes operator qualifications as well as controls on temperature and time measurements.

12.6.14 Calibration of measuring and test equipment

The controls in this section relate to the measuring and test equipment used by the fabricator as they relate to the QA system.

12.6.15 Records retention

This section relates to the types of documents to be stored and the retention period for each document.

12.6.16 Handling, storage, and shipping

Controls for material handling and storage locations at the fabricator facilities are described in this section. Shipping of completed vessels may also be included in this section where requirements such as those for tie-downs and shipping permits are detailed. Also included are the proper equipment used for shipping such as trucks, train cars, barges, or airplanes.

12.7 Additional Stamps Required for Pressure Vessels

The following stampings may be added to the basic ASME stamping requirements to comply with customer specifications or jurisdictional requirements.

12.7.1 National Board stamping, NB

Many jurisdictions require the pressure vessel to be registered with the National Board as well as ASME code stamped. The National Board of Boiler and Pressure Vessel Inspectors is a nonprofit organization comprised mainly of the Chief Inspectors of the various States and Jurisdictions in the United States and the Chief Inspectors of the various Canadian Provinces. The National Board provides various vital services to the pressure vessel industry such as

1) Maintaining U-1 data records of all boilers and pressure vessels registered with them including any alterations made to such vessels. These records can be obtained by any organization for a nominal fee. The National Board serves as a clearing house for these records when they are not available from the original fabricator or the user.
2) Serving as a Survey Team for qualifying fabricators.
3) Conducting various courses and seminars.
4) Qualifying authorized inspectors.
5) Certifying valves.
6) Maintaining copies of all previous editions of the ASME BPVC back to 1914.

Figure 12.2 shows an NB stamp at the top left-hand corner of the name plate.

12.7.2 Jurisdictional stamping

Many vessels require jurisdictional approval when their construction is not in strict accordance with the ASME code. A vessel constructed in accordance with a foreign code, or not to any code, will likewise require a special jurisdictional approval. In such situations, the Jurisdictional Board on Pressure Vessels will, if it approves the construction, require a name plate attached to the vessel with pertinent information.

12.7.3 User stamping

Users sometimes require an auxiliary name plate attached next to the ASME name plate showing various pieces of information regarding the operation of the pressure vessel.

12.7.4 Canadian Registration Numbers

A vessel shipped to Canada requires approval by the province where it will be used. When that approval is obtained, a Canadian Registration Number, CRN, is issued and must be shown on the name plate. Figure 12.2 shows such a CRN indicating the vessel may be shipped to Canada at a future date. The ".5" indicates that Ontario was the province of first approval, and the final "C" indicates that the vessel meets the requirements of all of the provinces.

12.8 Non-Code Jurisdictions

There are many jurisdictions around the world and a few jurisdictions in the United states such as the State of Texas and various governmental agencies that do not require vessels to be stamped with the ASME stamp. Users within these jurisdictions may specify process equipment built in accordance with the ASME code but not code stamped, or they may specify compliance with some other

Figure 12.2 Actual name plate

standard. This is usually done because the non-code vessel is lower in price. While a supplier may claim equivalence with code, significant compromises are typically made.

1) Welders may not be certified in accordance with Section IX of the ASME code, resulting in sub-par welding.
2) Material mill test reports may not be available.
3) Material properties may not be the same as those specified by the code.
4) Items such as minimum required manhole sizes and numbers may not be adhered to.
5) The design factor of safety and NDE testing may not be exactly the same as required by the ASME code.
6) Third-party inspections during fabrication may be reduced or nonexistent.
7) Following discovery of a defect during partial radiography, requirements for additional inspection may not be adhered to.

Users in jurisdictions not requiring ASME code stamping may need to specify additional requirements in their purchase specifications in order to avoid some of the above-listed issues. Additionally, third-party insurance companies may require users to purchase equipment with an ASME stamp even though the jurisdiction does not require it. And, finally, in the event of a vessel failure and lawsuits by injured parties, the user will most likely have a weaker defense in court since the primary standard for pressure vessels, almost universally accepted, has not been followed.

12.9 Temporary Shop Locations

Some fabricators use off-site fabrication facilities to reduce the cost of shipping large components or large pressure vessels from the fabrication plant to a distant job site. This can be justified when numerous parts or numerous pressure vessels are shipped to the same site. Temporary shop facilities may also be necessary when the job site is in a remote area where shipping of large components is not feasible. The temporary facility may consist of one of the following:

1) Renting or building a ware house and installing needed equipment for fabrication
2) Using the facility of an existing small manufacturing company and designating it, for a specific job and time period, as part of the parent company
3) Using indoor or outdoor space at or adjacent to the installation site

In order to qualify the fabrication in the temporary facility to the ASME code, the facility must be linked to the quality assurance program of the parent manufacturer. Detailed description is required on how to control welding PQRs and WPSs as well as other items such as PWHT and NDE. The program must also specify the technical ability of the supervisors needed for fabrication.

Reference

1 ASME. 2021. "*Boiler and Pressure Vessel Code, Rules for Construction of Pressure Vessels – Section VIII, Division 1*". American Society of Mechanical Engineers, New York.

13

Repair of Existing Equipment

13.1 Introduction

In-service pressure vessels inevitably require repairs or modification after operating for a time. Such repairs and alterations may consist of such items as

1) Overlaying corroded or eroded inside or outside surfaces
2) Repairing cracks that may occur in the welds and heat-affected zones
3) Adding nozzles or other items such as stiffening rings and lugs
4) Replacing portions of a vessel
5) Increasing or decreasing the length by adding or removing large sections of a vessel

Repairs of in-service pressure vessels are made in one of three ways depending on the extent of the repair and on various economic factors.

1) Repairs are made in the field after shutting down the vessel and isolating it from the system. Such repairs may require the use of cranes, scaffolding, weather protection enclosures, and portable equipment such as welding, drilling, and NDE machines. The cost of field labor is usually higher than that of shop labor. Weather may play a factor in the speed of the repair.
2) The vessel is shipped to a fabricator for repairs. This will require cranes and transportation systems such as on-the-road flat trailers, railroad cars, or barges. The cost of transporting the vessel back and forth must be weighed against repairing the vessel in the field.
3) Small nozzles are added in the field while the vessel is operating under restricted pressure and temperature. This operation is called hot tapping, and it requires fabricators familiar with this process and skilled operators.

Repairs of ASME code stamped vessels are done in accordance with the repair codes specified by the jurisdiction where the vessel is located or by the user in cases where there are no jurisdictions. Some of the repair codes used in the United States and Canada include the following:

1) National Board Inspection Code, NBIC, NB-23 [1]
2) ASME Post Construction Code, PCC-2 [2]
3) API 510 Pressure Vessel Inspection Code [3]
4) API 579 Fitness-For-Service Code [4]

This chapter briefly describes the fabrication requirements in these codes.

Fabrication of Metallic Pressure Vessels, First Edition. Owen R. Greulich and Maan H. Jawad.

13.2 National Board Inspection Code, NBIC, NB-23

The National Board Inspection Code, NB-23, instituted beginning in 1945, is the oldest and most widely used code for repairs of pressure vessels in the United States and Canada. The NB-23 is embedded in most State and other Jurisdictional laws and regulations for boilers and pressure vessels in the United States and in Provincial Laws in Canada. The National Board of Boiler and Pressure Vessel Inspectors is a nonprofit organization comprised mainly of the Chief Inspectors of the various States and Jurisdictions in the United States and the Chief Inspectors of the various Canadian Provinces. The NB-23 divides work affecting the structural integrity of pressure vessels into two categories: repairs and alterations. Alterations include reratings, which will be addressed separately here because they represent a category distinct from physical alterations.

13.2.1 Repairs

Repairs do not normally require the approval of the jurisdiction as long as the Authorized Inspector approves the repairs, but a record of the repairs (R1 Form) must be filed with the jurisdiction. Examples of repairs listed in NB-23 are as follows:

- Addition of welded attachments.
  - Studs for insulation or refractory lining.
  - Ladder clips.
- Corrosion resistant strip lining, or weld overlay.
- Weld buildup of wasted areas.
- Replacement of pressure-retaining parts identical to those existing on the pressure-retaining items such as.
  - Replacement of a shell or head.
  - Rewelding a circumferential or longitudinal seam in a shell or head.
  - Replacement of nozzles of a size where reinforcement is not a consideration.
- Installation of new nozzles where reinforcement is not a consideration.
- The addition of a nozzle where reinforcement is a consideration may be considered to be a repair provided the nozzle is identical to one in the original design and located in a similar part of the vessel.
- The replacement of a shell course in a cylindrical pressure vessel.
- Welding of wasted or distorted flange faces.
- The repair or replacement of a pressure part with a code-accepted material that has a nominal composition and strength that is equivalent to the original material and is suitable for the intended service.
- Replacement of a pressure-retaining part with a material of a different nominal composition but equal to or greater than allowable stress, from that used in the original design, provided the replacement material satisfies the material and design requirements of the original code of construction. The minimum required thickness shall be at least equal to the thickness stated in the original Manufacturer's Data Report.

In addition to following the repair standard specified by the jurisdiction, the fabricator performing the repairs of ASME code vessels must comply with the ASME code rules and requirements. These rules include design, fabrication details, welding procedures, etc.

13.2.2 Alterations

Alterations require the approval the Authorized Inspector, and typically must be distributed to the jurisdiction as well. A record of the alteration (R2 Form) must be filed with the jurisdiction as well as with the National Board. Examples of alterations are as follows:

- Increase in MAWP or increase in design temperature (Rerating)
- Decrease in MDMT (Rerating)
- The addition of new nozzles except those classified as repairs
- A change in the dimension of a pressure-retaining item
- The addition of a pressurized jacket to a pressure part
- The addition of a bracket that has substantial load

The fabricator performing the alterations must have design capability or access to it to perform alterations on an ASME code stamped vessel and must possess an “R” stamp.

13.2.3 Reratings

Rerating consists of increasing MAWP or design temperature or decreasing MDMT. Reratings require the approval of the jurisdiction and the Authorized Inspector. The following requirements must also be met for rerating of vessels:

- Revised calculations must be furnished.
- Rerating must be made to the original code of construction.
- Inspection records must be provided.
- The pressure-retaining item has been pressure tested for the proposed service conditions.

13.2.4 Post weld heat treating of repaired components

The rules for post weld heat treatment of repaired parts are detailed in NB-23. The following is a summary of the requirements:

1) PWHT subsequent to repair is to follow the requirements of ASME Section I [5] for boilers and ASME Section VIII [6] for pressure vessels, whenever possible. Some of these rules are detailed in Chapter 8.
2) When the rules of Sections I [5] and VIII [6] cannot practically be complied with, then the following methods provide examples of alternatives:

   a) Full PWHT of nozzles may be made locally around the nozzle rather than as a complete band around the vessel. In performing this operation, the following rules are to be followed:
      i) Weld soak area must be at full PWHT temperature.
      ii) A transition length of $4\sqrt{Rt}$ beyond the soak area is maintained where the temperature drop cannot exceed 0.5 times the maximum PWHT temperature. R is the outside radius of the component into which the nozzle is welded, and t is its nominal thickness.

   b) When PWHT cannot be performed, the following alternative welding procedure may be followed:
      i) The original material did not require impact testing.
      ii) This method is applicable to material group P1 Gr.1, 2, 3 or P3 Gr. 1, 2.

iii) Welding is limited to shielded metal-arc welding, gas metal-arc welding, flux-cored arc welding, and gas tungsten-arc welding.
iv) A preheat temperature of 300°F (150°C) must be maintained during welding. Maximum interpass temperature cannot exceed 450°F (230°C).

c) When a. or b. cannot be performed, the following temper bead welding procedure may be followed:

   i) The original material did not require impact testing.
   ii) For material group P1 Gr. 1, 2, and 3 or P3 Gr. 1, 2, and 3. A preheat temperature of 300°F (150°C) must be maintained during welding. Maximum interpass temperature cannot exceed 450°F (230°C).
   iii) For material group P4, the preheat temperature is 300°F (150°C) and for material group 5A, the preheat temperature is 400°F (205°C). The maximum interpass temperature is 800°F (425°C).
   iv) Welding is limited to shielded metal-arc welding, gas metal-arc welding, flux-cored arc welding, and gas tungsten-arc welding.

Examples of possible temper bead welding approaches provided in the 2013 edition of ASME Section XI [7] are shown in Figures 13.1 and 13.2.

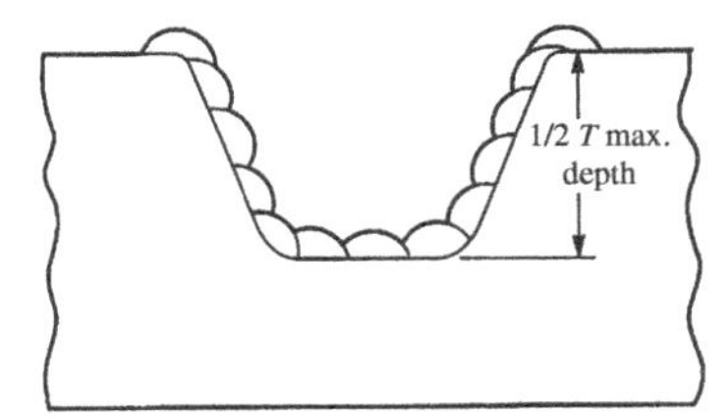

Step 1: Butter cavity with one layer of weld metal using 3/32 in. (2.5 mm) diameter coated electrode.

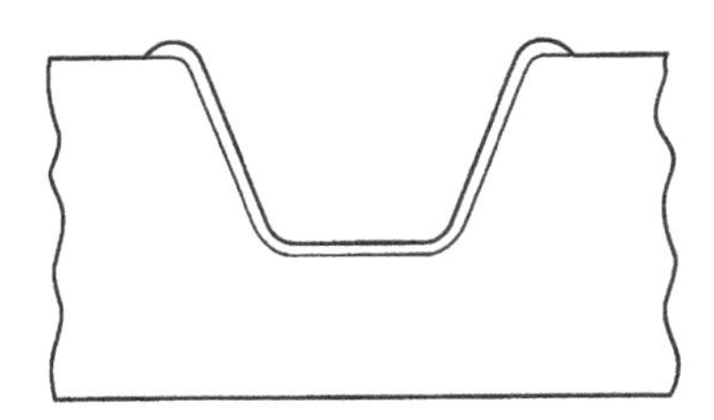

Step 2: Remove the weld bead crown of the first layer by grinding or machining.

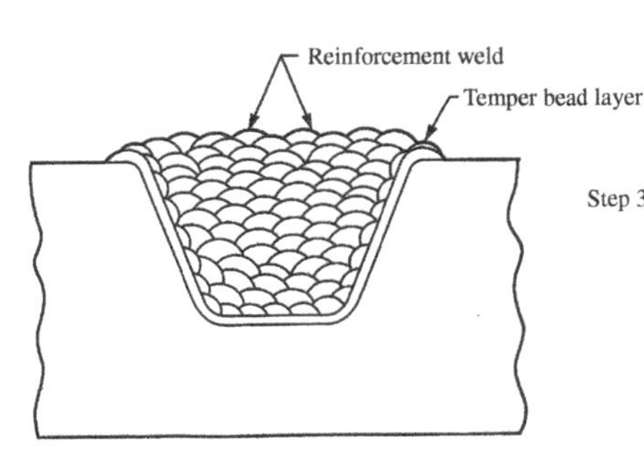

Step 3: The second layer shall be deposited with a 1/8 in. (3 mm) diameter electrode. subsequent layers shall be deposited with welding electrodes no larger than 5/32 in (4 mm) diameter. Bead deposition shall be performed in a manner as shown. Particular care shall be taken in the application of the temper bead reinforcement weld at the tie-in points as well as its removal to ensure that the heat affected zone of the base metal and the deposited weld metal is tempered and the resulting surface is substantially flush.

Figure 13.1 Temper bead manual welding (*Source:* ASME [7])

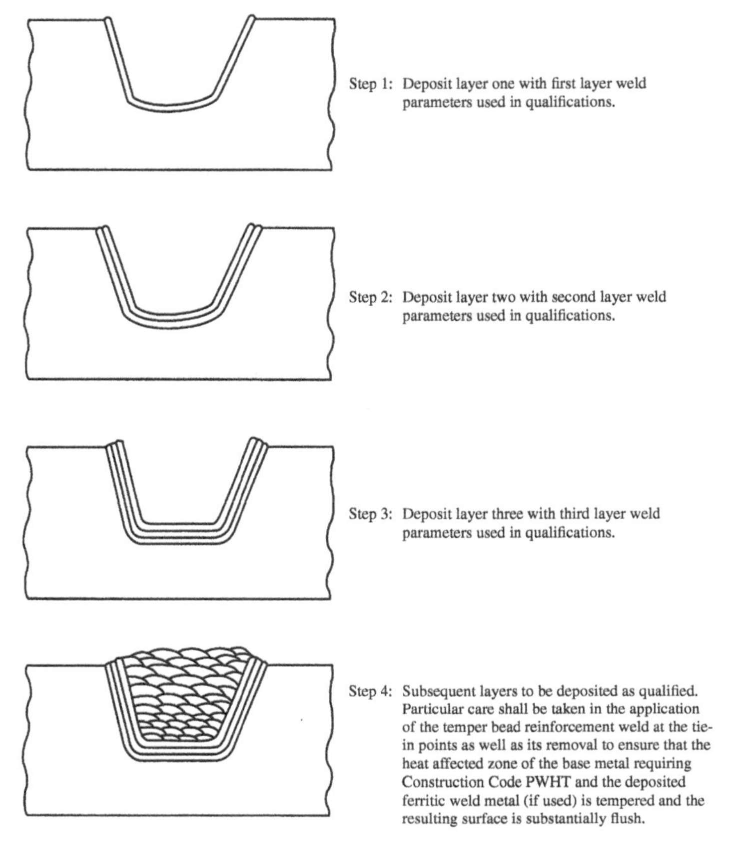

Figure 13.2 Temper bead automatic or machine welding (*Source:* ASME [7])

13.2.5 Hydrostatic or pneumatic testing of repaired vessels

The decision to hydrotest, pneumatic test, or to substitute NDE for the pressure test for repaired vessels depends on many factors. For example, the following conditions may prohibit a hydrotest:

1) The equipment, supports, or foundation cannot support the weight of the liquid.
2) The equipment cannot be dried and the process cannot tolerate moisture in the system.
3) The internal surface has refractory or other lining that would be affected.

Other factors may prohibit hydrotesting of the full vessel or otherwise make a partial hydro necessary or desirable. Some possible factors are as follows:

1) Other welds do not need to be checked.
2) Brittle fracture may be a consideration.
3) Need to minimize the amount of fluid.
4) Need to minimize the amount of flange disassembly.
5) Complete isolation of vessel is impractical.
6) Specified code NDE testing is not possible subsequent to hydrotest.

The fabricator, in conjunction with the user and authorized inspector, must evaluate the need for a hydrotest or pneumatic testing. Occasionally, NDE must be used instead of hydrotesting. Table 13.1 lists some available NDE methods

Table 13.1 Comparison of some selected NDE methods (*Source:* ASME [8])

Imperfection vs. Type of NDE Method

	Surface [Note (1)]		Subsurface [Note (2)]		Volumetric [Note (3)]				
	VT	PT	MT	ET	RT	UTA	UTS	AE	UTT
Service-Induced Imperfections									
Abrasive Wear (Localized)	◉	⊙	⊙	...	◉	⊙	⊙	...	⊙
Baffle Wear (Heat Exchangers)	◉	...	...	⊙	...	...	...	...	...
Corrosion-Assisted Fatigue Cracks	◎	⊙	◉	...	◎	◉	...	◉	...
Corrosion	...	...	...	...	...	...	...	...	...
-Crevice	◉	...	...	...	...	...	...	...	◎
-General / Uniform	...	...	...	◎	⊙	...	⊙	...	◉
-Pitting	◉	◉	◎	...	◉	◎	◎	⊙	◎
-Selective	◉	◉	◎	...	...	...	...	...	◎
Creep (Primary) [Note (4)]	...	...	...	...	...	...	...	...	...
Erosion	◉	...	...	...	◉	◎	⊙	...	⊙
Fatigue Cracks	◎	◉	◉	⊙	⊙	◉	...	◉	...
Fretting (Heat Exchanger Tubing)	⊙	...	...	⊙	...	...	...	...	⊙
Hot Cracking	...	⊙	⊙	...	⊙	◎	...	⊙	...
Hydrogen-Induced Cracking	...	⊙	⊙	...	◎	⊙	...	⊙	...
Intergranular Stress-Corrosion Cracks	...	...	...	...	...	◎	...	...	...
Stress-Corrosion Cracks (Transgranular)	◎	⊙	◉	◎	⊙	⊙	...	⊙	...
Welding Imperfections									
Burn Through	◉	...	...	...	◉	⊙	...	...	◎
Cracks	◎	◉	◉	⊙	⊙	◉	◎	◉	...
Excessive/Inadequate Reinforcement	◉	...	...	...	◉	⊙	◎	...	◎
Inclusions (Slag/Tungsten)	...	...	⊙	⊙	◉	⊙	◎	◎	...
Incomplete Fusion	⊙	...	⊙	⊙	⊙	◉	⊙	⊙	...
Incomplete Penetration	⊙	◉	◉	⊙	◉	◉	⊙	⊙	...
Misalignment	◉	...	...	...	◉	⊙	...	...	...
Overlap	⊙	◉	◉	◎	...	◎	...	...	...
Porosity	◉	◉	◎	...	◉	⊙	◎	◎	...
Root Concavity	◉	...	...	...	◉	⊙	◎	◎	◎
Undercut	◉	⊙	⊙	◎	◉	⊙	◎	◎	...
Product Form Imperfections									
Bursts (Forgings)	◎	◉	◉	⊙	⊙	⊙	⊙	◉	...
Cold Shuts (Castings)	◎	◉	◉	◎	◉	⊙	⊙	◎	...
Cracks (All Product Forms)	◎	◉	◉	⊙	⊙	⊙	◎	◉	...
Hot Tear (Castings)	◎	◉	◉	⊙	⊙	⊙	◎	◎	...
Inclusions (All Product Forms)	...	...	⊙	⊙	◉	⊙	◎	◎	...
Lamination (Plate, Pipe)	◎	⊙	⊙	...	...	◎	◉	◎	◉
Laps (Forgings)	◎	◉	◉	◎	⊙	...	◎	◎	...
Porosity (Castings)	◉	◉	◎	...	◉	◎	◎	◎	...
Seams (Bar, Pipe)	◎	◉	◉	⊙	◎	⊙	⊙	◎	...

Legend:

AE — Acoustic Emission
ET — Electromagnetic (Eddy Current)
MT — Magnetic Particle
PT — Liquid Penetrant
RT — Radiography
UTA — Ultrasonic Angle Beam
UTS — Ultrasonic Straight Beam
UTT — Ultrasonic Thickness Measurements
VT — Visual

◉ — All or most standard techniques will detect this imperfection under all or most conditions.

⊙ — One or more standard technique(s) will detect this imperfection under certain conditions.

◎ — Special techniques, conditions, and/or personnel qualifications are required to detect this imperfection.

GENERAL NOTE: Table A-110 lists imperfections and NDE methods that are capable of detecting them. It must be kept in mind that this table is very general in nature. Many factors influence the detectability of imperfections. This table assumes that only qualified personnel are performing nondestructive examinations and good conditions exist to permit examination (good access, surface conditions, cleanliness, etc.).

NOTES:

(1) Methods capable of detecting imperfections that are open to the surface only.
(2) Methods capable of detecting imperfections that are either open to the surface or slightly subsurface.
(3) Methods capable of detecting imperfections that may be located anywhere within the examined volume.
(4) Various NDE methods are capable of detecting tertiary (3rd stage) creep and some, particularly using special techniques, are capable of detecting secondary (2nd stage) creep. There are various descriptions/definitions for the stages of creep and a particular description/definition will not be applicable to all materials and product forms.

13.3 ASME Post Construction Code, PCC-2

This code addresses some topics covered by the NB-23 code, but it also covers some that are not detailed in that code, such as the following:

1) Full encirclement steel reinforcing sleeves for pipes in corroded areas
2) Welded hot taps in pressure equipment and pipelines

While NB-23 was first published in 1945 as a guide for chief inspectors, PCC-2 was first approved for publication in 2004 and is somewhat more limited in scope. Hence, PCC-2 is not as widely used as the NB-23 code. Fabricators using PCC-2 must ascertain whether the jurisdiction where the vessel will be used accepts the PCC-2 rules.

13.3.1 External weld buildup to repair internal thinning

Both PCC-2 and NB-23 have provisions for weld buildup on the outside surface to repair internal thinning of steel components (carbon steel, low-alloy steel, and stainless steel). This is done as shown in Figure 13.3.

The following limitations must be met:

1) The tensile strength of the weld shall match that of the pressure component.
2) The thickness w cannot exceed the thickness of the pressure component.
3) The weld shall extend full thickness past the repair area for a length $B = 0.75\sqrt{Rt_{\text{nom}}}$.
4) Two or more buildups cannot be closer than $0.75\sqrt{Rt_{\text{nom}}}$ between the toe of each buildup.
5) In order to use this procedure for pipes, a mock-up burst test is required.
   a) L/D and C/D of the mockup are the same as that of the actual pipe.
   b) The actual diameter is not less than 0.5 or greater than 2 times the mockup diameter.

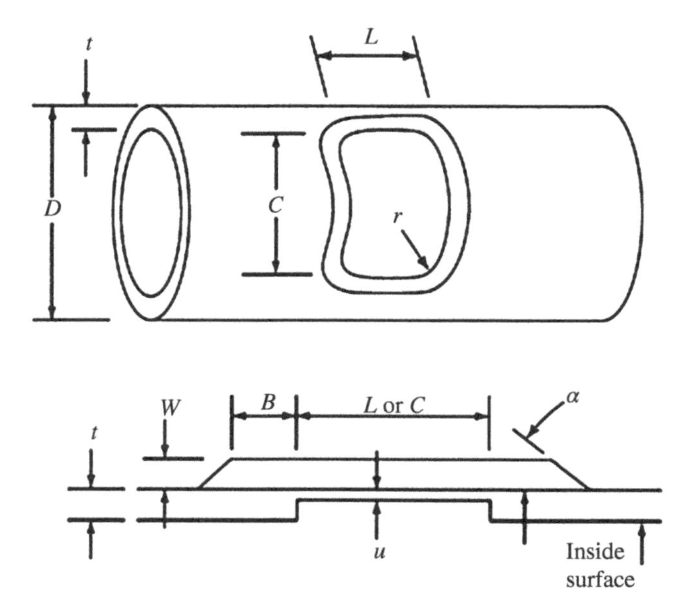

Figure 13.3 Outside weld buildup to repair inside thinning (*Source:* ASME [2])

c) The nominal t/D is not less than 0.5 or greater than 3 times the mockup t/d.
d) The burst pressure shall be greater than $P = 2St_{nom}/D$.

6) Instead of a mockup test and for large-diameter components, the weld buildup may be qualified by an Engineered Design that includes the following:
   a) The effect of weld shrinkage
   b) The effect of stress concentrations and section properties
   c) Stress concentration from internal surface configuration
   d) The effect of differing coefficients of thermal expansion between base material and weld material
   e) The effect of differing thermal mass of weld buildup area and adjacent area
   f) Possible creep degradation over 650°F (345°C).
7) The Engineered Design in piping sections may be considered prequalified if the following conditions are met:
   a) Design temperature is less than 650°F (345°C).
   b) The nominal thickness of the existing pipe is greater than schedule 40 or standard pipe, whichever is less.
   c) Length L does not exceed 0.5 times the diameter or 8 in. (205 mm), whichever is less.
8) PWHT must be performed in accordance with the applicable code of construction, post-construction code, or engineering design, as well as if required to meet service conditions or process requirements. Heat treatment may only be applied to equipment that is repaired while not in service.
9) Hydrotest must be performed in accordance with the applicable code if the component can be isolated.

13.3.2 Full encirclement steel reinforcing sleeves for pipes in corroded areas

When an eroded or corroded area results in a pipe or shell thickness that is less than the minimum required thickness but without a leak, then a sleeve, Figure 13.4, can be placed over the area until a more permanent repair can be done.

The required thickness of the sleeve is a function of

1) Pressure, radius, allowable stress
2) Requirements of the applicable code
3) Weld details shown in Figure 13.4
4) Amount of NDE performed
5) Other requirements specified in PCC-2

When an eroded or corroded area springs a leak in a pipe or shell, then a sleeve, Figure 13.5, can be placed over the area until a more permanent repair can be done.

The required thickness of the sleeve is a function of

1) Pressure, radius, allowable stress
2) Requirements of the applicable code
3) Weld details shown in Figure 13.5
4) Amount of NDE performed
5) Other requirements specified in PCC-2

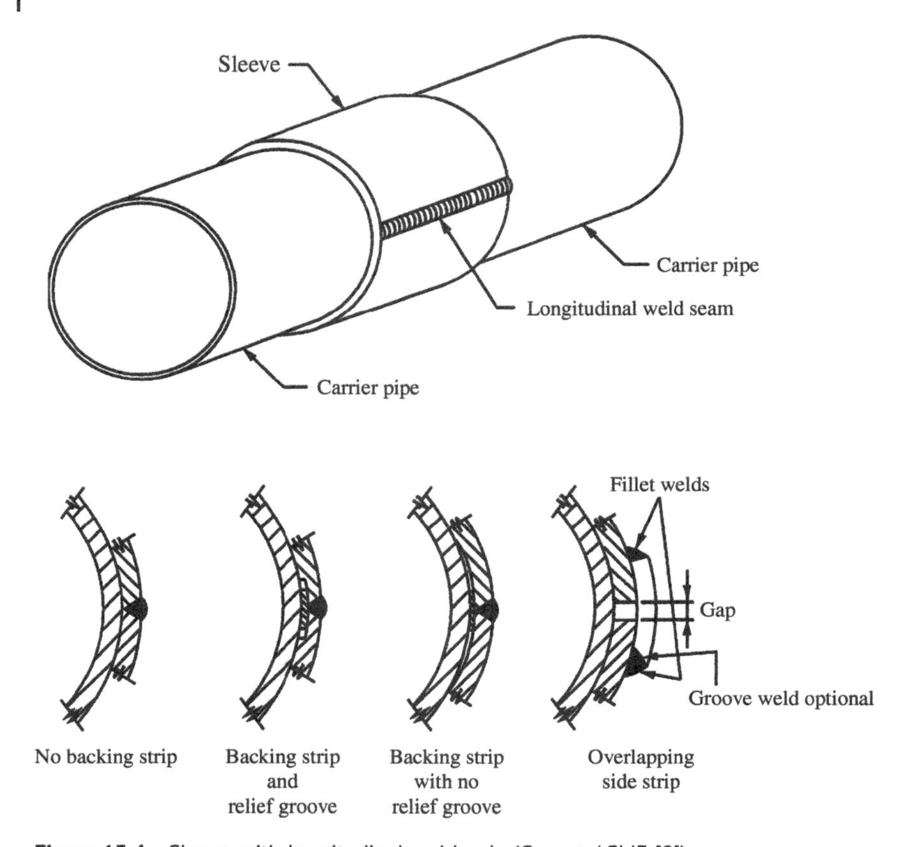

Figure 13.4 Sleeve with longitudinal weld only (*Source:* ASME [2])

13.3.3 Welded hot taps

Hot tapping is the process of drilling a small hole in an operating pipe or vessel for extracting a small amount of the product on a temporary basis. Later, at shutdown, a permanent nozzle is attached to the hole.

Hot tapping must be performed by experienced people due to the danger involved in welding and drilling on a pipe or vessel operating at pressure and temperature.

The procedure for hot tapping is as follows:

1) Provide a sleeve with a thickness equal to the required thickness of the pipe or shell.
2) Add a nozzle to the top part of the sleeve with proper reinforcement as shown in Figure 13.6. The nozzle size must be large enough to accommodate the hot tap valve.
3) Prior to welding the longitudinal seam, the pressure must be adjusted as follows:
   a) Measure the actual wall thickness. t = measured wall thickness – 3/32 in. (2.5 mm). The 3/32 number is the approximate amount of outside wall thickness that is affected by the welding process.
   b) Adjust the outside diameter. D = measured outside diameter – 3/16 in. (4.8 mm).
   c) Let T_1 = operating temperature, D_1 = inside diameter, and T_2 = temperature in the area of weld penetration and is equal to 1380°F (750°C).
   d) Calculate average temperature of remaining metal available for internal pressure containment.

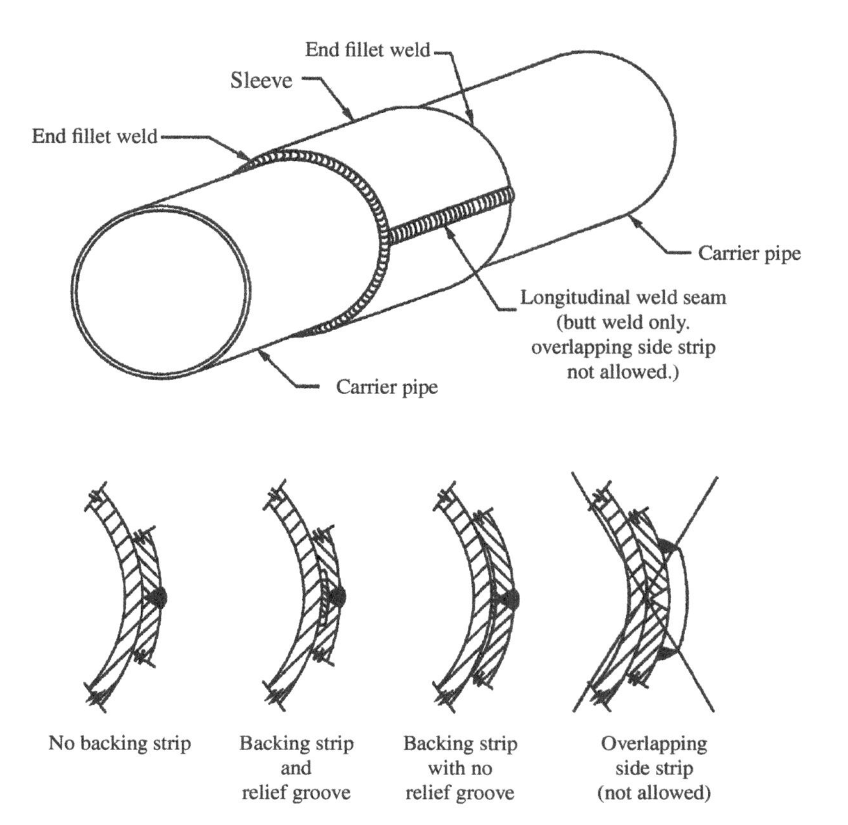

Figure 13.5 Sleeve with circumferential and longitudinal welds (*Source:* ASME [2])

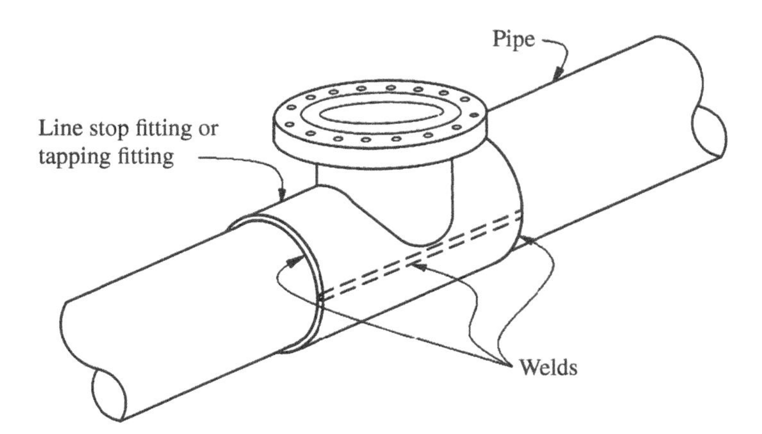

Figure 13.6 Hot tapping sleeve (*Source:* ASME [2])

$$T_m = T_2 - (T_1 - T_2)\{[\ln(D/D_1)]^{-1} - [D/(D - D_1)]\} \tag{13.1}$$

e) Determine the yield stress S_y at T_m.
f) The allowable stress $S_a = 2/3S_y$.
g) The maximum pressure during hot tapping shall be determined in accordance with the applicable code of construction and the additional provisions of PCC-2.

4) Weld the longitudinal seams as shown in Figure 13.6.
5) Weld the circumferential seams as shown in Figure 13.6.
6) Attach the hot tap valve and drilling machine, Figure 13.7.
7) Drill the hole, extract the disk, close the valve, and remove the drilling machine, Figure 13.7.
8) Attach the new pipe to the flanged hot tap.

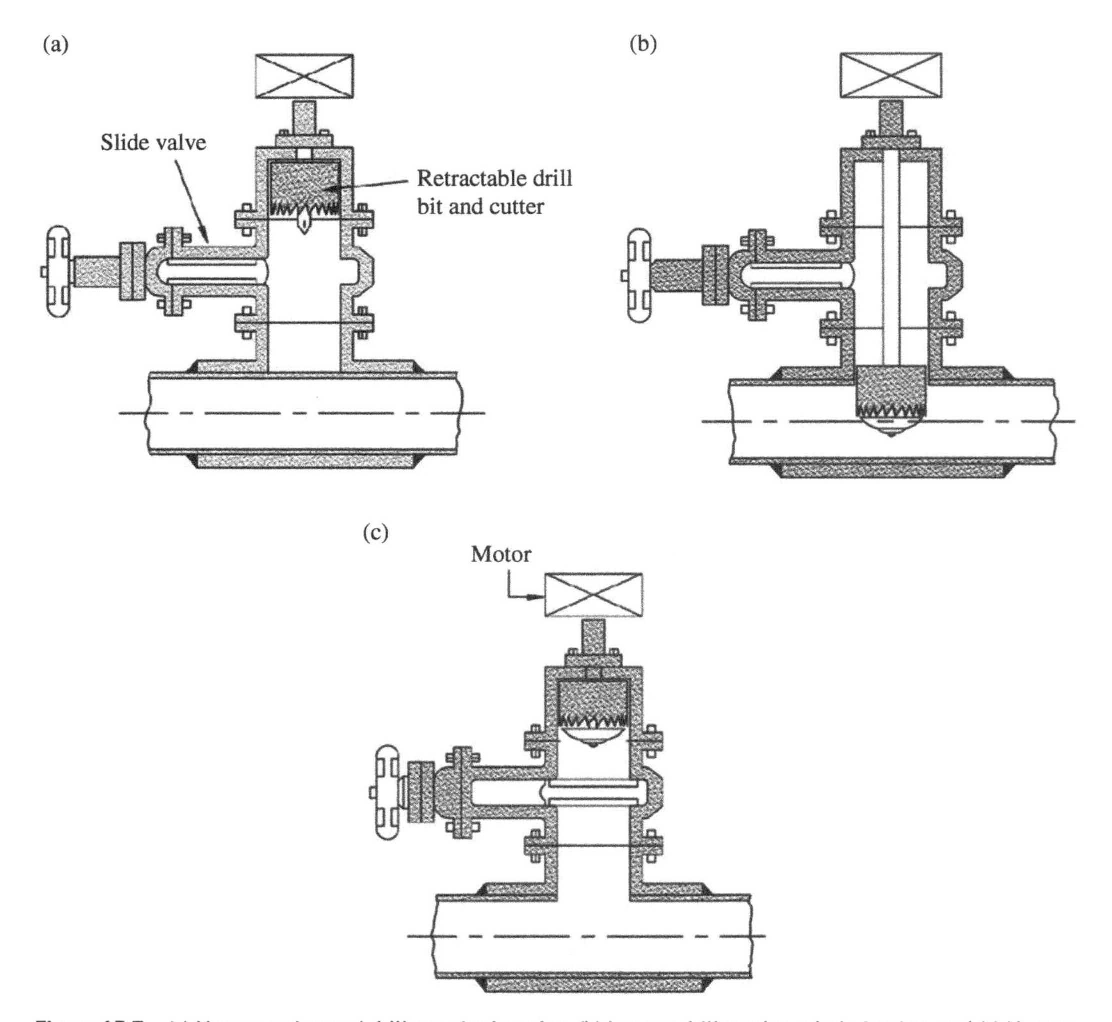

Figure 13.7 (a) Hot tap valve and drill attached to pipe, (b) hot tap drill cutting a hole in pipe, and (c) Hot tap drill extracting cut-piece from pipe and isolating it

Example 13.1

A pressure vessel has the following design data:

Material: SA 516-70
Inside diameter = 24 in.
Thickness = 0.375 in.
Allowable stress = 20,000 psi
Design pressure = 615 psi
Operating pressure = 600 psi
Design temperature = 410°F
Operating temperature = 400°F

A 2-in. diameter hole is to be hot tapped into the vessel. How low must the pressure be dropped during hot tapping?

The yield stresses of SA 516-70 for various temperatures are given as follows:

Value, ksi	Temperature, °F
38.0	100
34.8	200
33.6	300
32.5	400
31.0	500
29.1	600
27.2	700
25.5	800
24.0	900
22.6	1000

Solution

Effective $t = 0.375 - 0.094 = 0.281$ in.

Effective outside diameter, $D = 24.75 - 0.1875 = 24.563$ in.

$T_1 = 400^\circ F$

$D_1 = 24.00$ in.

$T_2 = 1380^\circ F$

$T_m = 1380 - (400 - 1380)\{[\ln(24.563/24)]^{-1} - [24.563/(24.563 - 24)]\}$

$T_m = 888°F$

Interpolating in the above table, $S_y = 24{,}180$ psi

$S_a = (2/3)(24{,}180) = 16{,}120$ psi

$P = (16{,}120)(0.281)/[12.0 + 0.6(0.281)]$

$P = 372$ psi, which is substantially less than the operating pressure of 600 psi

13.4 API Pressure Vessel Inspection Code, API-510

The API-510 code is used by refineries in the United States for repairing pressure vessels. The owner/user must exercise overall control of activities relating to the in-service inspection, repair, and alteration and rerating of pressure vessels and pressure-relieving devices. Some of the requirements of the owner/user organization are as follows:

- An owner/user organization is responsible for developing, documenting, implementing, executing, and assessing pressure vessel inspection systems and inspection/repair procedures that meet the requirements of API 510.
- These systems and procedures will be contained and maintained in a quality assurance inspection/repair management system.
- The engineer is responsible to the owner/user for activities involving design, engineering review, analysis, or evaluation of pressure vessels.
- All repairs and alterations shall be performed by a repair organization possessing an R stamp from the National Board of Boiler and Pressure Vessel Inspectors. The repair organization is responsible to the owner/user and shall provide the materials, equipment, quality control, and workmanship that are necessary to maintain and repair the vessel.
- The inspector is responsible to the owner/user to assure that the inspection, NDE, and pressure testing activities meet the requirements of API 510. Inspectors shall have the following experience in order to qualify for certification:
  - A bachelor degree in engineering plus one year experience, or
  - A two-year engineering degree plus two years' experience, or
  - A high school diploma plus three years' experience, or
  - A minimum of five years' experience.
- An API 510 authorized pressure vessel inspector certificate may be issued when the inspector passes the NBIC inspector examination and meets one of the above-listed requirements.
- The NDE examiners need not be API 510 inspectors or employees of the owner/user.

13.5 API 579/ASME FFS-1 Fitness-For-Service Code

This code has comprehensive rules for evaluating various conditions in a pressure vessel to determine whether the vessel can be operated under these conditions before making any repairs. Some of the conditions considered are as follows:

1) Methods for establishing remaining strength
2) Assessment of existing equipment for brittle fracture
3) Assessment of general metal loss
4) Assessment of local metal loss
5) Assessment of pitting corrosion
6) Assessment of hydrogen blisters and hydrogen damage
7) Assessment of crack-like flaws
8) Assessment of fire damage
9) Assessment of fatigue damage

API 579/ASME FFS-1 lists three levels for assessing various conditions. These levels provide for a graduated evaluation sequence in which a basic approach is tried, followed by more sophisticated evaluations if needed.

13.6 Miscellaneous Repairs

There are many instances where the repair is performed based on good engineering principles and good judgment due to lack of applicable rules or guidance. A few of such repairs are given next.

Figure 13.8 Typical hydraulic nut-cutter

13.6.1 Removal of seized nuts

Corroded or seized nuts are often encountered in the field. Using penetrating lubricant is the first step in trying to remove jammed nuts. Otherwise, a nut-cutter, Figure 13.8, is often used to split small-diameter nuts. For large-diameter nuts, such as 4 in. (100 mm) or 6 in. (150 mm) in size, the removal becomes more complicated. Wrenches with large handles are sometimes fabricated to use extra leverage to remove the nuts. Heating torches are sometime used. Portable grinders are another means of removing nuts. A last resort is to drill the bolt or stud away from the nut and replace it.

13.6.2 Structural supports and foundation

Most pressure vessels are supported by structural members or foundations. Repairing such supports can present unique problems. Figure 13.9 shows a crack in a deep steel structural girder supporting

Figure 13.9 A crack in a deep structural steel vessel support girder (*Source:* Nooter Construction)

a tall and heavy pressure vessel. The crack occurred when the temperature dropped precipitately, causing a brittle fracture in the girder. Repairing it required removing the loads from the girder by supporting the vessel by some other means while the repairs were performed. Careful planning is needed in such a project since it is not possible to provide published rules for every eventuality.

References

1 NBBI. 2019. "*National Board Inspection Code, NBIC, NB-23.*" National Board of Boiler and Pressure Vessel Inspectors, Columbus, OH.
2 ASME. 2018. "*Repair of Pressure Equipment and Piping, ASME PCC-2*". American Society of Mechanical Engineers, New York, NY.
3 ASME. 2014. "*Pressure Vessel Inspection Code, API 510*". American Petroleum Institute, Washington, DC.
4 ASME. 2016. *Fitness-For-Service, API 579/ASME FFS-1*. American Petroleum Institute. Washington, DC.
5 ASME. 2021. "*Boiler and Pressure Vessel Code, Rules for Construction of Power Boilers – Section I*". American Society of Mechanical Engineers, New York, NY.
6 ASME. 2019. "*Boiler and Pressure Vessel Code, Rules for Construction of Pressure Vessels – Section VIII*". 2019. American Society of Mechanical Engineers, New York, NY.
7 ASME. 2013. "*Boiler and Pressure Vessel Code, Rules for Inservice Inspection of Nuclear Power Plant Components – Section XI*". American Society of Mechanical Engineers, New York, NY.
8 ASME. 2019. "*Boiler and Pressure Vessel Code, Nondestructive Examination – Section V.*" American Society of Mechanical Engineers, New York, NY.

Appendix A

Units and Conversion Factors

A.1 Some Customary Units

Item	U.S. Customary units	Metric SI units
Plate thickness	Inches and fractions (in.)	Millimeters (mm)
Pipe diameters	Fractional inches (small sizes) and inches (in.)	Millimeters (mm)
Pipe walls	Inches and decimals (in.)	Millimeters (mm)
Lathe capacity	Inches × inches (in.)	Millimeters (mm)
Mill capacity	Inches × inches × inches (in.)	Millimeters × millimeters × millimeters (mm)
Plate shear capacity	Inches (in.) × feet (ft)	Millimeters × millimeters (mm)
Grinding wheel diameter	Inches (in.)	Millimeters (mm)
Flame, plasma, and laser cutting speeds	Inches per minute (in./min)	Meters per minute (m/min)
Machining feeds	Decimal inches per minute (in./min), or decimal inches per revolution (in./rev)	Millimeters per minute (mm/min) or millimeters per revolution (mm/rev)
Machining speeds	Feet per minute (ft/min)	Meters per minute (m/min)
Machining dimensions	Inches and decimals (in.)	Millimeters and decimals (mm)
Machining material removal rates	Pounds per hours	Cubic centimeters per minute (cm^3/min)
Surface finishes	Ra, microinches (μin.)	Ra, micrometers (μm)
Weld deposition rates	Pounds per hour (lb/hr)	Kilograms/hr

Fabrication of Metallic Pressure Vessels, First Edition. Owen R. Greulich and Maan H. Jawad.

A.2 Conversion Factors

Multiply	By factor	To get units
bar	14.504	psi
kg	2.205	lb
kg/mm^2	14.22	psi
ksi	6.895	MPa
lb	0.454	kg
MPa	0.1450	ksi
MPa	10	bars
MPa	1	MN/mm^2
psi	0.06895	bar
psi	0.0704	kg/mm^2

A.3 Length Conversions

Length units	1 mm	1 cm	1 in.	1 ft	1 m	1 km	1 mile
Millimeter	1	10	25.4	304.8	1000		
Centimeter	0.1	1	2.54	30.48	100		
Inch	0.0394	0.3937	1	12	39.37		
Foot	0.00328	0.0328	0.0833	1	3.281	3281	5280
Meter	0.001	0.01	0.0254	0.3048	1	1000	1609
Kilometer				0.000305	0.001	1	1.609
Mile				0.000189	0.000621	0.621	1

Example: Multiply 1 ft by 304.8 to obtain mm.

A.4 Miscellaneous Unit Conversions

1 metric ton = 1 tonne = 1.1023 U.S. Tons = 2205 lb
1 in. of water = 0.0361 psi
1 ft of water = 0.433 psi
$°K = °C + 273.15$
$°R = °F + 459.67$
$°R = 1.8°K$
$°F = 1.8°C + 32$
$°C = (°F - 32)/1.8$

Appendix B

Welding Symbols

The following weld symbols are commonly used (AWS [1]).

Basic welding symbols their location significance

Location significance	Fillet	Plug or slot	Spot or projection	Stud	Seam	Back or backing	Surfacing	Edge
Arrow side								
Other side				Not used			Not used	
Both sides		Not used	Not used	Not used	Not used	Not used	Not used	
No arrow side or other side significance	Not used	Not used		Not used		Not used	Not used	Not used

Location significance	Groove							Scarf for brazed joint
	Square	V	Bevel	U	J	Flare-V	Flare-bevel	
Arrow side								
Other side								
Both sides								
No arrow side or other side significance		Not used	Not used	Not used	Not used	Not used	Not used	Not used

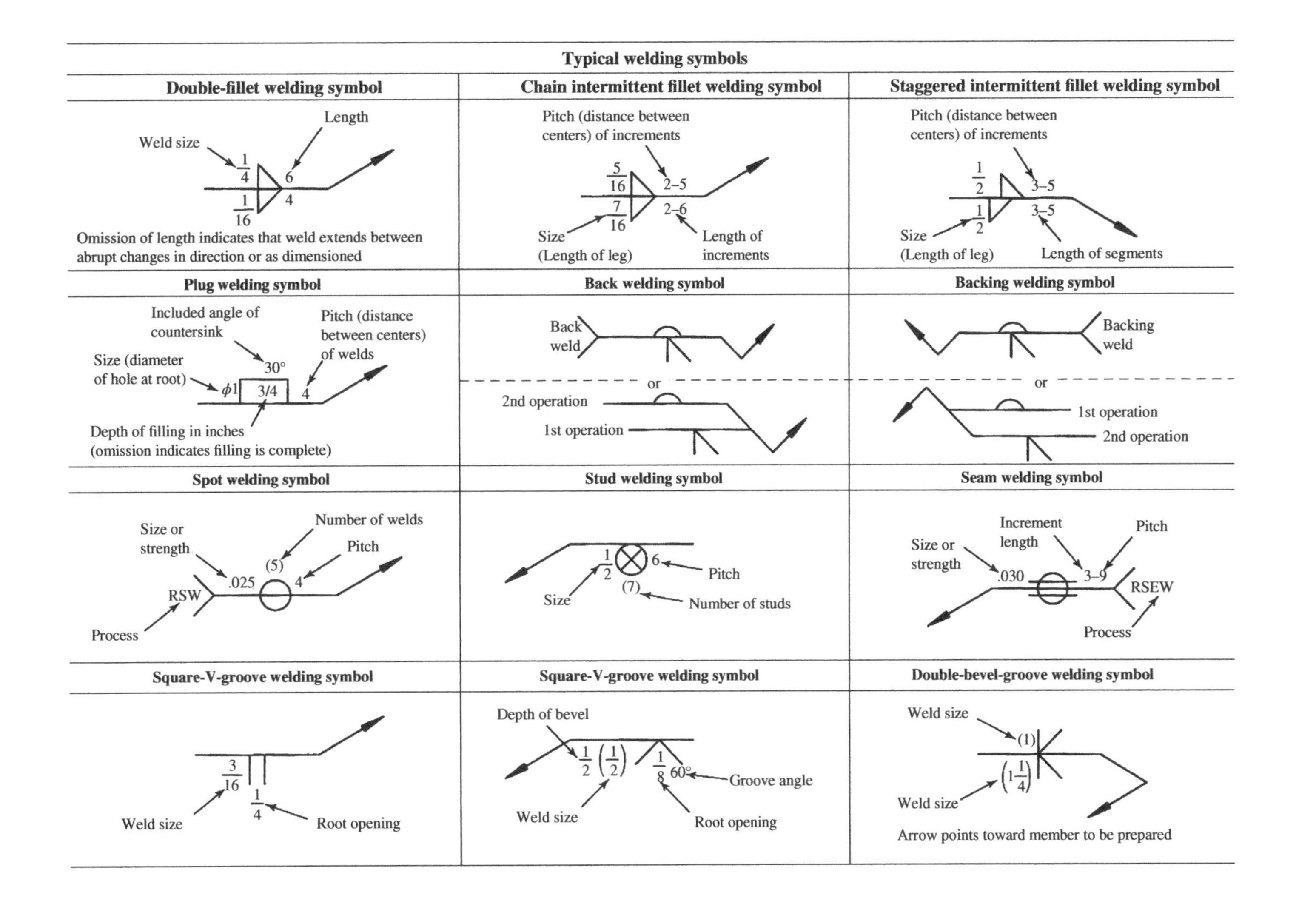
Typical welding symbols
Double-fillet welding symbol
Length
Weld size
1/4
6
1/16
4
Omission of length indicates that weld extends between abrupt changes in direction or as dimensioned
Chain intermittent fillet welding symbol
Pitch (distance between centers) of increments
5/16
2–5
7/16
2–6
Size (Length of leg)
Length of increments
Staggered intermittent fillet welding symbol
Pitch (distance between centers) of increments
1/2
3–5
1/2
3–5
Size (Length of leg)
Length of segments
Plug welding symbol
Included angle of countersink
Pitch (distance between centers) of welds
Size (diameter of hole at root)
30°
ϕ1
3/4
4
Depth of filling in inches (omission indicates filling is complete)
Back welding symbol
Back weld
or
2nd operation
1st operation
Backing welding symbol
Backing weld
or
1st operation
2nd operation
Spot welding symbol
Size or strength
Number of welds
Pitch
.025
(5)
4
RSW
Process
Stud welding symbol
1/2
6
Pitch
(7)
Number of studs
Size
Seam welding symbol
Increment length
Pitch
Size or strength
.030
3–9
RSEW
Process
Square-V-groove welding symbol
3/16
1/4
Weld size
Root opening
Square-V-groove welding symbol
Depth of bevel
1/2 (1/2)
1/8
60°
Groove angle
Weld size
Root opening
Double-bevel-groove welding symbol
Weld size
(1)
(1 1/4)
Weld size
Arrow points toward member to be prepared

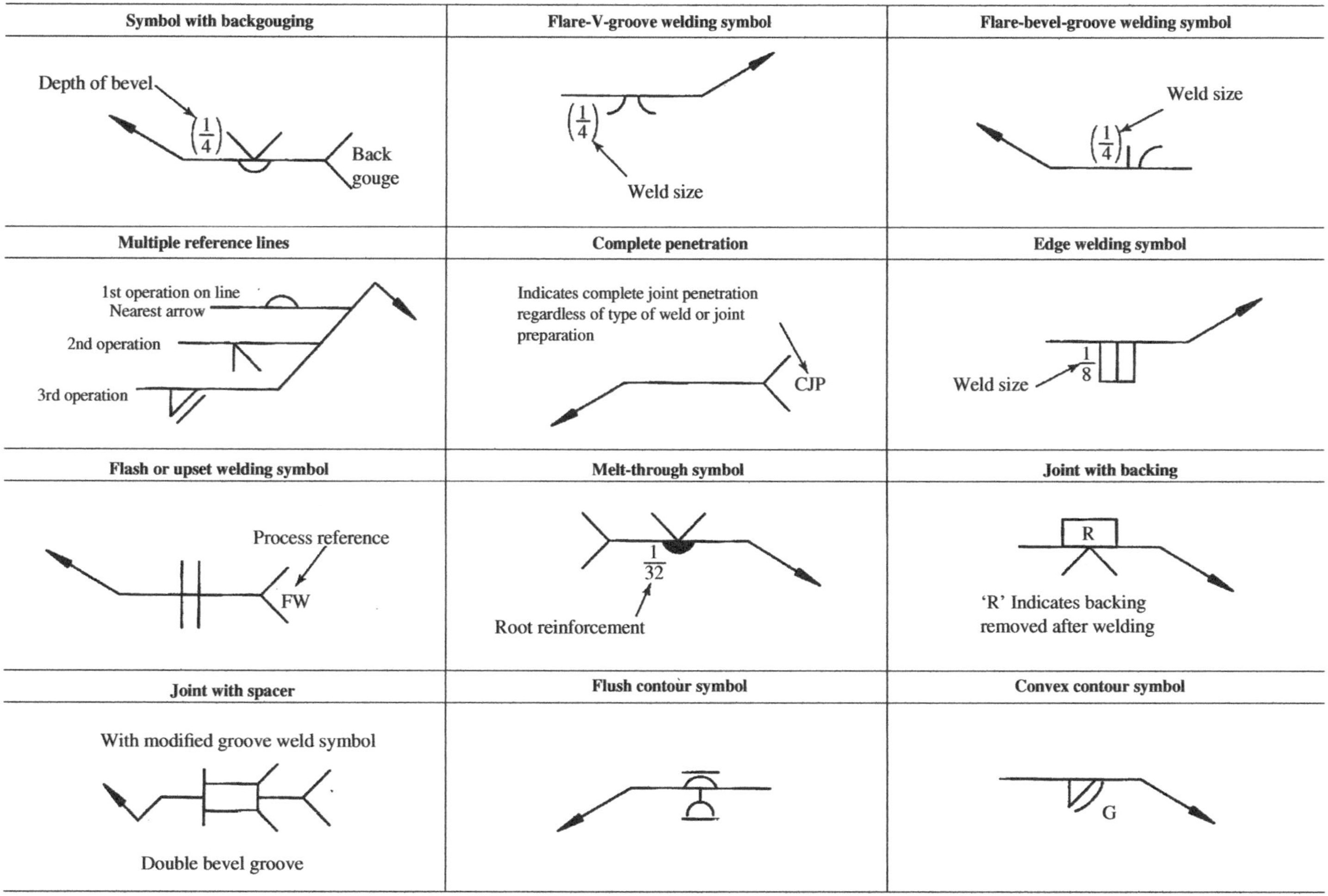

Symbol with backgouging	Flare-V-groove welding symbol	Flare-bevel-groove welding symbol
Depth of bevel $\left(\frac{1}{4}\right)$ Back gouge	$\left(\frac{1}{4}\right)$ Weld size	Weld size $\left(\frac{1}{4}\right)$
Multiple reference lines	**Complete penetration**	**Edge welding symbol**
1st operation on line Nearest arrow 2nd operation 3rd operation	Indicates complete joint penetration regardless of type of weld or joint preparation CJP	Weld size $\frac{1}{8}$
Flash or upset welding symbol	**Melt-through symbol**	**Joint with backing**
Process reference FW	$\frac{1}{32}$ Root reinforcement	R 'R' Indicates backing removed after welding
Joint with spacer	**Flush contour symbol**	**Convex contour symbol**
With modified groove weld symbol Double bevel groove		G

Reference

1 AWS. 2020. "*American Welding Society Welding Symbol Chart A2.1-DC:2020*". American Welding Society, Miami, FL.

Appendix C

Weld Process Characteristics

C.1 Weld Process Advantages and Disadvantages

	Advantages																								
Advantage / Welding Process	No HAZ	Similar and Dissimilar materials	Reactive alloys such as Ti and Zr	No Flux	No shielding gas	No Filler	Complex Geometry	Very clean	Small HAZ	High depth/width ratio	Large unit can weld 0.1" to 4"	Wide range of thickness	Deposition rates(#/hr)	High production rates	Weld heavy sections in single pass	Weld across full thickness at once	Low joint prep cost	No starts and stops	Good small to large jobs	Good in windy conditions	Fine control	No slag inclusions	Slow cooling prevents martensite and cold cracking of HAZ	Easily portable	Easy training for experienced welders
Diffusion (DFW)	×	×	×	×		×	×	×		×			n/a						×			×			
Electron beam (EBW)		×	×	×	×	×	×	×	×	×	×	×	n/a	×	4″	×		×	×		×	×			
Elextrogas (EGW)				×						×			35–60	×	14″	×	×	×				×	×		
Electroslag (ESW)					×					×			35–60	×		×	×	×					×		
Flux cored arc welding (FCAW)		×			(1)							×	3–8					×	×	×				×	×
Flash		×		×	×	×	×			×		×	n/a	×		×	×	n/a				×			×
Friction Stir (FSW)		×	×	×	×	×		×	(2)				n/a	×			×	×				×			
Gas metal arc (GMAW)		×		×								×	3–8					×	×					×	×
Gas tungsten arc (GTAW)		×	×	×		(3)		×					.5–3						×			×			
Laser beam (LBW)		×		×	(4)	×		×	×	×			n/a	×		×		×	×		×	×			
Orbital			×	×		×		×					n/a	×			×	×	×	×	×	×		×	
Oxyfuel (OFW)				×	×								.5–3						×			×		×	×
Plasma arc (PAW)		×	×	×				×	×	×			.5–5					×				×			
Resistance Spot (RSW)		×		×	×	×			×				n/a	×			×	n/a	×	×		×			
Resistance Seam (RSEW)		×		×	×	×			×				n/a	×			×	n/a		×		×			
Submerged arc (SAW)		×			×							×	3–20	×				×		×					
Shielded metal arc (SMAW)		×	×		×							×	3–8						×	×				×	
Stud welding		×		×	×	×			×				n/a				×	n/a	×	×		×		×	

Notes: (1) Shielding gas sometimes used for arc stability and to control heat transfer.
(2) HAZ effectively non-existent.
(3) May be used either with or without filler.
(4) Shielding gas often used to protect the workpiece from oxidation, disperse the plasma shield produced by high power laser welding, and protect the focusing lens.

		Disadvantages															
Disadvantage / Welding Process		May not be suited for mass production	Not suited tor some materials due to long term high temperature exposure	High heat makes cooling provisions desirable	High heat leads to low toughness weld and HAZ and high transition temps	Time consuming	Weld size limited by equipment	Equipment is expensive	Requires vacuum or controlled atmosphere	Requires vacuum	Requires containment of x-rays	Mass production only for small products	Flux or flux coat ea rods must be stored in oven after opening containers	Flux must be chipped after welding	May require cleanup after welding	Flux may not be continuous in core	Sections joined must be similar cross section
Diffusion (DFW)		×	×			×	×	×	×								
Electron beam (EBW)			×				×	×	×	×	×	×					
Electrogas (EGW)			×	×	×												
Electroslag (ESW)				×	×								×				
Flux cored arc Welding (FCAW)		×					×		(1)					×		×	
Flash							×	×							×		×
Friction Stir (FSW)		×					×										×
Gas metal arc (GMAW)															×		
Gas tungsten arc (GTAW)						×						×					
Laser beam (LBW)							×	×	(2)								
Orbital							×		×			×					
Oxyfuel (OFW)		×				×						×			×		
Plasma arc (PAW)								×									
Resistance Spot (RSW)							×										×
Resistance Seam (RSEW)							×										×
Submerged arc (SAW)													×				
Shielded metal arc (SMAW)		×											×	×			
Stud welding																	

Notes: (1) Shielding gas sometimes used for arc stability and to control heat transfer.
(2) Shielding gas often used to protect the workpiece from oxidation, disperse the plasma shield produced by high power laser welding, and protect the focusing lens.

C.2 Weld Process Applications

Position, Process, or Application / Welding Process		Process			Applications											
	Weld positions	Manual	Semi-automatic	Automatic	Cladding	Fuel housings, impellers, valves	Ends of long tubes, pipes, structurals	Repairs, small sections	Heat exchanger plates	Flex hose collars	Tube sheet welds	Pressure vessels	Shipping industry	Aerospace	High purity applications	Very thick parts
Diffusion (DFW)	All			×	×							×		×		
Electron beam (EBW)	All			×		×		×		×		×		×	s	
Electrogas (EGW)	1G			×								×	×			×
Electroslag (ESW)	1G			×								×	×			
Flux cored arc Welding (FCAW)	1F, 1G, 2F, 2G,	×	×	×			×					×	×			
Flash	All		×	×			×									
Friction Stir (FSW)	All			×								×		×		
Gas metal arc (GMAW)	All	×	×	×								×				
Gas tungsten arc (GTAW)	All	×	×	×		×		×		×	×	×		×	×	
Laser beam (LBW)	All			×								×		×	×	
Orbital	All			×			×				×			×	×	
Oxyfuel (OFW)	All	×						×								
Plasma arc (PAW)	All	×	×	×	×	×								×	×	
Resistance Spot (RSW)	All	×	×	×					×							
Resistance Seam (RSEW)	All	×	×	×					×	×						
Submerged arc (SAW)	1F, 1G,		×	×	×	×						×	×	×	×	
Shielded metal arc (SMAW)	All	×			×	×		×				×	×	×		
Stud welding	All	×	×													

Appendix D

Weld Deposition

Weld Deposition - Pounds Per Foot of Weld - Carbon Steel																	
Size	Fillet	Square Butt (opening half of plate thickness)	Single Vee 60 included, no back gouge	Single Vee 45 included no back gouge	Single Vee 30 included, no back gouge	Single Vee 60 included, back gouge	Single Vee 45 included, back gouge	Single Vee 30 included, back gouge	Single J 30 included, no back gouge	Single J 15 included, no back gouge	Single J 30 w/ back gouge	Single J 15 include w/ back gouge	Double J 30 deg included w/ backgouge	Double J 30 included, one side w/ back gouge	Double Vee 30 included w/ back gouge	Double Vee 45	Double Vee 60
1/8	0.03	0.03	0.04	0.03	0.02	0.06	0.06	0.04	x	x	x	x	x	x	x	x	x
3/16	0.06	0.07	0.08	0.07	0.04	0.15	0.14	0.10	x	x	x	x	x	x	x	x	x
1/4	0.11	0.11	0.20	0.18	0.12	0.28	0.26	0.20	x	x	x	x	x	x	x	x	x
5/16	0.17	0.18	0.30	0.27	0.18	0.46	0.43	0.34	x	x	x	x	x	x	x	x	x
3/8	0.24	0.27	0.44	0.39	0.25	0.71	0.66	0.52	0.81	0.76	1.08	1.03	x	x	0.54	0.62	0.76
7/16	0.33	0.37	0.57	0.51	0.32	0.84	0.78	0.59	0.99	0.93	1.25	1.19	x	x	0.60	0.72	0.90
1/2	0.42	0.48	0.79	0.70	0.46	1.05	0.97	0.73	1.17	1.09	1.43	1.36	x	x	0.67	0.83	1.07
5/8	0.66	x	1.16	1.03	0.66	1.42	1.30	0.92	1.55	1.44	1.82	1.71	x	x	0.83	1.07	1.46
3/4	0.96	x	1.60	1.42	0.88	1.87	1.69	1.15	1.97	1.81	2.24	2.07	x	x	1.01	1.37	1.92
7/8	1.30	x	2.22	1.98	1.25	2.49	2.24	1.52	2.43	2.23	2.69	2.49	3.41	4.83	1.14	1.57	2.24
1	1.70	x	2.79	2.48	1.54	3.06	2.75	1.81	2.90	2.62	3.17	2.89	3.89	5.30	1.34	1.88	2.75
1-1/4	2.66	x	4.09	3.62	2.21	4.36	3.89	2.47	3.91	3.44	4.18	3.71	5.63	6.25	1.75	2.58	3.89
1-1/2	3.82	x	5.64	4.98	2.98	5.91	5.24	3.25	5.04	4.31	5.31	4.58	6.79	7.27	2.22	3.39	5.23
1-3/4	5.21	x	7.43	6.54	3.87	7.70	6.81	4.13	6.28	5.24	6.54	5.51	8.01	8.35	2.76	4.32	6.77
2	6.80	x	9.47	8.32	4.87	9.74	8.59	5.14	7.63	6.23	7.90	6.49	9.29	9.51	3.36	5.36	8.52
2-1/4	8.61	x	11.75	10.31	5.99	12.02	10.58	6.25	9.10	7.27	9.36	7.53	10.63	10.73	4.03	6.53	10.47
2-1/2	10.62	x	14.28	12.52	7.22	14.61	12.85	7.55	10.68	8.36	11.01	8.69	12.11	12.08	4.82	7.88	12.69
2-3/4	12.85	x	17.06	14.93	8.56	17.39	15.26	8.89	12.38	9.51	12.71	9.85	13.58	13.44	5.62	9.28	15.05

(Continued)

3	15.30	x	20.07	17.56	10.02	20.41	17.89	10.35	14.18	10.72	14.52	11.05	15.11	14.86	6.47	10.80	17.62
3-1/4	17.95	x	23.34	20.40	11.59	23.67	20.73	11.92	16.11	11.99	16.44	12.32	16.71	16.36	7.39	12.43	20.39
3-1/2	20.82	x	26.85	23.46	13.28	27.18	23.79	13.61	18.14	13.31	18.47	13.64	18.37	17.92	8.37	14.19	23.36
3-3/4	23.90	x	30.60	26.72	15.08	30.94	27.05	15.41	20.29	14.68	20.63	15.02	20.09	19.55	9.41	16.06	26.54
4	27.20	x	34.60	30.20	16.99	34.94	30.53	17.32	22.56	16.12	22.89	16.45	21.88	21.25	10.52	18.02	29.93

Multiply the values above by 1.483 to obtain kg/m.
For other metals, multiply by the following factors:

Stainless steel	1.045	Nickel	1.134	Aluminum	0.346	Titanium	0.573	Magnesium	0.221

This chart provides expected weld deposition poundage for carbon steel for a number of common weld configurations and thicknesses. A number of assumptions were made in the development of this chart, and its applicability will depend on choices regarding weld configuration as well as on various things controlled by the welder in the field, some of which are given as follows:

1) A number of welds are not calculated because it is considered that the weld configuration would be inefficient for practical applications. For example, the square butt weld in a thick plate requires a large mass of weld and high heat input compared to some of the other weld designs, and the J-grooves are not used for relatively thin designs because they can cost more in both material and labor.
2) All double welds are based on a one-third, two-thirds design because this design has been found to work well for back-gouging and minimizing distortion of the piece.
3) Double J-groove welds in which both plates are beveled are assessed based on a 5/16 in. radius, while those with only one plate beveled use a 5/8 in. radius. These radii are selected to permit access for the welding head and may need to be adjusted depending on the particular equipment and skills available.
4) Back-gouge dimensions vary somewhat with the thickness of the weld. The size of this portion of the weld may vary significantly depending on actual field practice. A gouge deep enough to remove defects at the root of the weld is needed, but an excessively large gouge will take additional weld to fill and will probably result in greater distortion.
5) A crown averaging 3/32 in. is generally assumed on each face of the weld. Field practice may result in a crown that is either larger or smaller than this and must be considered accordingly.
6) Weld designs are selected to enhance the quality of the weld and to minimize the cost of fabrication and distortion of the weldment. If all else is equal, then minimizing the weight of weld will result in the least cost, since less weld metal is needed and the welding will take less time. Some of the lightest welds in thick materials are those with a steep-walled J-groove. When selecting one of these, the extra cost of machining the weld prep must be considered, as well as the greater difficulty in penetrating the sidewall to ensure a quality weld.
7) In some cases, such as nozzle welds, partial penetration groove welds may be appropriate, and a fillet weld will sometimes be required on top of a full or partial penetration weld. Groove welds may be selected to represent the appropriate configuration, and a fillet weld added to arrive at the total weight of weld.

Appendix E

Shape Properties

E.1 Properties of Cross Sections

Shape	Properties
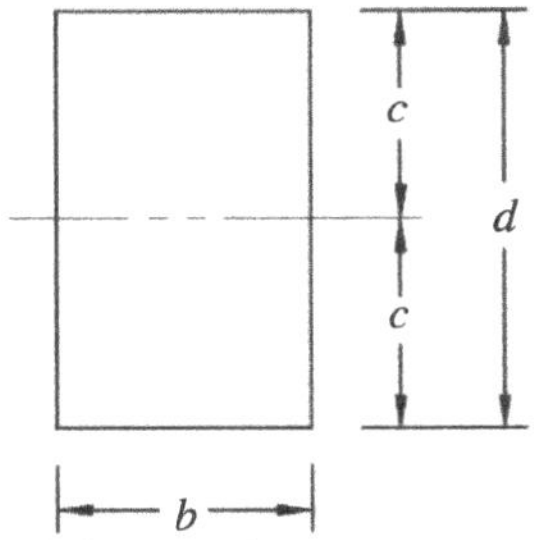 Solid rectangle	$c = d/2$ $A = bd$ $I = bd^3/12$ $r = d/(12)^{0.5}$ $S = bd^2/6$
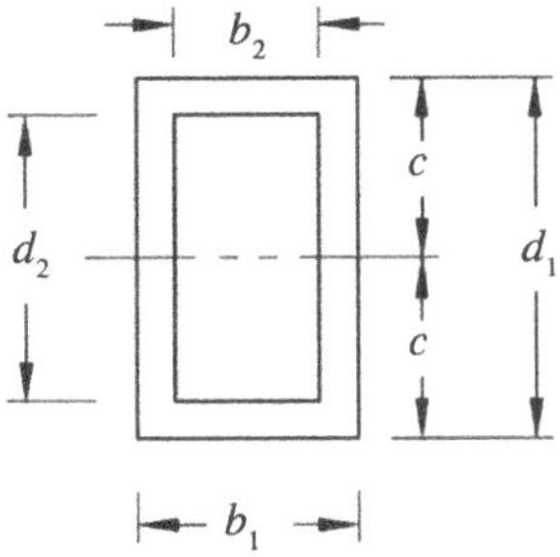 Hollow rectangle	$c = d_1/2$ $A = b_1d_1 - b_2d_2$ $I = (b_1d_1^3 - b_2d_2^3)/12$ $r = (I/A)^{0.5}$ $S = (b_1d_1^2 - b_2d_2^3/d_1)/6$

(*Continued*)

Fabrication of Metallic Pressure Vessels, First Edition. Owen R. Greulich and Maan H. Jawad.

Shape	Properties
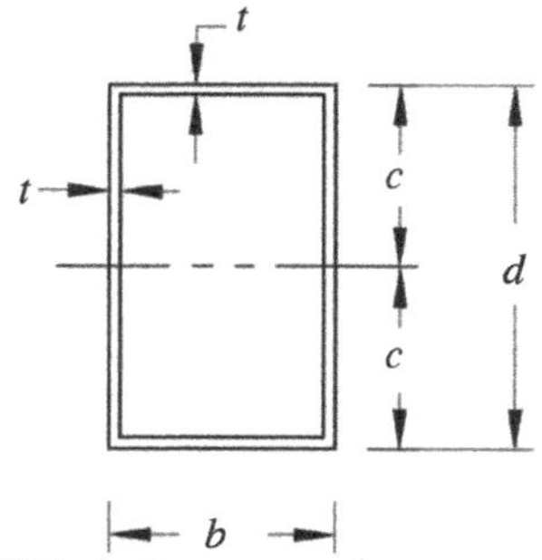 Thin hollow rectangle	$c = d/2$ $A = 2(b+d)t$ $I = td^2(d+3b)/6$ $r = \frac{d\sqrt{3}}{2}\sqrt{\frac{d+3b}{d+b}}$ $S = td(d+3b)/3$
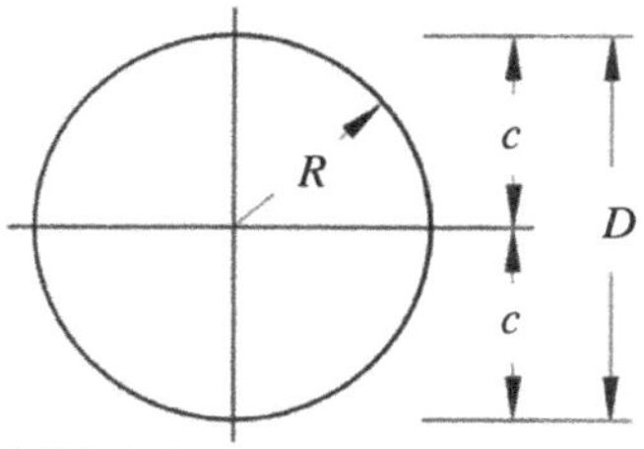 Solid circle	$c = R = D/2$ $A = \pi R^2 = \pi D^2/4$ $I = \pi R^4/4 = \pi D^4/64$ $r = R/2 = D/4$ $S = \pi R^3/4 = \pi D^3/32$
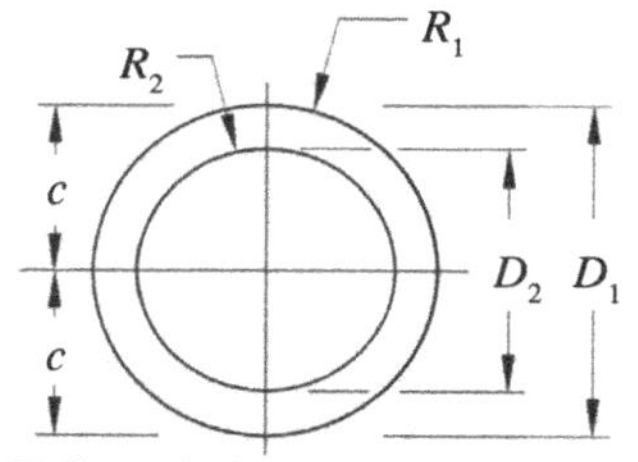 Hollow circle	$c = R_1 = D_1/2$ $R_2 = D_2/2$ $A = \pi(R_1^2 - R_2^2)$ $= \pi(D_1^2 - D_2^2)/4$ $I = \pi(R_1^4 - R_2^4)/4$ $= \pi(D_1^4 - D_2^4)/64$ $r = \frac{1}{2}\sqrt{R_1^2 + R_2^2}$ $r = \frac{1}{4}\sqrt{D_1^2 + D_2^2}$ $S = \pi(R_1^4 - R_2^4)/4R_1$ $= \pi(D_1^4 - D_2^4)/32D_1$
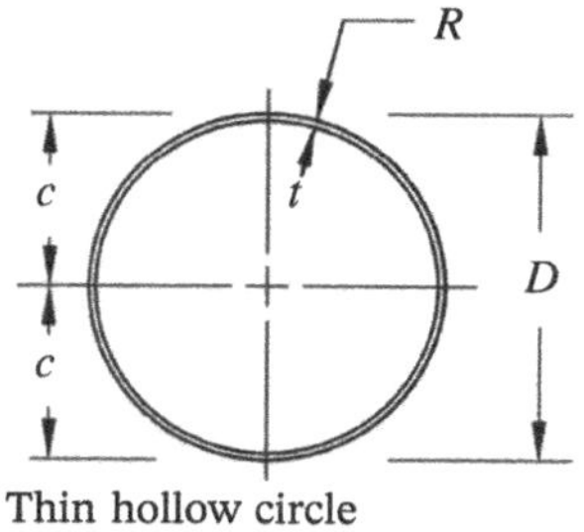 Thin hollow circle	$c = R = D/2$ $A = 2\pi Rt = \pi Dt$ $I = \pi tR^3 = \pi tD^3/8$ $r = R/1.414 = D/2.828$ $S = \pi tR^2 = \pi tD^2/4$

Shape	Properties
Triangular section	$c_1 = d/3$ $c_2 = 2d/3$ $A = bd/2$ $I = bd^3/36$ $r = d/(18)^{0.5}$ $S_1 = bd^2/12$ at bottom of triangle $S_2 = bd^2/24$ at top of triangle

Notation: A = area, b = width, c = distance from neutral axis to edge, d = height, D = diameter of circle, I = moment of inertia, r = radius of gyration, R = radius of circle, S = section modulus.

E.2 Properties of Solids

Cylinder	Convex surface = πDH Volume = $\pi HD^2/4$ Centroid of mass above base = $H/2$ Projected area = HD Centroid of projected area above base = $H/2$
Cone	Convex surface = $(\pi D/4)(D^2 + 4H^2)^{0.5}$ Volume = $\pi HD^2/12$ Centroid of mass above base = $H/4$ Projected area = $HD/2$ Centroid of projected area above base = $H/3$
Frustum of cone	Convex surface = $(\pi/4)(D_1 + D_2)[(D_1 - D_2)^2 + 4H^2]^{0.5}$ Volume = $(\pi H/12)(D_1^2 + D_1D_2 + D_2^2)$ Centroid of mass above base = $(H/4)[(D_1^2 + 2D_1D_2 + 3D_2^2)/(D_1^2 + D_1D_2 + D_2^2)]$ Projected area = $H(D_1 + D_2)/2$ Centroid of projected area above base = $(H/3)[(D_1+2D_2)/(D_1+D_2)]$

(Continued)

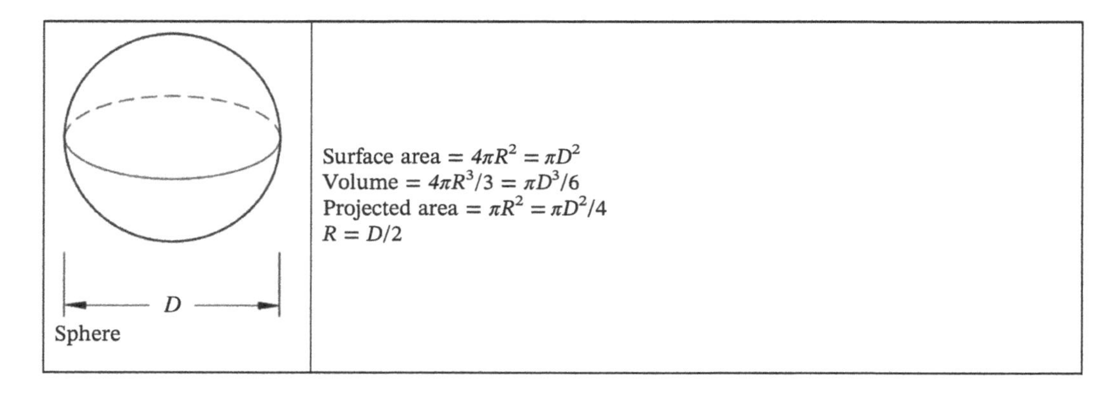

Surface area = $4\pi R^2 = \pi D^2$
Volume = $4\pi R^3/3 = \pi D^3/6$
Projected area = $\pi R^2 = \pi D^2/4$
$R = D/2$

E.3 Properties of Hemispheres and Spherical Segments

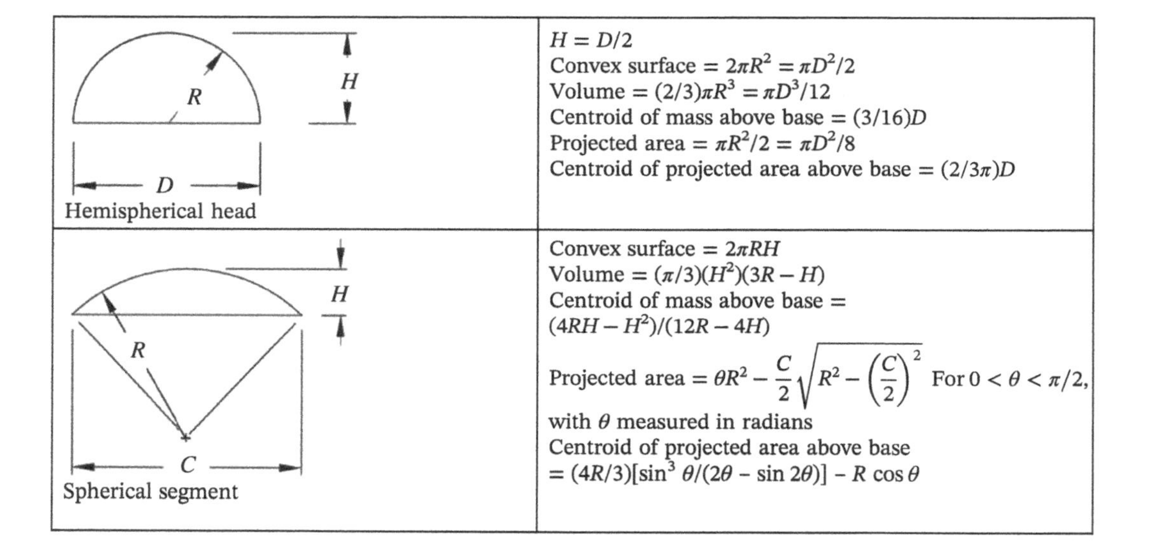

Hemispherical head:

$H = D/2$
Convex surface = $2\pi R^2 = \pi D^2/2$
Volume = $(2/3)\pi R^3 = \pi D^3/12$
Centroid of mass above base = $(3/16)D$
Projected area = $\pi R^2/2 = \pi D^2/8$
Centroid of projected area above base = $(2/3\pi)D$

Spherical segment:

Convex surface = $2\pi RH$
Volume = $(\pi/3)(H^2)(3R - H)$
Centroid of mass above base =
$(4RH - H^2)/(12R - 4H)$
Projected area = $\theta R^2 - \frac{C}{2}\sqrt{R^2 - \left(\frac{C}{2}\right)^2}$ For $0 < \theta < \pi/2$,
with θ measured in radians
Centroid of projected area above base
= $(4R/3)[\sin^3 \theta/(2\theta - \sin 2\theta)] - R\cos\theta$

E.4 Properties of Commonly Used Ellipsoidal Heads

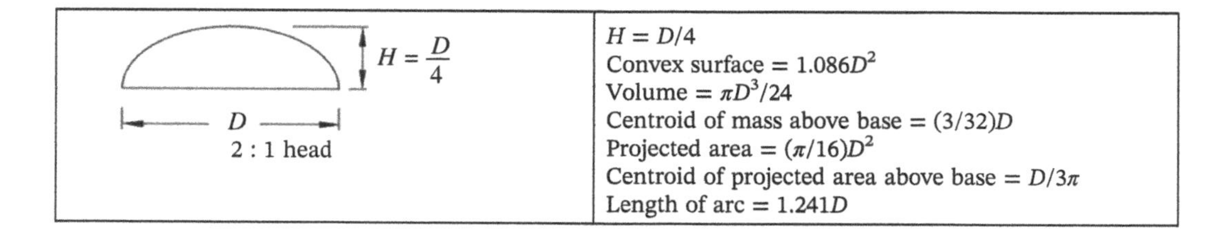

$H = D/4$
Convex surface = $1.086D^2$
Volume = $\pi D^3/24$
Centroid of mass above base = $(3/32)D$
Projected area = $(\pi/16)D^2$
Centroid of projected area above base = $D/3\pi$
Length of arc = $1.241D$

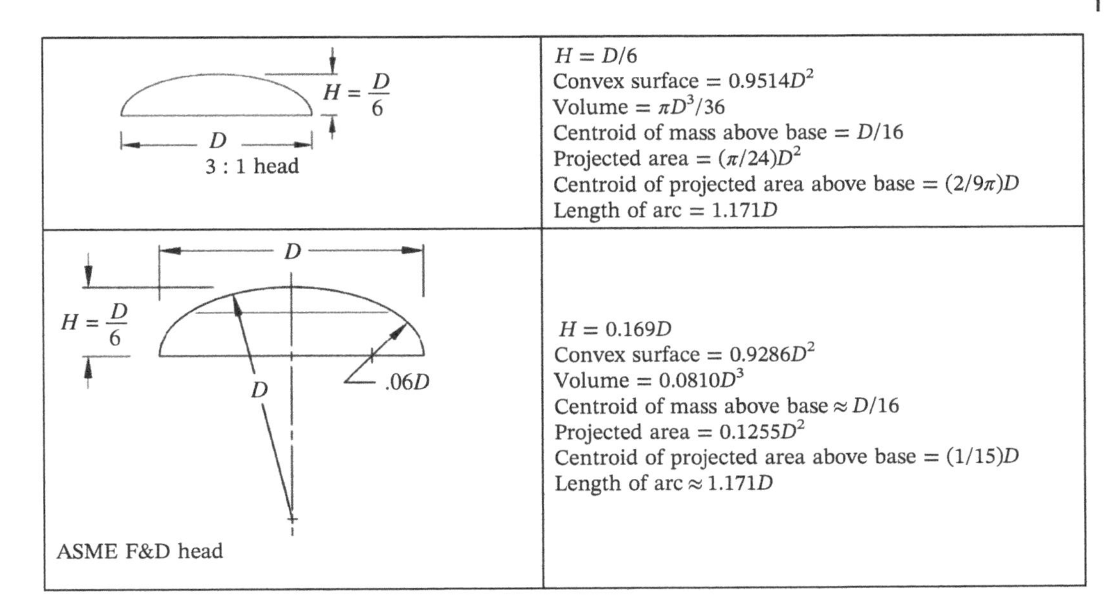

Shape	Properties
$H = \frac{D}{6}$, D, 3 : 1 head	$H = D/6$ Convex surface = $0.9514D^2$ Volume = $\pi D^3/36$ Centroid of mass above base = $D/16$ Projected area = $(\pi/24)D^2$ Centroid of projected area above base = $(2/9\pi)D$ Length of arc = $1.171D$
D, $H = \frac{D}{6}$, D, $.06D$, ASME F&D head	$H = 0.169D$ Convex surface = $0.9286D^2$ Volume = $0.0810D^3$ Centroid of mass above base $\approx D/16$ Projected area = $0.1255D^2$ Centroid of projected area above base = $(1/15)D$ Length of arc $\approx 1.171D$

E.5 Ellipsoidal Head General Formulas

The volume and surface area of an ellipsoidal head, Figure E.1, are expressed as

$$V = (\pi/6)(D^2H) \tag{E.1}$$

$$S = (2\pi)\left\{\left[\left(D^2/4\right)^{1.6075} + 2(DH/2)^{1.6075}\right]/3\right\}^{1/1.6075} \tag{E.2}$$

or, in a different format

$$S = 2\pi D^2\left(1 + \frac{H^2}{eD^2}\tanh^{-1}e\right)$$

where

$$e = \sqrt{1 - \left(\frac{H}{D}\right)^2}$$

where

D = diameter
H = height
S = surface area
V = volume

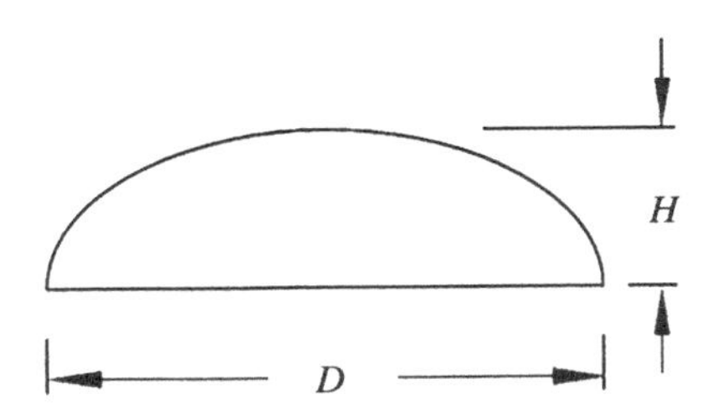

Figure E.1 Ellipsoidal head

For ease of manufacturing, large-diameter ellipsoidal heads are sometimes constructed by using two constant radii to approximate an ellipsoidal shape. The heads are

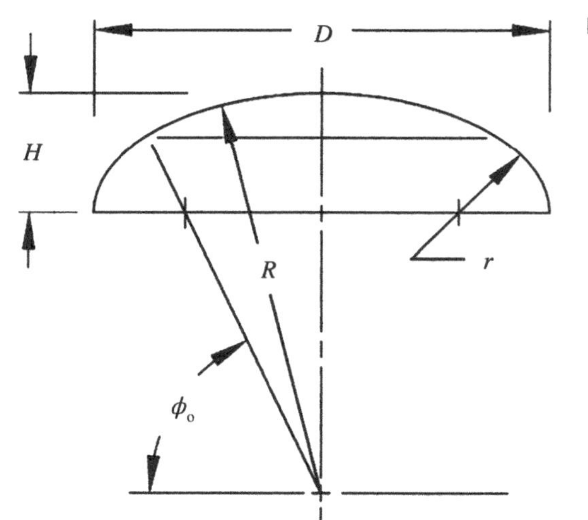

Figure E.2

constructed by forming a knuckle section and a spherical section as shown in Figure E.2. The height of the head, H, is obtained as follows:

Calculate angle ϕ_o from the equation

$$(R - r)\cos\phi_o + r = D/2 \tag{E.3}$$

Then determine H from the equation

$$H = R - (R - r)\sin\phi_o \tag{E.4}$$

A common practice in the United States for approximating large 2 : 1 ellipsoidal heads ($D/H = 4$) is to use $R = 0.9D$ and $r = 0.17D$. From Eq. (E.3), the angle $\phi_o = 63.12°$ and from Eq. (E.4) the height is calculated as $H = 0.25D$. This profile simulates closely a 2 : 1 ellipsoidal head.

Shallow heads are sometimes used where $R = D$ and $r = 0.06D$. In such cases, $\phi_o = 62.09°$ and $H = 0.169D$. This ratio of $D/H \approx 6$ closely approximates a 3 : 1 ellipsoidal head.

Appendix F

Pipe and Tube Dimensions and Weights

Dimensions and weights for some commonly used pipe sizes and schedules are provided, followed by those for tubes.

The weight/ft is in lb for carbon steel. For other metal alloys, multiply the value found in Table F.1 by the following factors:

Material	Factor
Steel	1.0
Aluminum	0.35
Copper alloys	1.14
Nickel alloys	1.02
Stainless steel	1.04
Titanium alloys	0.57
Zirconium alloys	0.81

For SI weights (kg) of pipe or tube, use the following conversion factors:

Material	Factor
Steel	0.454
Aluminum	0.158
Copper alloys	0.517
Nickel alloys	0.462
Stainless steel	0.472
Titanium alloys	0.258
Zirconium alloys	0.367

For SI weight (kg) of water, multiply by 0.454.

Fabrication of Metallic Pressure Vessels, First Edition. Owen R. Greulich and Maan H. Jawad.

Table F.1 Pipe properties

Nominal pipe size	Outside diameter (in.)	Schedule		Wall thickness (in.)	Inside diameter (in.)	Weight (lb/ft)	Water Weight (lb/ft)
1/8	0.405	10	...	0.049	0.307	0.19	0.032
	0.405	30	...	0.057	0.291	0.21	0.029
	0.405	40	Std	0.068	0.269	0.24	0.025
	0.405	80	XS	0.095	0.215	0.31	0.016
1/4	0.540	10	...	0.065	0.41	0.33	0.057
	0.540	30	...	0.073	0.394	0.36	0.053
	0.540	40	Std	0.088	0.364	0.43	0.045
	0.540	80	XS	0.119	0.302	0.54	0.031
3/8	0.675	10	...	0.065	0.545	0.42	0.101
	0.675	30	...	0.073	0.529	0.47	0.095
	0.675	40	Std	0.091	0.493	0.57	0.083
	0.675	80	XS	0.126	0.423	0.74	0.061
1/2	0.840	5	...	0.065	0.71	0.54	0.172
	0.840	10	...	0.083	0.674	0.67	0.155
	0.840	30	...	0.095	0.65	0.76	0.144
	0.840	40	Std	0.109	0.622	0.85	0.132
	0.840	80	XS	0.147	0.546	1.09	0.101
	0.840	160	...	0.188	0.464	1.31	0.073
	0.840	...	XXS	0.294	0.252	1.72	0.022
3/4	1.050	5	...	0.065	0.92	0.69	0.288
	1.050	10	...	0.083	0.884	0.86	0.266
	1.050	30	...	0.095	0.86	0.97	0.252
	1.050	40	Std	0.113	0.824	1.13	0.231
	1.050	80	XS	0.154	0.742	1.48	0.187
	1.050	160	...	0.219	0.612	1.95	0.127
	1.050	...	XXS	0.308	0.434	2.44	0.064

Table F.1 (Continued)

Nominal pipe size	Outside diameter (in.)	Schedule		Wall thickness (in.)	Inside diameter (in.)	Weight (lb/ft)	Water Weight (lb/ft)
1	1.315	5	...	0.065	1.185	0.87	0.478
	1.315	10	...	0.109	1.097	1.41	0.409
	1.315	30	...	0.114	1.087	1.46	0.402
	1.315	40	Std	0.133	1.049	1.68	0.374
	1.315	80	XS	0.179	0.957	2.17	0.312
	1.315	160	...	0.250	0.815	2.85	0.226
	1.315	...	XXS	0.358	0.599	3.66	0.122
1¼	1.660	5	...	0.065	1.53	1.11	0.796
	1.660	10	...	0.109	1.442	1.81	0.707
	1.660	30	...	0.117	1.426	1.93	0.692
	1.660	40	Std	0.140	1.38	2.27	0.648
	1.660	80	XS	0.191	1.278	3.00	0.556
	1.660	160	...	0.250	1.16	3.77	0.458
	1.660	...	XXS	0.382	0.896	5.22	0.273
1½	1.900	5	...	0.065	1.77	1.28	1.07
	1.900	10	...	0.109	1.682	2.09	0.96
	1.900	30	...	0.125	1.65	2.37	0.93
	1.900	40	Std	0.145	1.61	2.72	0.88
	1.900	80	XS	0.200	1.5	3.63	0.77
	1.900	160	...	0.281	1.338	4.86	0.61
	1.900	...	XXS	0.400	1.1	6.41	0.41

(Continued)

Table F.1 (Continued)

Nominal pipe size	Outside diameter (in.)	Schedule		Wall thickness (in.)	Inside diameter (in.)	Weight (lb/ft)	Water Weight (lb/ft)
2	2.375	5	...	0.065	2.245	1.61	1.71
	2.375	10	...	0.109	2.157	2.64	1.58
	2.375	30	...	0.125	2.125	3.01	1.54
	2.375	40	Std	0.154	2.067	3.66	1.45
	2.375	80	XS	0.218	1.939	5.03	1.28
	2.375	160	...	0.344	1.687	7.47	0.97
	2.375	...	XXS	0.436	1.503	9.04	0.77
2½	2.875	5	...	0.083	2.709	2.48	2.50
	2.875	10	...	0.120	2.635	3.53	2.36
	2.875	30	...	0.188	2.499	5.40	2.12
	2.875	40	Std	0.203	2.469	5.80	2.07
	2.875	80	XS	0.276	2.323	7.67	1.84
	2.875	160	...	0.375	2.125	10.02	1.54
	2.875	...	XXS	0.552	1.771	13.71	1.07
3	3.500	5	...	0.083	3.334	3.03	3.78
	3.500	10	...	0.120	3.26	4.34	3.62
	3.500	30	...	0.188	3.124	6.66	3.32
	3.500	40	Std	0.216	3.068	7.58	3.20
	3.500	80	XS	0.300	2.9	10.26	2.86
	3.500	160	...	0.438	2.624	14.34	2.34
	3.500	...	XXS	0.600	2.3	18.60	1.80
3½	4.000	5	...	0.083	3.834	3.48	5.00
	4.000	10	...	0.120	3.76	4.98	4.81
	4.000	30	...	0.188	3.624	7.66	4.47
	4.000	40	Std	0.226	3.548	9.12	4.28
	4.000	80	XS	0.318	3.364	12.52	3.85

Table F.1 (Continued)

Nominal pipe size	Outside diameter (in.)	Schedule		Wall thickness (in.)	Inside diameter (in.)	Weight (lb/ft)	Water Weight (lb/ft)
4	4.500	5	...	0.083	4.334	3.92	6.39
	4.500	10	...	0.120	4.26	5.62	6.17
	4.500	30	...	0.188	4.124	8.67	5.79
	4.500	40	Std	0.237	4.026	10.80	5.51
	4.500	80	XS	0.337	3.826	15.00	4.98
	4.500	120	...	0.438	3.624	19.02	4.47
	4.500	160	...	0.531	3.438	22.53	4.02
	4.500	...	XXS	0.674	3.152	27.57	3.38
5	5.563	5	...	0.109	5.345	6.36	9.72
	5.563	10	...	0.134	5.295	7.78	9.54
	5.563	40	Std	0.258	5.047	14.63	8.67
	5.563	80	XS	0.375	4.813	20.80	7.88
	5.563	120	...	0.500	4.563	27.06	7.08
	5.563	160	...	0.625	4.313	32.99	6.33
	5.563	...	XXS	0.750	4.063	38.59	5.62
6	6.625	5	...	0.109	6.407	7.59	14.0
	6.625	10	...	0.134	6.357	9.30	13.7
	6.625	40	Std	0.280	6.065	18.99	12.5
	6.625	80	XS	0.432	5.761	28.60	11.3
	6.625	120	...	0.562	5.501	36.43	10.30
	6.625	160	...	0.719	5.187	45.39	9.15
	6.625	...	XXS	0.864	4.897	53.21	8.16

(*Continued*)

Table F.1 (Continued)

Nominal pipe size	Outside diameter (in.)	Schedule		Wall thickness (in.)	Inside diameter (in.)	Weight (lb/ft)	Water Weight (lb/ft)
8	8.625	5	...	0.109	8.407	9.92	24.0
	8.625	10	...	0.148	8.329	13.41	23.6
	8.625	20	...	0.250	8.125	22.38	22.5
	8.625	30	...	0.277	8.071	24.72	22.2
	8.625	40	Std	0.322	7.981	28.58	21.7
	8.625	60	...	0.406	7.813	35.67	20.8
	8.625	80	XS	0.500	7.625	43.43	19.8
	8.625	100	...	0.594	7.437	51.00	18.8
	8.625	120	...	0.719	7.187	60.77	17.6
	8.625	140	...	0.812	7.001	37.82	16.7
	8.625	...	XXS	0.875	6.875	72.49	16.1
	8.625	160	...	0.906	6.813	74.76	15.8
10	10.750	5	...	0.134	10.482	15.21	37.4
	10.750	10	...	0.165	10.42	18.67	36.9
	10.750	20	...	0.250	10.25	28.06	35.7
	10.750	30	...	0.307	10.136	34.27	35.0
	10.750	40	Std	0.365	10.02	40.52	34.2
	10.750	60	XS	0.500	9.75	54.79	32.3
	10.750	80	...	0.594	9.562	64.49	31.1
	10.750	100	...	0.719	9.312	77.10	29.5
	10.750	120	...	0.844	9.062	89.38	27.9
	10.750	140	XXS	1.000	8.75	104.23	26.0
	10.750	160	...	1.125	8.5	115.75	24.6

Table F.1 (Continued)

Nominal pipe size	Outside diameter (in.)	Schedule		Wall thickness (in.)	Inside diameter (in.)	Weight (lb/ft)	Water Weight (lb/ft)
12	12.750	5	...	0.156	12.438	21.00	52.6
	12.750	10	...	0.180	12.39	24.19	52.2
	12.750	20	...	0.250	12.25	33.41	51.1
	12.750	30	...	0.330	12.09	43.81	49.7
	12.750	...	Std	0.375	12	49.61	49.0
	12.750	40	...	0.406	11.938	53.57	48.5
	12.750	...	XS	0.500	11.75	65.48	47.0
	12.750	60	...	0.562	11.626	73.22	46.0
	12.750	80	...	0.688	11.374	88.71	44.0
	12.750	100	...	0.844	11.062	107.42	41.6
	12.750	120	XXS	1.000	10.75	125.61	39.3
	12.750	140	...	1.000	10.75	139.81	39.3
	12.750	160	...	1.312	10.126	160.42	34.9
14	14.000	5	...	0.156	13.688	23.09	63.7
	14.000	10	...	0.250	13.5	36.75	62.0
	14.000	20	...	0.312	13.376	45.66	60.9
	14.000	30	Std	0.375	13.25	54.63	59.7
	14.000	40	...	0.438	13.124	63.51	58.6
	14.000	...	XS	0.500	13	72.17	57.5
	14.000	60	...	0.594	12.812	85.14	55.8
	14.000	80	...	0.750	12.5	106.25	53.2
	14.000	100	...	0.938	12.124	130.99	50.0
	14.000	120	...	1.094	11.812	150.95	47.5
	14.000	140	...	1.250	11.5	170.40	45.0
	14.000	160	...	1.406	11.188	189.32	42.6

(Continued)

Table F.1 (Continued)

Nominal pipe size	Outside diameter (in.)	Schedule		Wall thickness (in.)	Inside diameter (in.)	Weight (lb/ft)	Water Weight (lb/ft)
16	16.000	5		0.165	15.67	27.93	83.5
	16.000	10	...	0.250	15.5	42.10	81.7
	16.000	20	...	0.312	15.376	52.33	80.4
	16.000	30	Std	0.375	15.25	62.65	79.1
	16.000	40	XS	0.500	15	82.86	76.6
	16.000	60	...	0.656	14.688	107.62	73.4
	16.000	80	...	0.844	14.312	136.76	69.7
	16.000	100	...	1.031	13.938	165.00	66.1
	16.000	120	...	1.219	13.562	192.64	62.6
	16.000	140	...	1.438	13.124	223.88	58.6
	16.000	160	...	1.594	12.812	245.51	55.8
18	18.000	5	...	0.165	17.67	31.46	106
	18.000	10	...	0.250	17.5	47.44	104
	18.000	20	...	0.312	17.376	59.00	103
	18.000	...	Std	0.375	17.25	70.66	101
	18.000	30	...	0.438	17.124	82.24	100
	18.000		XS	0.500	17	93.55	98.3
	18.000	40	...	0.562	16.876	104.78	96.9
	18.000	60	...	0.750	16.5	138.32	92.6
	18.000	80	...	0.938	16.124	171.11	88.5
	18.000	100	...	1.156	15.688	208.18	83.7
	18.000	120	...	1.375	15.25	244.40	79.1
	18.000	140	...	1.562	14.876	274.52	75.3
	18.000	160	...	1.781	14.438	308.83	70.9

Table F.1 (Continued)

Nominal pipe size	Outside diameter (in.)	Schedule		Wall thickness (in.)	Inside diameter (in.)	Weight (lb/ft)	Water Weight (lb/ft)
20	20.000	5	...	0.188	19.624	39.82	131
	20.000	10	...	0.250	19.5	52.79	129
	20.000	20	Std	0.375	19.25	78.68	126
	20.000	30	XS	0.500	19	104.24	123
	20.000	40	...	0.594	18.812	123.24	120
	20.000	60	...	0.812	18.376	166.58	115
	20.000	80	...	1.031	17.938	209.09	109
	20.000	100	...	1.281	17.438	256.37	103
	20.000	120	...	1.500	17	296.69	98
	20.000	140	...	1.750	16.5	341.46	93
	20.000	160	...	1.969	16.062	379.58	88
22	22.000	5	...	0.188	21.624	43.84	159
	22.000	10	...	0.250	21.5	58.13	157
	22.000	20	Std	0.375	21.25	86.70	154
	22.000	30	XS	0.500	21	114.93	150
	22.000	60	...	0.875	20.25	197.63	140
	22.000	80	...	1.125	19.75	251.08	133
	22.000	100	...	1.375	19.25	303.20	126
	22.000	120	...	1.625	18.75	353.99	120
	22.000	140	...	1.875	18.25	403.44	113
	22.000	160	...	2.125	17.75	451.55	107

(Continued)

Table F.1 (Continued)

Nominal pipe size	Outside diameter (in.)	Schedule		Wall thickness (in.)	Inside diameter (in.)	Weight (lb/ft)	Water Weight (lb/ft)
24	24.000	10	...	0.250	23.5	63.48	188
	24.000	20	Std	0.375	23.25	94.72	184
	24.000	...	XS	0.500	23	125.63	180
	24.000	30	...	0.562	22.876	140.83	178
	24.000	40	...	0.688	22.624	171.48	174
	24.000	60	...	0.969	22.062	238.60	166
	24.000	80	...	1.219	21.562	296.90	158
	24.000	100	...	1.531	20.938	367.79	149
	24.000	120	...	1.812	20.376	429.85	141
	24.000	140	...	2.062	19.876	483.64	134
	24.000	160	...	2.344	19.312	542.72	127
26	26.000	10	...	0.312	25.376	85.69	219
	26.000	...	Std	0.375	25.25	102.74	217
	26.000	20	XS	0.500	25	136.32	213
28	28.000	10	...	0.312	27.376	92.36	255
	28.000	...	Std	0.375	27.25	110.76	253
	28.000	20	XS	0.500	27	147.01	248
	28.000	30	...	0.625	26.75	182.92	243
30	30.000	5	...	0.250	29.5	79.52	296
	30.000	10	...	0.312	29.376	99.03	294
	30.000	...	Std	0.375	29.25	118.78	291
	30.000	20	XS	0.500	29	157.70	286
	30.000	30	...	0.625	28.75	196.29	281

Table F.1 (Continued)

Nominal pipe size	Outside diameter (in.)	Schedule		Wall thickness (in.)	Inside diameter (in.)	Weight (lb/ft)	Water Weight (lb/ft)
32	32.000	10	...	0.312	31.376	105.70	335
	32.000	...	Std	0.375	31.25	126.79	332
	32.000	20	XS	0.500	31	168.39	327
	32.000	30	...	0.625	30.75	209.65	322
	32.000	40	...	0.688	30.624	230.32	319
34	34.000	10	...	0.312	33.376	112.37	379
	34.000	...	Std	0.375	33.25	134.81	376
	34.000	20	XS	0.500	33	179.08	371
	34.000	30	...	0.625	32.75	223.02	365
	34.000	40	...	0.688	32.624	245.03	362
36	36.000	10	...	0.312	35.376	119.05	426
	36.000	...	Std	0.375	35.25	142.83	423
	36.000	20	XS	0.500	35	189.77	417
	36.000	30	...	0.625	34.75	236.38	411
	36.000	40	...	0.750	34.5	282.66	405
42	42.000	...	Std	0.375	41.25	166.89	579
	42.000	...	XS	0.500	41	221.85	572

Table F.2 Tube properties

Tube outside diameter	Birmingham wire gauge (BWG)	Wall thickness (in.)	Inside diameter (in.)	Weight (lb/ft)	Water weight (lb/ft)
1/4	27	0.016	0.218	0.33	0.094
	26	0.018	0.214	0.36	0.093
	24	0.022	0.206	0.43	0.089
	22	0.028	0.194	0.54	0.084
3/8	24	0.022	0.331	0.42	0.143
	22	0.028	0.319	0.47	0.138
	20	0.035	0.305	0.57	0.132
	18	0.049	0.277	0.74	0.120
1/2	22	0.028	0.444	0.54	0.192
	20	0.035	0.43	0.67	0.186
	18	0.049	0.402	0.76	0.174
	16	0.065	0.37	0.85	0.160
5/8	20	0.035	0.555	0.69	0.240
	19	0.042	0.541	0.86	0.234
	18	0.049	0.527	0.97	0.228
	17	0.058	0.509	1.13	0.220
	16	0.065	0.495	1.48	0.214
	15	0.072	0.481	1.95	0.208
	14	0.083	0.459	2.44	0.199
	13	0.095	0.435	0.87	0.188
	12	0.109	0.407	1.41	0.176

Table F.2 (Continued)

Tube outside diameter	Birmingham wire gauge (BWG)	Wall thickness (in.)	Inside diameter (in.)	Weight (lb/ft)	Water weight (lb/ft)
3/4	20	0.114	0.522	1.46	0.226
	18	0.133	0.484	1.68	0.210
	17	0.179	0.392	2.17	0.170
	16	0.250	0.25	2.85	0.108
	15	0.358	0.034	3.66	0.015
	14	0.065	0.62	1.11	0.269
	13	0.109	0.532	1.81	0.230
	12	0.117	0.516	1.93	0.224
	11	0.140	0.47	2.27	0.204
	10	0.191	0.368	3.00	0.159
7/8	20	0.114	0.647	3.77	0.280
	18	0.133	0.609	5.22	0.264
	17	0.179	0.517	1.28	0.224
	16	0.250	0.375	2.09	0.162
	15	0.358	0.159	2.37	0.069
	14	0.065	0.745	2.72	0.323
	13	0.109	0.657	3.63	0.285
	12	0.117	0.641	4.86	0.278
	11	0.140	0.595	6.41	0.258
	10	0.191	0.493	1.61	0.214

(Continued)

Table F.2 (Continued)

Tube outside diameter	Birmingham wire gauge (BWG)	Wall thickness (in.)	Inside diameter (in.)	Weight (lb/ft)	Water weight (lb/ft)
1	20	0.035	0.93	2.64	0.403
	18	0.049	0.902	3.01	0.391
	16	0.065	0.87	3.66	0.377
	15	0.072	0.856	5.03	0.371
	14	0.083	0.834	7.47	0.361
	13	0.095	0.81	9.04	0.351
	12	0.109	0.782	2.48	0.339
	11	0.120	0.76	3.53	0.329
	10	0.134	0.732	5.40	0.317
	8	0.165	0.67	5.80	0.290
1¼	20	0.035	1.18	7.67	0.511
	18	0.049	1.152	10.02	0.499
	16	0.065	1.12	13.71	0.485
	14	0.083	1.084	3.03	0.470
	13	0.095	1.06	4.34	0.459
	12	0.109	1.032	6.66	0.447
	11	0.120	1.01	7.58	0.438
	10	0.134	0.982	10.26	0.425
	8	0.165	0.92	14.34	0.399
	7	0.180	0.89	18.60	0.386
1½	16	0.065	1.37	3.48	0.593
	14	0.083	1.334	4.98	0.578
	12	0.109	1.282	7.66	0.555
	10	0.134	1.232	9.12	0.534

Table F.2 (Continued)

Tube outside diameter	Birmingham wire gauge (BWG)	Wall thickness (in.)	Inside diameter (in.)	Weight (lb/ft)	Water weight (lb/ft)
2	14	0.083	1.834	27.57	0.794
	13	0.095	1.81	6.36	0.784
	12	0.109	1.782	7.78	0.772
	11	0.12	1.76	14.63	0.762
2½	14	0.083	2.334	20.80	1.01
	12	0.109	2.282	27.06	0.989
	10	0.134	2.232	32.99	0.967
3	14	0.083	2.834	7.59	1.23
	12	0.109	2.782	9.30	1.21
	10	0.134	2.732	18.99	1.18

Appendix G

Bending and Expanding of Pipes and Tubes

Pipes and tubes are extensively used in boilers, pressure vessels, and heat exchangers. Bending of these pipes and tubes or expanding them into tubesheet holes alters their thickness. Different codes address the changes in thickness in different ways. The following is a summary of various applications and their effect on thickness.

G.1 TEMA Requirements for U-Tube Thinning

TEMA calculates the thinning of a tube and adds it to the calculated thickness of a straight tube. Figure G.1 shows a U-tube bend. The relationship between stress and strain in the bend is obtained from the theory of elasticity. It assumes the tube wall thickness to be small relative to the tube diameter and the bend radius.

$$\varepsilon_L = (1/E)(S_L - \mu S_r - \mu S_c) \tag{G.1}$$

$$\varepsilon_r = (1/E)(-\mu S_L + S_r - \mu S_c) \tag{G.2}$$

$$\varepsilon_c = (1/E)(-\mu S_L - \mu S_r + S_c) \tag{G.3}$$

where

E = modulus of elasticity
ε_c = strain in the circumferential direction
ε_L = strain in the longitudinal direction
ε_r = strain in the radial direction
μ = Poisson's ratio
S_c = stress in the circumferential direction
S_L = stress in the longitudinal direction
S_r = stress in the radial direction

Bending of the tube from straight to a radiused configuration results in a force in the longitudinal directions and thus, S_L. The forces in the other two directions are zero. Hence, $S_c = S_r = 0$. Also, because the deformation is plastic, $\mu = 0.5$, rather than 0.3 as is generally used for elastic deformation of metals. Equations (G.1), (G.2), and (G.3) become

$$\varepsilon_L = (1/E)(S_L) \tag{G.4}$$

$$\varepsilon_r = (1/E)(-\mu S_L) = -0.5\varepsilon_L \tag{G.5}$$

Fabrication of Metallic Pressure Vessels, First Edition. Owen R. Greulich and Maan H. Jawad.

Figure G.1 U-tube bend

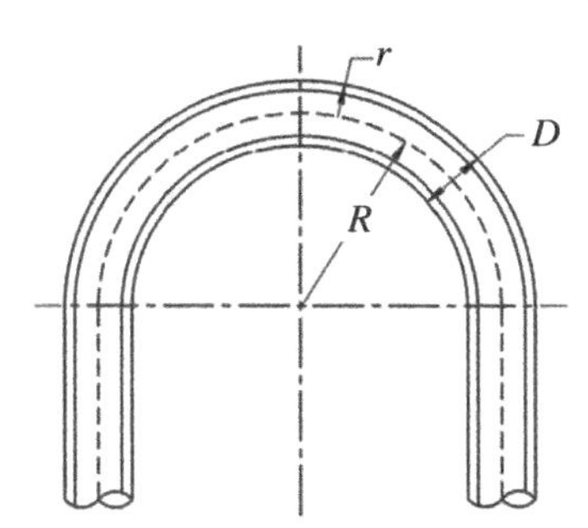

$$\varepsilon_c = (1/E)(-\mu S_L) = -0.5\varepsilon_L \tag{G.6}$$

Equations (G.5) and (G.6) show the radial and circumferential strain to be half the magnitude of the longitudinal strain in a U-tube bend.

From geometry, the longitudinal strain due to stretching of the outer surface of the bend is given by

$$\varepsilon_L = \frac{r}{R} = \frac{D/2}{R} \tag{G.7}$$

where

D = outside diameter of the tube
R = bend radius as shown in Figure G.1
r = outside radius of the tube

From Eqs. (G.5) and (G.7),

$$\varepsilon_t = -\frac{D/4}{R} \tag{G.8}$$

Equation (G.8) shows that the tube thickness reduces at the outside surface by the following amount due to forming:

$$\Delta T = -\frac{(T)(D/4)}{R} \tag{G.9}$$

where

ΔT = reduction in tube wall at the outside of the bend
T = thickness

Hence, the original tube thickness T_o required prior to forming is

$$T_o = T_i + \Delta T \tag{G.10}$$

where

T_i = required minimum wall thickness of a straight tube
T_o = required minimum purchased thickness

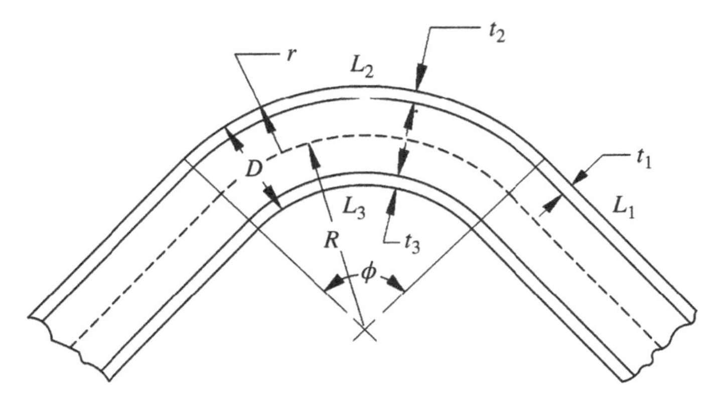

Figure G.2 U-Tube bend

Substituting Eq. (G.9) into Eq. (G.10) gives

$$T_o = T_i\left(1 + \frac{D/4}{R}\right) \tag{G.11}$$

Equation (G.11) is identical to the TEMA Equation in RCB-2.31 (TEMA [1]).

It is of interest to note that the increase in wall thickness at the inside surface of the tube due to forming is given by

$$\Delta T = +\frac{(T)(D/4)}{R} \tag{G.12}$$

Example G.1

A U-tube heat exchanger requires 1 in. diameter tubes. The centerline bend radius for the inner bank of the tubes is $R = 2.5$ in. Determine the minimum required starting thickness for the inner bank if $P = 1470$ psi and $S = 15{,}000$ psi.

Solution

The required thickness of the straight tubes is obtained from the equation $T = \text{PD}/2\text{S}$.

$$\begin{aligned} T &= (1470)(1.0)/2(15{,}000) \\ &= 0.049 \text{ in.} \end{aligned}$$

The increased thickness required due to the tube bends is obtained from Eq. (G.11) as

$$\begin{aligned} T_o &= (0.049)\left(1 + \frac{(1.0)/4}{2.5}\right) \\ &= 0.054 \text{ in.} \end{aligned}$$

Hence, the minimum required starting tube thickness is 0.054 in.

G.2 ASME B31 Code Requirements for Tube Thickness at U-Bends

The required thicknesses at the outside and inside surfaces of the bend are usually obtained from the pressure-area (Jawad [2]). The formulation is as follows:

G.2.1 Thickness of straight tubes

From Figure F.2, for a unit length, L_1, of pipe and internal pressure P,

Force due to pressure = force in the pipe thickness

$$(L_1)(P)(r) = (L_1)(t_1)(S)$$

or

$$t_1 = Pr/S \tag{G.13}$$

G.2.2 Thickness of outside of bend

Force due to pressure in the outside portion of the bend = force in the outside thickness

$$P\left[\pi(R+r)^2(\theta/2\pi) - \pi R^2(\theta/2\pi)\right] = (L_2)(t_2)(S)$$

$$P(\theta/2)\left[(R+r)^2 - R^2\right] = \theta(R+r)(t_2)(S)$$

$$(P/2)\left[2Rr + r^2\right] = (t_2)(S)(R+r)$$

$$t_2 = \frac{Pr(2R+r)}{S(2R+2r)} = \frac{PD(2R+D/2)}{2S(2R+D)} \tag{G.14}$$

$$t_2 = \frac{PD(4R+D)}{S(4R+2D)} = \frac{PD(4R/D+1)}{2S(4R/D+2)}$$

$$t_2 = \frac{PD}{2S}\left[\frac{4R/D+1}{4R/D+2}\right]$$

This equation is equivalent to that given in Paragraph 102.4.5 of the Power Piping code (ASME B31.1) and Paragraph 304.2.1 of the Process Piping code (ASME B31.3) except for the assumption of thin wall and use of

$$t = \frac{PD}{2S} \ \textit{instead of} \ t = \frac{PD}{2(SEW + PY)}$$

The actual thickness at the outside surface is obtained from Eq. (G.9):

$$T_o = T_i\left(1 - \frac{D/4}{R}\right) \tag{G.15}$$

G.2.3 Thickness of inside of bend

Force due to pressure in the inside portion of the bend = force in the inside thickness

$$P\left[\pi R^2(\theta/2\pi) - \pi(R-r)^2(\theta/2\pi)\right] = (L_3)(t_3)(S)$$

$$P(\theta/2)\left[R^2 - (R-r)^2\right] = \theta(R-r)(t_3)(S)$$

$$(P/2)\left[2Rr = r^2\right] = (t_3)(S)(R-r)$$

$$t_3 = \frac{P\,r(2R-r)}{S(2R-2r)} = \frac{PD(2R-D/2)}{2S(2R-D)} \tag{G.16}$$

$$t_3 = \frac{P\,D(4R-D)}{S(4R-2D)} = \frac{PD(4R/D-1)}{2S(4R/D-1)}$$

$$t_3 = \frac{PD}{2S}\left[\frac{4R/D-1}{4R/D-2}\right]$$

This equation is also equivalent to that given in Paragraph 102.4.5 of the Power Piping code (ASME B31.1) and Paragraph 304.2.1 of the Process Piping code (ASME B31.3), with the same simplification. Note that this derivation applies to external pressure and is slightly conservative for internal pressure.

The actual thickness at the outside surface is obtained from Eq. (G.12):

$$T_\mathrm{o} = T_\mathrm{i}\left(1 + \frac{D/4}{R}\right) \tag{G.17}$$

Example G.2

U-tubes with 1 in. diameter are used in a power plant, ASME Section I. The minimum centerline bend radius for the inner bank of tubes is $R = 2.5$ in. Determine the minimum required thickness for the inner bank if $P = 1470$ psi and $S = 15{,}000$ psi.

Solution

Cycle 1

The required thickness of the straight tubes is obtained from Eq. (G.13) as

$$T = (1470)(1.0)/2(15,000)$$
$$= 0.049 \text{ in.}$$
$$4R/D = 5(2.5)/1 = 10$$

The required thickness at the outside surface is obtained from Eq. (G.14) as

$$T_\mathrm{o} = (0.049)\left[\frac{10+1}{10+2}\right]$$
$$= 0.045 \text{ in.}$$

The required thickness at the inside surface is obtained from Eq. (G.16) as

$$T_\mathrm{i} = (0.049)\left[\frac{10-1}{10-2}\right]$$
$$= 0.055 \text{ in.}$$

The above calculations show that the tube thickness is governed by the inside bend surface.

The actual thickness of the tube bends is given by Eqs. (G.15) and (G.17) as

$$T_o' = (0.049)\left(1 - \frac{1/4}{2.5}\right)$$
$$= 0.044 \text{ in.}$$

and,

$$T_i' = (0.049)\left(1 - \frac{1/4}{2.5}\right)$$
$$= 0.054 \text{ in.}$$

A comparison of T_o with T_o' and T_i with T_i' shows the tube thickness needs to be increased slightly.

Cycle 2

Try $T = 0.050$ in.

From Cycle 1, $T_o = 0.045$ in. and $T_i = 0.055$ in.

The actual thickness of the tube bends is given by Eqs. (G.15) and (G.17) as

$$T_o' = (0.050)\left(1 - \frac{1/4}{2.5}\right)$$
$$= 0.045 \text{ in.}$$

$$T_i' = (0.050)\left(1 + \frac{1/4}{2.5}\right)$$
$$= 0.055 \text{ in.}$$

A comparison of T_o with T_o' and T_i with T_i' shows the tube thickness of 0.050 is adequate.

Hence, the minimum required beginning tube thickness is 0.050 in.

A comparison of the results in Examples G.1 and G.2 show a 7% saving in thickness is achieved by using thickness equations for the outside and inside surfaces of the bends.

It should be pointed out that Eqs. (G.13) through (G.17) are based on theoretical formulation. Actual results may vary, and the fabricator needs to establish a design standard that considers these variations. These variations are due in part to

1) Nonhomogeneous material properties
2) Friction during the forming process
3) Out-of-roundness in the tube/pipe during forming that affects the thickness changes
4) Strain hardening of the material

G.3 Expansion of Tubes into Tubesheet Holes

The following equations determine the change in length due to roller expansion of the tubes into the tubesheet.

Since the volume of metal in the tube is constant before and after forming, the manufacturer can calculate the shortening of the tube from the following equation.

$$L' = L[(2r_i + t)/(2r_i + t + 2c)] \tag{G.18}$$

where

c = radial clearance between the tube outside diameter and tube hole surface
L = length of rolled section
L' = reduced length due to rolling
r_i = inside radius of the tube
t = thickness of the tube

Subsequent to the tube touching the tube hole surface, additional rolling will cause the tube wall to thin and the tube to increase in length, Figure G.3, as the tube wall is forced against the tubesheet metal. This additional rolling is used to ensure a preload and associated friction between the tube and the tubesheet, holding the tube in place. The amount of tube elongation can be calculated by the fabricator as

$$L' = L[(2r_o - t)(t)/(2r_o - t')(t')] \tag{G.19}$$

where

L = length of rolled section
L' = reduced length due to rolling
r_o = outside radius of the tube
t = thickness of the tube
t' = reduced thickness of the tube

Equations (G.18) and (G.19) are used to determine the change in length of the rolled tube, if such a dimension is desired.

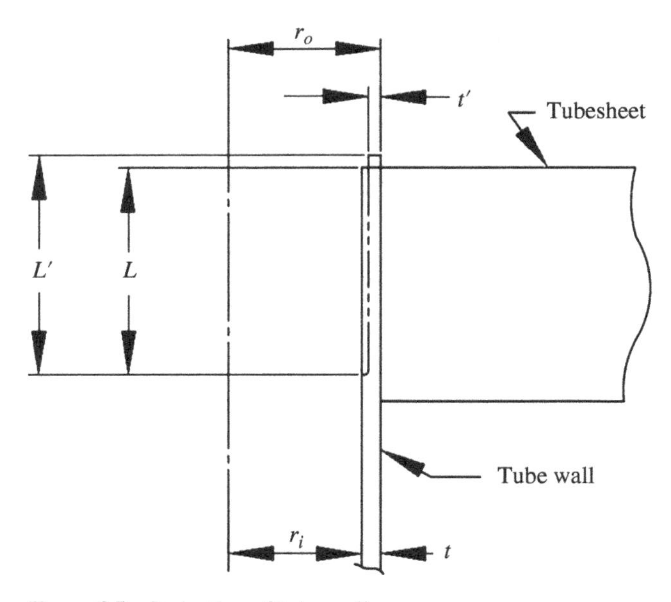

Figure G.3 Reduction of tube wall

Example G.3 A heat exchanger has 1.50-in.-OD tubes with $t = 0.035$ in. The tubesheet thickness is 4.0 in. The tube holes are 1.531 in. in diameter and the rolling length is 3.0 in. A tube thinning of 4% is required in this case, based on experience, to secure the tubes to the tubesheet. Determine the change in the length of the tubes.

Solution
$r_o = 1.50/2 = 0.75$ in.
$r_i = 0.75 - 0.035 = 0.715$ in.
$c = 0.5(1.53125 - 1.50) = 0.0156$ in.
$t' = 0.035(0.96) = 0.0336$ in.
$L = 3.0$ in.

From Eq. (G.18), the reduced length of the tube due to expansion until it touches the hole surface is

$$L' = (3.0)[(2(0.715) + 0.035t)/(2(0.715) + 0.035 + 2(0.0156))]$$
$$= 2.937 \text{ in.}$$

So the shrinkage due to expansion is 0.0630 in.

From Eq. (G.19), the increased length due to 4% thinning is

$$L' = (3.0)[(2(0.75) - 0.035)(0.035)/(2(0.75) - 0.0336)(0.0336)]$$
$$= 3.122 \text{ in.}$$

So, the increase in length due to radial compression is 0.122 in.

Hence, the net change in the length of the tube is an increase of 0.056 in. (0.122 – 0.063) during the rolling process.

References

1 TEMA. 2007. "*Standards of the Tubular Exchanger Manufacturers Association*". Tubular Exchanger Manufacturers Association, Tarrytown, New York.
2 Maan Jawad. 2018. "*Stress in ASME Pressure Vessels, Boilers, and Nuclear Components*". John Wiley, Hoboken, New Jersey.

Appendix H

Dimensions of Some Commonly Used Bolts and Their Required Minimum Spacing

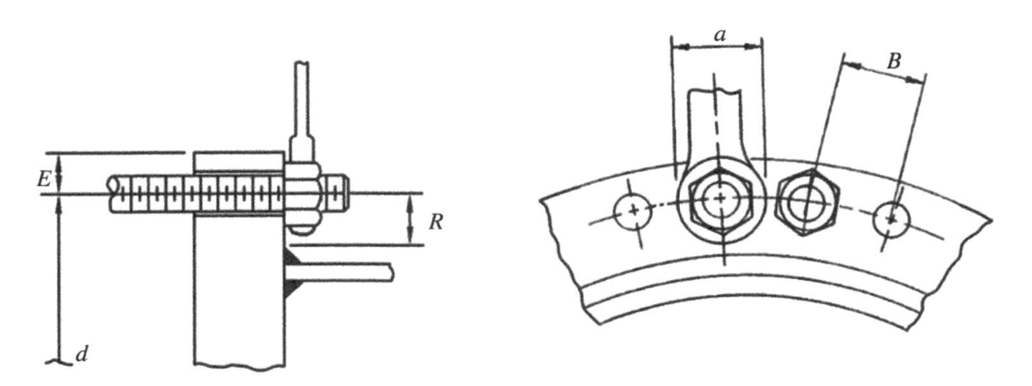

Bolt size	Threads		Nut dimensions					
	No. of threads	Root area, in.^{2}	Across flats	Across corners	Bolt spacing, *B*	Radial distance, *R*	Edge distance, *E*	Wrench diameter *a*
1/2	13	0.126	7/8	0.969	1¼	13/16	5/8	1½
5/8	11	0.202	1¹⁄₁₆	1.175	1½	15/16	3/4	1¾
3/4	10	0.302	1¼	1.383	1¾	1⅛	13/16	2¹⁄₁₆
7/8	9	0.419	1⁷⁄₁₆	1.589	2¹⁄₁₆	1¼	15/16	2⅜
1	8	0.551	1⅝	1.796	2¼	1⅜	1¹⁄₁₆	2⅝
1⅛	8	0.728	1¹³⁄₁₆	2.002	2½	1½	1⅛	2⅞
1¼	8	0.929	2	2.209	2¹³⁄₁₆	1¾	1¼	3¼
1⅜	8	1.155	2³⁄₁₆	2.416	3¹⁄₁₆	1⅞	1⅜	3½
1½	8	1.405	2⅜	2.622	3¼	2	1½	3¾
1⅝	8	1.608	2⁹⁄₁₆	2.828	3½	2⅛	1⅝	4
1¾	8	1.98	2¾	3.035	3¾	2¼	1¾	4¼
1⅞	8	2.304	2¹⁵⁄₁₆	3.242	4	2⅜	1⅞	4½

Fabrication of Metallic Pressure Vessels, First Edition. Owen R. Greulich and Maan H. Jawad.

Bolt size	Threads		Nut dimensions						
	No. of threads	Root area, in.^{2}	Across flats	Across corners	Bolt spacing, *B*	Radial distance, *R*	Edge distance, *E*	Wrench diameter *a*
2	8	2.652	3⅛	3.449	4¼	2½	2	4¾
2¼	8	3.423	3½	3.862	4¾	2¾	2¼	5¼
2½	8	4.292	3⅞	4.275	5¼	3 1/16	2⅜	5⅞
2¾	8	5.259	4¼	4.688	5¾	3⅜	2⅝	6½
3	8	6.324	4⅝	5.105	6¼	3⅝	2⅞	7
3¼	8	7.487	5	5.515	6¾	3⅞	3	7½
3½	8	8.749	5⅜	5.928	7⅛	4⅛	3¼	8
3¾	8	10.108	5¾	6.341	7⅝	4⅜	3½	8½
4	8	11.566	6⅛	6.755	8⅛	4⅝	3⅝	9

Appendix I

Shackles

Shackles come in two shapes. They are either chain type or anchor type as shown [1] in Figure I.1. Chain-type shackles are intended to be used for straight lift, while anchor type may be used for angle lift and are usually intended for heavier lifts.

Shackles also come with a round pin for light loads, a screw pin for intermediate loads, or a bolt pin for heavy loads as shown [1] in Figure I.2.

The rating for a given shackle is specified by the manufacturer and is normally stamped on the shackle. Samples of the dimensions and the associated ratings of chain shackles manufactured by Crosby [1] are shown in Figure I.3 and Table I.1. Comparable values for samples of anchor shackles are provided in Figure I.4 and Table I.2.

The loads shown in Table I.2 are adequate for maximum pickup angle of 120° as shown [1] in Figure I.5.

The load capacity of anchor shackles shown in Table I.2 must be reduced when the shackle is side loaded as shown [1] in Figure I.6. The reduction factors are given in Tables I.3 and I.4.

Chain-type shackle

Anchor type shackle

Figure I.1 Types of shackles (*Source:* Crosby General Catalog [1])

Fabrication of Metallic Pressure Vessels, First Edition. Owen R. Greulich and Maan H. Jawad.

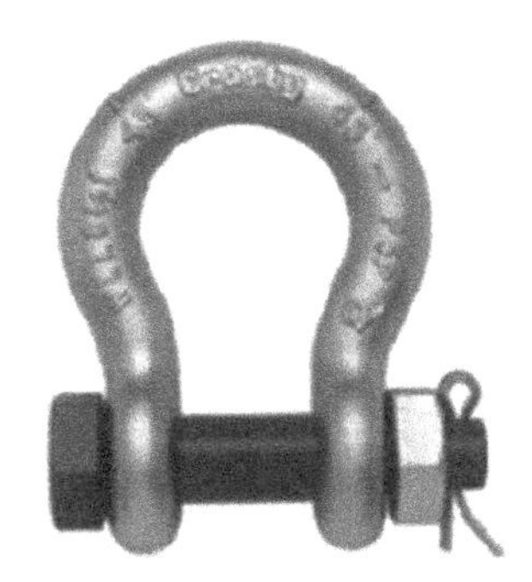

Figure I.2 Types of pins (*Source:* Crosby General Catalog [1])

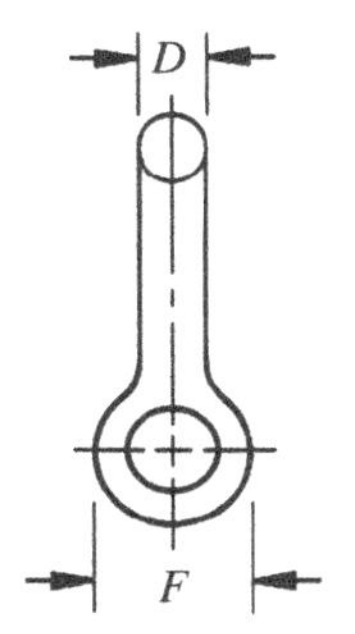

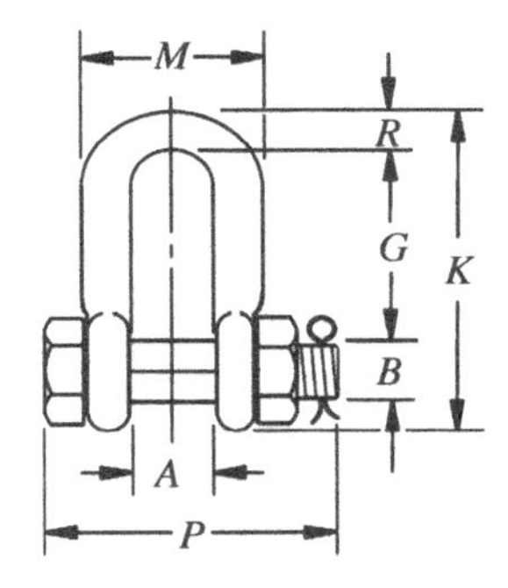

Figure I.3 Dimensions of chain shackle (*Source:* Crosby General Catalog [1])

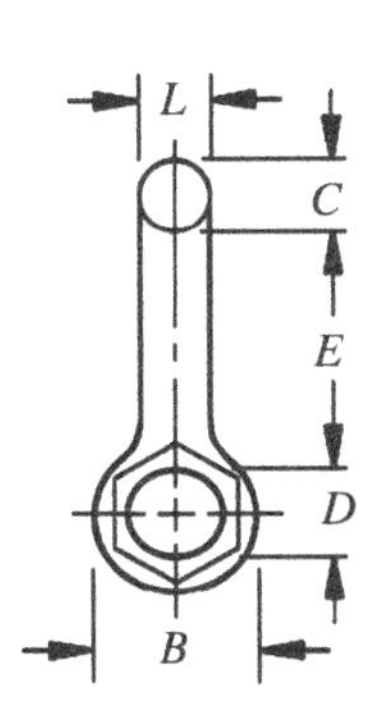

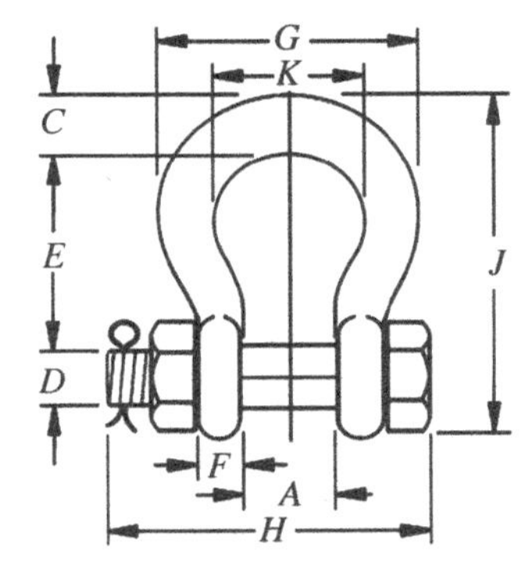

Figure I.4 Dimensions of anchor shackle (*Source:* Crosby General Catalog [1])

Table I.1 Sample Loads and Dimensions for Chain Shackles (*Source:* Crosby General Catalog [1])

Nom. size, in.	Load limit, tons*	Shackle weight, lb	Dimensions, in.								
			A	*B*	*D*	*F*	*G*	*K*	*M*	*P*	*R*
1/4	0.5	0.13	0.47	0.31	0.25	0.62	0.91	1.59	0.97	1.56	0.25
5/16	0.75	0.23	0.53	0.38	0.31	0.75	1.07	1.91	1.15	1.82	0.31
3/8	1	0.33	0.66	0.44	0.38	0.92	1.28	2.31	1.42	2.17	0.38
7/16	1.5	0.49	0.75	0.5	0.44	1.06	1.48	2.67	1.63	2.51	0.44
1/2	2	0.75	0.81	0.64	0.50	1.18	1.66	3.03	1.81	2.80	0.50
5/8	3.25	1.47	1.06	0.77	0.63	1.50	2.04	3.76	2.32	3.56	0.63
3/4	4.75	2.52	1.25	0.89	0.75	1.81	2.40	4.53	2.75	4.15	0.81
7/8	6.5	3.85	1.44	1.02	0.88	2.10	2.86	5.33	3.20	4.82	0.97
1	8.5	5.55	1.69	1.15	1.00	2.38	3.24	5.94	3.69	5.39	1.00
1⅛	9.5	7.60	1.81	1.25	1.13	2.68	3.61	6.78	4.07	5.90	1.25
1¼	12	10.81	2.03	1.4	1.25	3.00	3.97	7.50	4.53	6.69	1.38
1⅜	13.5	13.75	2.25	1.53	1.38	3.31	4.43	8.28	5.01	7.21	1.50
1½	17	18.50	2.38	1.66	1.50	3.62	4.87	9.05	5.38	7.73	1.62
1¾	25	31.40	2.88	2.04	1.75	4.19	5.82	10.97	6.38	9.33	2.12
2	35	46.75	3.25	2.3	2.10	5.00	6.82	12.74	7.25	10.41	2.36
2½	55	85.00	4.12	2.8	2.63	5.68	8.07	14.85	9.38	13.58	2.63
3	85	124.25	5	3.25	3.00	6.50	8.56	16.87	11.00	15.13	3.50

*Metric tons (1 metric ton = 1.1023 U.S. tons = 2205 lb).
Proof load is 2 times the load limit.
Ultimate strength is 6 times the load limit.

Table I.2 Sample Loads and Dimensions for Anchor Shackles (*Source:* Crosby General Catalog [1])

Nom. size, in.	Load limit, tons*	Shackle weight, lb	Dimensions, in.										
			A	*B*	*C*	*D*	*E*	*F*	*G*	*H*	*J*	*K*	*L*
3/8	2	0.33	0.66	0.91	0.38	0.44	1.44	0.38	1.78	2.17	2.49	1.03	0.38
7/16	2.67	0.49	0.75	1.06	0.44	0.50	1.69	0.41	2.03	2.51	2.91	1.16	0.44
1/2	3.33	0.79	0.81	1.19	0.50	0.64	1.88	0.46	2.31	2.80	3.28	1.31	0.50
5/8	5	1.68	1.06	1.5	0.69	0.77	2.38	0.58	2.94	3.56	4.19	1.69	0.63
3/4	7	2.72	1.25	1.81	0.81	0.89	2.81	0.69	3.50	4.15	4.97	2.00	0.75
7/8	9.5	3.95	1.44	2.09	0.97	1.02	3.31	0.81	4.03	4.82	5.83	2.28	0.88
1	12.5	5.66	1.69	2.38	1.06	1.15	3.75	0.92	4.69	5.39	6.56	2.69	1.00
1⅛	15	8.27	1.81	2.69	1.25	1.25	4.25	1.04	5.16	5.90	7.47	2.91	1.13
1¼	18	11.7	2.03	3	1.38	1.40	4.69	1.16	5.75	6.69	8.25	3.25	1.29
1⅜	21	15.8	2.25	3.31	1.50	1.53	5.25	1.28	6.38	7.21	9.16	3.63	1.42
1½	30	18.8	2.38	3.62	1.62	1.63	5.75	1.39	6.88	7.73	10.00	3.88	1.53
1¾	40	33.8	2.88	4.19	2.25	2.00	7.00	1.75	8.81	9.33	12.34	5.00	1.84
2	55	49.9	3.25	4.81	2.40	2.25	7.75	2.00	10.16	10.41	13.68	5.75	2.08
2½	85	103	4.12	5.81	3.12	2.75	10.50	2.62	12.75	13.58	17.90	7.25	2.71
3	120	162	5	6.5	3.63	3.25	13.00	3.00	14.62	15.13	21.50	7.88	3.12

(*Continued*)

Table I.2 (Continued)

Nom. size, in.	Load limit, tons*	Shackle weight, lb	Dimensions, in.										
			A	*B*	*C*	*D*	*E*	*F*	*G*	*H*	*J*	*K*	*L*
3½	150	327	5.25	8	4.38	3.75	14.63	3.75	17.02	20.33	24.88	9.00	3.62
4	175	318	5.5	9	4.56	4.25	14.50	4.00	18.00	21.20	25.68	10.00	4.00
4¾	200	461	7.25	10.5	5.00	4.75	15.19	4.58	20.84	24.04	27.81	11.00	4.75
5	250	608	8.5	12	5.62	5.00	18.50	4.85	23.62	24.87	32.61	13.00	5.00
6	300	797	8.38	13	6.06	6.00	18.72	4.89	24.76	26.22	34.28	13.00	5.88
7	400	1289	8.25	14	7.25	7.00	22.50	6.50	26.00	29.66	40.25	13.00	6.00

*Metric tons (1 metric ton = 1.1023 U.S. tons = 2205 lb).
Proof load is 2 times the load limit.
Ultimate strength is 5 times the load limit for 2 through 21 tons.
Ultimate strength is 5.4 times the load limit for 30 through 175 tons.
Ultimate strength is 4.0 times the load limits for 200 through 400 tons.

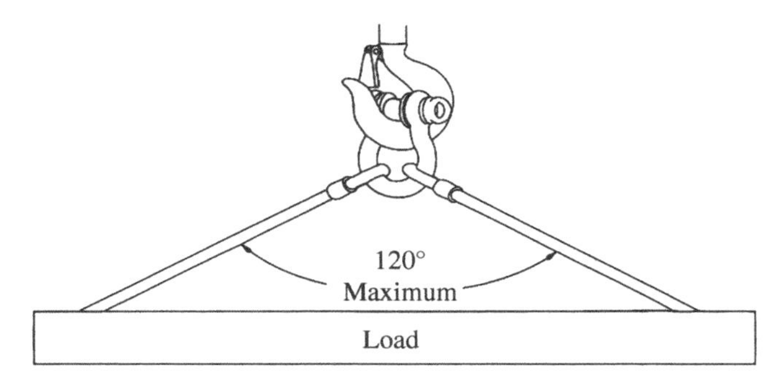

Figure I.5 Maximum pickup angle (*Source:* Crosby General Catalog [1])

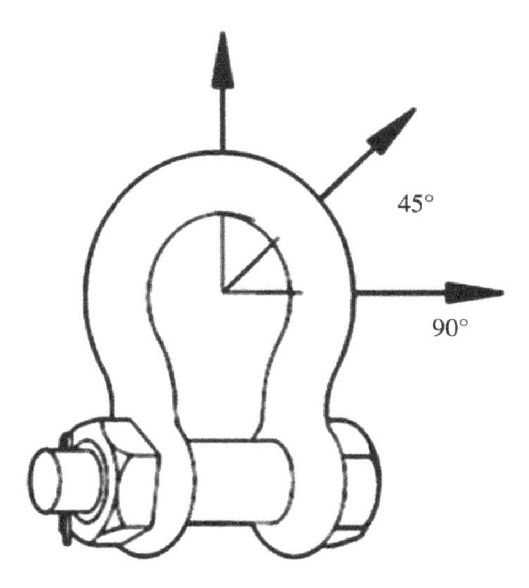

Figure I.6 Side-loaded shackle (*Source:* Crosby General Catalog [1])

Table I.3 Load Reduction Factors for Side-Loaded Shackles with Size Up to 3 in. (*Source:* Crosby General Catalog [1])

Angle of side load from vertical in-line of shackle	Reduction of load limit
0–10° in-line	0% of rated load limit
11–20° from in-line	15% of rated load limit
21–30° from in-line	25% of rated load limit
31–45° from in-line	30% of rated load limit
46–55° from in-line	40% of rated load limit
56–70° from in-line	45% of rated load limit
71–90° from in-line	50% of rated load limit

Table I.4 Load Reduction Factors for Side Loaded Shackles with Size Larger Than 3 in. (*Source:* Crosby General Catalog [1])

Angle of side load from vertical in-line of shackle	Reduction of load limit
0–5° in-line	0% of rated load limit
6–10° from in-line	15% of rated load limit
>10° from in-line	Analysis is required

Reference

1 General Catalog, 2019. The Crosby Group, LLC, Tulsa, Oklahoma.

Appendix J

Shears, Moments, and Deflections of Beams

Nomenclature

- E modulus of elasticity
- F concentrated load
- I moment of inertia of beam
- L length of beam between supports
- L_1 length of beam overhang
- w uniformly distributed load

Case 1. Simply supported beam – concentrated load at center

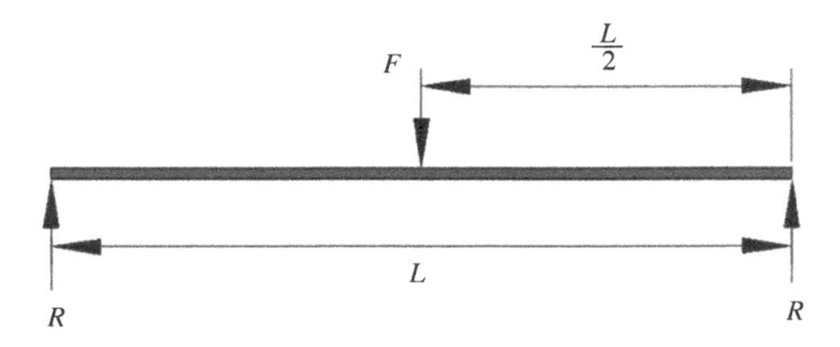

Reaction at support $= \dfrac{F}{2}$

Maximum shear $= \dfrac{F}{2}$

Maximum bending moment at mid-span $= \dfrac{FL}{4}$

Maximum slope at end $= \dfrac{FL^2}{16EI}$

Maximum deflection at mid-span $= \dfrac{FL^3}{48EI}$

Fabrication of Metallic Pressure Vessels, First Edition. Owen R. Greulich and Maan H. Jawad.

Case 2. Simply supported beam – concentrated load at any point

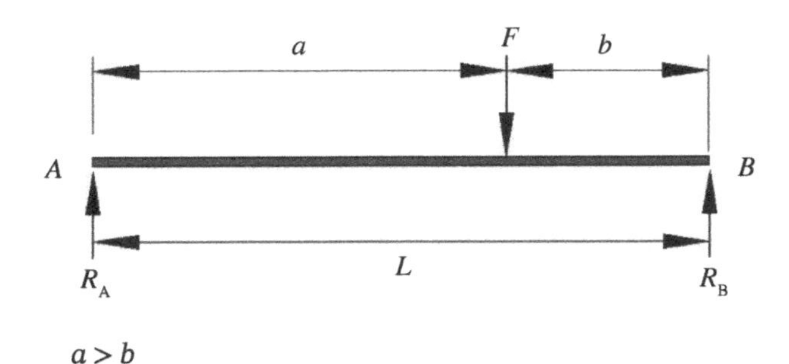

$a > b$

Reaction $R_A = \frac{Fb}{L}$ reaction $R_B = \frac{Fa}{L}$

Shear to the left of the load $= \frac{Fb}{L}$

Shear to the right of the load $= \frac{Fa}{L}$

Maximum bending moment at the load $= \frac{Fab}{L}$

Maximum slope at point $A = \theta_A = \frac{Fab(L + b)}{6LEI}$

Maximum slope at point $B = \theta_B = \frac{Fab(L + a)}{6LEI}$

Maximum deflection $= \frac{Fab}{27LEI}(a + 2b)[3a(a + 2b)]^{0.5}$ at a distance of $[a(a + 2b)/3]^{0.5}$ from left support

Case 3. Cantilever beam – concentrated load at free end

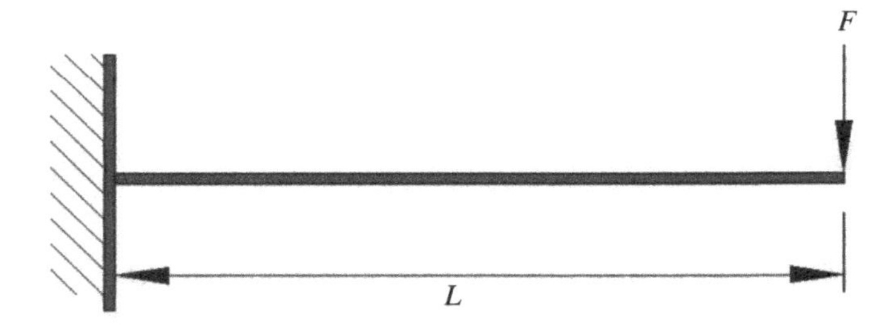

Reaction at support $= F$

Maximum shear at support $= F$

Maximum bending moment at support $= FL$

Maximum slope at end $= \frac{FL^2}{6EI}$

Maximum deflection at end $= \frac{FL^3}{3EI}$

Case 4. Cantilever beam – concentrated load at any point

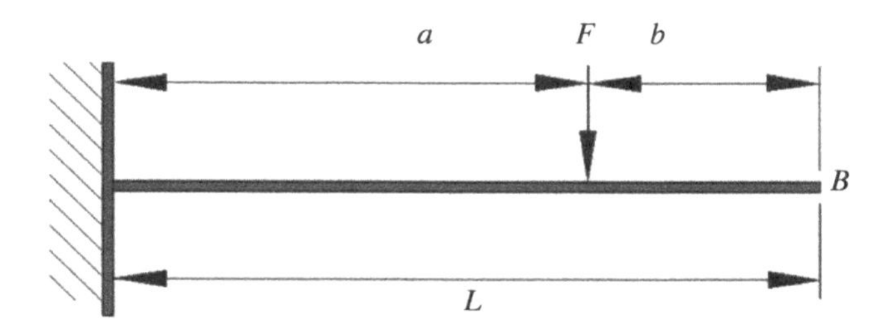

Reaction at support = F
Maximum shear at support = F
Maximum bending moment at support = Fa

Maximum slope at free end = $\dfrac{Fa^2}{2EI}$

Maximum deflection at free end = $\dfrac{(Fa^2)(2L + b)}{6EI}$

Case 5. Simply supported beam – uniformly distributed load

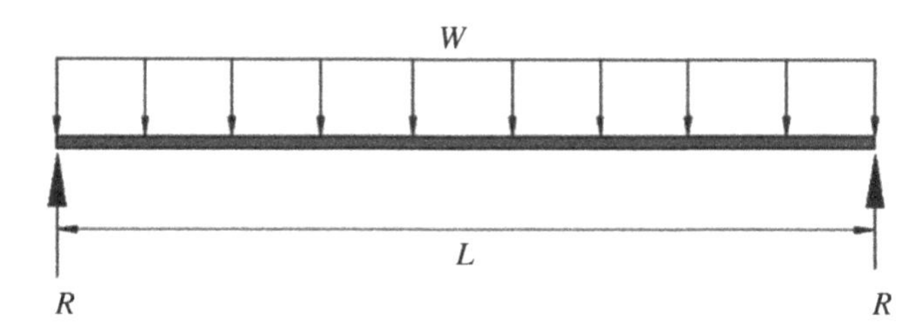

Reactions at supports = $\dfrac{wL}{2}$

Maximum shear at reaction = $\dfrac{wL}{2}$

Maximum bending moment at mid-span = $\dfrac{wL^2}{8}$

Maximum slope at end = $\dfrac{wL^3}{24EI}$

Maximum deflection at mid-span = $\dfrac{5wL^4}{384EI}$

Case 6. Cantilever beam – uniformly distributed load

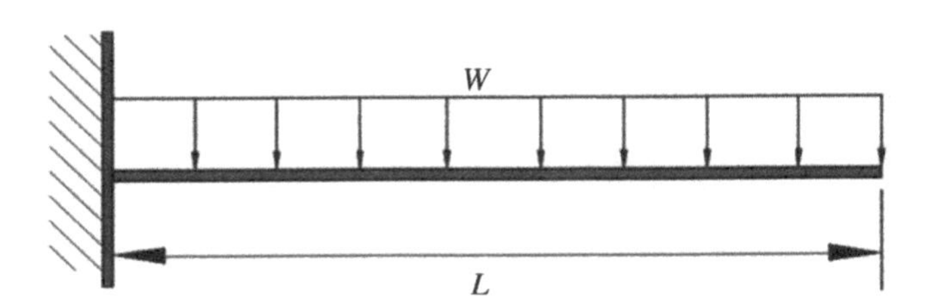

Reaction at support $= wL$

Maximum shear at support $= wL$

Maximum bending moment at support $= \dfrac{wL^2}{2}$

Maximum slope at free end $= \dfrac{wL^3}{6EI}$

Maximum deflection at free end $= \dfrac{wL^4}{8EI}$

Case 7. Beam overhanging two supports-uniformly distributed load

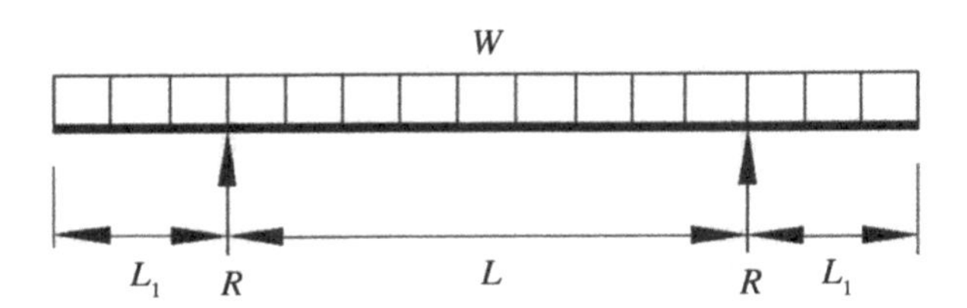

Reactions at support $= \left(\dfrac{w}{2}\right)(L + 2L_1)$

Maximum shear at support = larger of $\dfrac{wL}{2}$ or wL_1

Maximum bending moment at mid-span $= \dfrac{wL^2}{8} - \dfrac{wL_1^2}{2}$

Maximum bending moment at support $= \dfrac{wL_1^2}{2}$

Maximum slope at free end $= \dfrac{wL^3}{24EI} - \dfrac{wL_1^2}{12EI}(3L + 2L_1)$

Maximum deflection at free end $= \dfrac{wL^3L_1}{24EI} - \dfrac{wL_1^3}{8EI}(2L + L_1)$

Maximum deflection at mid-span $= \dfrac{5wL^4}{384EI} - \dfrac{wL_1^2L^2}{16EI}$

Appendix K

Commonly Used Terminology

K.1 Glossary

Air bend	A forming process, usually in a press brake, using a punch (top die) and a vee or rectangular bottom die, in which a plate is formed by pressing it into the bottom die, but with space ("air") remaining between the bottom of the plate and the bottom die. Air bending requires much less force than bending in which the die is "bottomed."
Annealing	A heat treatment process that alters the metallic microstructure, causing changes in such properties as strength and ductility.
Austenite	A face-centered cubic form of gamma iron. It is typically soft and ductile and is the main constituent in the 300 series austenitic stainless steels (304, 316, 321, etc.)
Boring mill	A machine tool used for accurate placement of holes and cutting using a rotating cutting tool. Generally classified as either vertical or horizontal, depending on the orientation of the axis of the cutter. Sometimes referred to simply as "mill." Such mills range from small ones using a single small diameter cutter, to large units capable of driving cutting heads up to two or more feet in diameter with multiple cutting inserts. A wide range of cutting styles and configurations are available. The term "boring mill" is also sometimes used to refer to a vertically oriented lathe, which if equipped with a tool changing turret, is also called a vertical turret lathe, or VTL.
Brake	A metalworking machine for bending of metal sheet or plate. See press brake.
Bump rolling	The process of rolling plate to a radius using a series of successive bends produced using a press brake. Especially common for forming cones and transitions.
C-press	A forming press with a "C" type frame. Commonly used for forming the crown radius of pressure vessel heads.
Cap	A cover for the end of a pipe or tube. Sometimes used for the ends of small pressure vessels.
Carbon steel	An iron alloy with a carbon content of 0.05–2.0%, with small amounts of other elements such as manganese, phosphorus, silicon, and sulfur.

Fabrication of Metallic Pressure Vessels, First Edition. Owen R. Greulich and Maan H. Jawad.

Chill bar	A fixed or removable bar or part of a fixture used to control interpass temperature, shape a weld root pass, or contain a weld, as at the bottom of a V-groove weld in which blow through would otherwise be likely. Sometimes referred to as a backing bar that is either removed or left in place after welding.
Code	The American Society of Mechanical Engineers (ASME) Boiler and Pressure Vessel Code.
Cold spring	The amount that a component is elastically deflected in order to make it fit during fabrication fit up or assembly.
Coupling	A fitting, typically encompassing two axially aligned female threads or openings, used for joining pipes.
Course	A section of a vessel shell between two circumferential seams. Also referred to as a "can."
Cribbing	Materials used to support a product during fabrication or shipping. Same as dunnage.
Crimp	Same as pinch. Pre-bend the beginning of a plate for rolling, either in the roll machine or a press break.
Crown radius	The spherical forming radius of a head in the area of the axis of symmetry.
Fitter	Fabrication shop person who specializes in assembling components and tack welding them together in preparation for final welding.
Flanged and dished	Describes a head produced by forming a spherical radius in the center of a circular plate followed by forming the edges to a toroidal radius, usually with a straight cylindrical section for welding to a shell course.
Flat head	A flat end cover for a pressure vessel. May be round, rectangular, or another configuration, and may be permanently installed or removable.
Handhole	In process equipment, an inspection opening large enough for some visual examination and for reaching into a vessel, but not large enough for a person to enter.
Head	The end cap or cover on a pressure vessel. Available in a number of configurations including flat, conical, flanged and dished, hemispherical, etc.
Heat-affected zone	The unmelted portion of the parent material adjacent to a weld, which has undergone changes in material properties and grain structure from exposure to high temperatures.
Horizontal boring mill	A machine tool whose work spindle and cutting tool rotate around a horizontal axis. Typically, at least three axes of relative motion are available between the tool and the workpiece, and often as many as five or six axes.
Hydrogen embrittlement	Also known as hydrogen assisted cracking, a reduction in ductility, toughness, and tensile strength associated with certain metals in the presence of hydrogen and stress, particularly at elevated temperatures.
Inspection opening	A manhole, handhole, or other opening in a pressure vessel, suitable for examination and cleaning.
Integrally reinforced	Having all required reinforcing material encompassed in the shell, the penetrating component, and the associated welds, without the addition of other material in the form of reinforcing pads, etc.

Internals	Components such as trays, cyclones, demister pads and associated hardware within a pressure vessel but not part of it.
Interpass	Refers to the time after one weld pass and immediately before the next, at which there is often a limit, either minimum or maximum, on temperature.
Knuckle radius	The radius of the toroidal portion of a flanged only or flanged and dished head.
Layer out	The person trained and qualified to perform layout functions.
Layout	The process of marking plates or other materials for cutting and/or forming, including associated mathematical calculations, allowances for bends, etc.
Long seam	A longitudinal weld seam in a shell course.
Long welding neck flange	A flange with an integral long heavy walled neck used to create integrally reinforced pressure vessel nozzles.
Longitudinal seam	A shell course weld seam oriented in the axial direction of a vessel.
Manipulator	A piece of equipment used to hold a welding head in position with respect to the workpiece.
Manway	A port to provide personnel access to the interior of a pressure vessel.
Nozzle	A pipe, flange, or other opening in a pressure vessel.
Nozzle insert	A component, such as a studding outlet, used to create a nozzle in a shell without a pipe or tube between.
Passivate	To treat a metallic surface to make it less chemically active.
Pi-tape®	A precision tape graduated in units of pi inches or millimeters, used to determine the diameter accurately based on a measurement of the circumference of a part such as a cylinder or a head.
Pickle	To chemically treat a metal surface, most often with a strong acid, but sometimes with an alkaline solution, to remove surface impurities and inclusions.
Pinch	Same as crimp. Pre-bend the beginning of a plate for rolling, either in the roll machine or a press break.
Preheat	A minimum parent material temperature required before welding. Used to drive off moisture, control distortion, and reduce the cooling rate of the weld metal.
Press brake	A pressing tool used for forming metal sheet and plate. Typical press brake operation involves top and bottom dies arranged to provide needed bending moment. Press brakes are mechanical or hydraulic, with hydraulics predominating for thick plates.
Procedure qualification record	A record documenting the essential and supplementary essential variables used for production of the test coupon, the ranges of variables qualified, and the results of required testing and/or nondestructive examinations.
Process equipment	Expression related to pressure vessels, heat exchangers, and other equipment used in the petrochemical and power industries.
Reinforcing pad	An internal or external plate added to a shell/nozzle intersection to provide additional strength.
Semielliptical head	A flanged and dished head in which the cross section is one half of an ellipse.

Shell	A cylindrical section of a pressure vessel including one or more courses.
Slag	A vitreous coating created by the melting of welding flux and used to protect the weld from oxidation prior to cooling.
Solution anneal	A heat treatment process in which the alloying elements of a metal are brought into a solid solution.
Spider	A structure using radial arms to achieve or maintain roundness of a shell or head.
Spring back	The degree to which a part returns to its original configuration when the load of a forming process is removed.
Strain harden	Also referred to as work hardening, the strengthening of a metal achieved through plastic deformation.
Studding outlet	A fitting designed for welding into a pipe or pressure vessel and including drilled and tapped holes for installation of threaded studs for joining with a mating flange.
Taping	The circumferential measurement, usually on the outside, of a formed head, shell, or flat head.
Thredolet®	An integrally reinforced fitting for welding to a pipe or pressure vessel, with internal threads for installation of threaded pipe.
Weld prep	The edge or end of material prepared with a configuration favorable for welding.
Welder performance qualification	A record demonstrating the ability of a welder to produce acceptable welds using a particular welding procedure specification (WPS).
Welding neck flange	A flange with an integral neck designed to facilitate welding and withstand mechanical loading.
Welding procedure specification	A document defining welding parameters to ensure the quality of the weld.
Weldolet®	An integrally reinforced weld fitting for joining a pipe to the side of another pipe or a vessel.

K.2 Acronyms and Other Letter Designations

ASME	American Society of Mechanical Engineers
ASTM	Formerly, the American Society for Testing Materials, now ASTM International
BPQ	Brazing performance qualification
BPVC	Boiler and Pressure Vessel Code
DFW	Diffusion welding
EBW	Electron beam welding
EGW	Electrogas welding
ESW	Electroslag welding
ET	Eddy current testing
FCAW	Flux-cored arc welding
FSW	Friction stir welding
GMAW	Gas metal-arc welding
GTAW	Gas tungsten-arc welding
HAZ	Heat-affected zone

HRAP	Hot rolled, annealed, and pickled
HSLA	High strength low alloy
ICR	Inside crown radius
ID	Inside diameter (dimension)
IKR	Inside knuckle radius
LBW	Laser beam welding
MAWP	Maximum allowable working pressure
MT	Magnetic particle testing
MTR	Mill test report
NB	National Board of Boiler and Pressure Vessel Inspectors
NDE	Nondestructive evaluation (examination)
NDT	Nondestructive testing
OD	Outside diameter (dimension)
OFW	Oxyfuel welding
OP	Operating pressure
PAW	Plasma-arc welding
PQR	Performance qualification record
PT	Penetrant testing
PWHT	Post weld heat treatment
QA	Quality assurance
QQS	A federal specification for low carbon steel sheet and strip
RoHS	Restriction of Hazardous Substances regulations of the European Union
RSEW	Resistance seam welding
RSW	Resistance spot welding
RT	Radiographic testing
SA (material designation)	ASME prefix designating ferrous metals
SAW	Submerged-arc welding
SB (material designation)	ASME prefix designating nonferrous metals
SE	Semielliptical
SFM	Surface feet per minute
SFPM	Surface feet per minute
SMAW	Shielded metal-arc welding
UNS	Unified numbering system
VTL	Vertical turret lathe
WPQ	Welding performance qualification

Index